The use of ion implantation for the modification and control of optical properties of a wide range of materials has revolutionised the fabrication of many optical systems, and is being employed in an increasing number of applications. This book is the first to give a detailed description of the factors and processes which govern the optical properties of ion implanted materials, as well as an overview of the variety of devices which can be produced in this way.

Ion implantation is already a well-established technique for changing surface properties of semiconductors and metals, and the book commences with a survey of the basic physics and practical methods involved, which are equally valid for insulating materials. The topics of optical absorption and luminescence are then discussed, before moving on to a chapter describing optical waveguide analysis techniques. An understanding of waveguides provides the background for particular optical devices, such as waveguide lasers and non-linear waveguide systems. The book concludes with a survey of the exciting potential applications which range from fields as diverse as lasers and new phosphors to car mirrors.

Combining both theoretical and practical aspects of the subject, the book will be invaluable to graduate students, scientists and technologists in the fields of solid state physics, lasers, quantum electronics, materials science and surface engineering.

CAMBRIDGE STUDIES IN MODERN OPTICS : 13

Series Editors

P. L. KNIGHT
Optics Section, Imperial College of Science and Technology

A. MILLER
Department of Physics and Astronomy, University of St Andrews

Optical effects of ion implantation

TITLES IN PRINT IN THIS SERIES

Interferometry (second edition)
W. H. Steel

Optical Holography – Principles, Techniques and Applications
P. Hariharan

Fabry–Perot Interferometers
G. Hernandez

Holographic and Speckle Interferometry (second edition)
R. Jones and C. Wykes

Laser Chemical Processing for Microelectronics
edited by K. G. Ibbs and R. M. Osgood

The Elements of Nonlinear Optics
P. N. Butcher and D. Cotter

Optical Solitons – Theory and Experiment
edited by J. R. Taylor

Particle Field Holography
C. S. Vikram

Ultrafast Fiber Switching Devices and Systems
M. N. Islam

Optical Effects of Ion Implantation
P. D. Townsend, P. J. Chandler and L. Zhang

Optical effects of ion implantation

P. D. TOWNSEND, P. J. CHANDLER and L. ZHANG

School of Mathematical and Physical Sciences
University of Sussex

Published by the Press Syndicate of the University of Cambridge
The Pitt Building, Trumpington Street, Cambridge CB2 1RP
40 West 20th Street, New York, NY 10011-4211, USA
10 Stamford Road, Oakleigh, Melbourne 3166, Australia

© Cambridge University Press 1994

First published 1994

Printed in Great Britain at the University Press, Cambridge

A catalogue record of this book is available from the British Library

Library of Congress cataloguing in publication data

Townsend, P. D. (Peter David)
  Optical effects of ion implantation / P. D. Townsend,
P. J. Chandler, and L. Zhang.
    p.   cm. – (Cambridge studies in modern optics; 13)
  Includes index..
  ISBN 0 521 39430 9
  1. Ion implantation.   2. Materials – Optical properties.
I. Chandler, P. J. II. Zhang, L. III. Title. IV. Series.
QC702.7.I55T673   1994
621.36–dc20   93–44923   CIP

ISBN 0 521 39430 9 hardback

Contents

Preface		*page* xiii
1	**An overview of ion implantation**	1
1.1	Development of ion implantation	1
1.2	Properties influenced by ion implantation	4
1.2.1	Mechanical and chemical properties	5
1.2.2	Electrical properties	6
1.2.3	Optical properties	8
1.2.4	Optical properties controlled by surface layers	10
1.3	Processes occurring during ion implantation	11
1.3.1	Nuclear collisions and high defect densities	12
1.3.2	Point defects and electronic interactions	14
1.3.3	Synergistic effects	14
1.3.4	Radiation enhanced diffusion	14
1.3.5	Thermal effects	16
1.3.6	Compositional effects	17
1.4	A summary of the advantages of ion beam processing	18
1.5	Pattern definition	20
1.6	Energy and dose requirements	21
1.7	Summary of implantation effects	21
References		22
2	**Ion ranges, damage and sputtering**	24
2.1	Predictions of range distributions	24
2.1.1	Nuclear collisions	27
2.1.2	Differential cross-section	29
2.1.3	Electronic stopping	31
2.1.4	Summary of nuclear and electronic stopping	33
2.1.5	Ion range distributions	34
2.2	Damage distributions	35

2.2.1	Electronic defect formation	38
2.2.2	Displacement threshold effects	39
2.2.3	Diffusion, relaxation and amorphisation	39
2.2.4	Stability of amorphised layers	47
2.2.5	Amorphisation of semiconductors	47
2.2.6	Stability of point and cluster defects	49
2.2.7	Defect diffusion and crystallography	51
2.2.8	Structural and compositional changes	52
2.2.9	Conclusions on damage distributions	54
2.3	Channelling	55
2.4	Sputtering	57
2.5	Computer simulations	62
2.5.1	Simulation approaches	63
2.5.2	Molecular dynamics	65
2.5.3	Boltzmann transport equation	66
2.5.4	State of simulation programs	66
References		67
3	**Optical absorption**	**70**
3.1	Analysis methods using absorption, ESR and RBS	70
3.2	*In situ* optical absorption	71
3.3	Crystallographic effects on stress and defect motion	75
3.4	Sapphire	77
3.5	Alkali halides	83
3.5.1	F and F_2 centres	85
3.5.2	F_3, F_2' and F_3' bands	88
3.5.3	Other features	89
3.6	Defect complexes	90
3.7	Growth curves	93
3.8	Molecular beam effects on absorption	95
3.9	Isotopic and ion species effects	100
3.10	Measurement of oscillator strength	101
3.11	ESR and ENDOR	103
3.12	High dose effects	103
3.12.1	Amorphisation	103
3.12.2	Colloids	105
3.12.3	Precipitate phases	108
3.13	Summary of problems in interpretation	109
References		112
4	**Luminescence**	**115**
4.1	Luminescence processes	115

4.2	Luminescence during ion implantation	116
4.3	Effects of implantation temperature	117
4.4	*In situ* luminescence	119
4.4.1	Alkali halides – excitons	119
4.4.2	Alkali halides – a search for bi-excitons	122
4.4.3	CaF_2	123
4.4.4	Silica	123
4.4.5	Sapphire	127
4.4.6	$LiNbO_3$ – impurity and stoichiometric effects	130
4.4.7	$LiNbO_3$ – excitons	132
4.4.8	Surface impurity emission	133
4.4.9	Solid argon	135
4.4.10	Summary of *in situ* luminescence effects	135
4.5	Photoluminescence	135
4.5.1	Luminescence of NaF	136
4.5.2	Synthesis of new semiconductor alloys	136
4.5.3	Divacancies in sapphire	137
4.6	Waveguide lasers	137
4.7	Thermoluminescence	140
4.7.1	Silica and quartz	140
4.7.2	CaF_2	143
4.7.3	LiF dosimeters	145
4.8	Impurity doping of CaO	145
4.9	Cathodoluminescence	146
4.10	Depth effects	147
References		148
5	**Ion implanted waveguide analysis**	**151**
5.1	Characteristics of ion implanted waveguides	151
5.2	Waveguide mode theory	152
5.2.1	Maxwell equation approach	154
5.2.2	Quantum mechanics analogy	158
5.3	Waveguide coupling	160
5.3.1	End coupling	160
5.3.2	Prism coupling	163
5.4	Index profile determination	167
5.4.1	WKB approximation for a graded index profile	167
5.4.2	Ion implanted optical barrier waveguides	168
5.4.3	Reflectivity calculation method (RCM)	169
5.4.4	Index profile characterisation	175
5.4.5	Examples of refractive index profile fitted by using RCM	179

5.4.6	Thin film reflectivity method	183
5.5	Planar waveguide attenuation	189
5.5.1	Prism methods	190
5.5.2	Insertion loss	192
References		194
6	**Ion implanted optical waveguides**	**196**
6.1	Practical waveguide structures	197
6.1.1	Conventional fabrication methods	197
6.1.2	Fabrication by ion implantation structural effects	198
6.1.3	Chemically formed ion implanted waveguides	200
6.2	Summary of effects of ion implantation on index	201
6.3	Materials exhibiting index changes	202
6.4	Crystalline quartz	202
6.5	Niobates	207
6.5.1	Lithium niobate	207
6.5.2	Optical damage in lithium niobate	213
6.5.3	Other niobates	215
6.6	Tantalates	217
6.7	Bismuth germanate	219
6.8	Laser hosts	222
6.8.1	Garnets	222
6.8.2	Other laser substrates	226
6.9	Non-linear materials	228
6.10	Other crystalline materials	232
6.11	Non-crystalline materials	233
6.12	Combination with conventional techniques	238
6.13	Ion implanted chemical waveguides	238
6.14	Summary of progress so far	241
References		242
7	**Applications of ion implanted waveguides**	**247**
7.1	Waveguide construction techniques	248
7.1.1	Channel waveguides	248
7.1.2	Optical writing	251
7.1.3	Double barrier implants	253
7.2	Ion implanted waveguide lasers	255
7.2.1	Spectroscopic effects	259
7.2.2	Planar waveguide laser performance	260
7.2.3	Channel waveguide lasers	263
7.3	Frequency doubling	264
7.3.1	Quartz	265

7.3.2	Potassium niobate	266
7.3.3	Potassium titanyl phosphate	270
7.4	Photorefractive effects	272
7.5	Future and related applications	275
References		277
Index		279

Preface

Ion implantation is a superb method for modifying surface properties of materials since it offers accurate control of dopant composition and structural modification at any selected temperature. In the field of semiconductor technology there was a time lag of some 20 years from the initial development of ion implantation to its becoming a cornerstone of production technology. A similar delay in the acceptance time occurred for metal surface treatments. For insulating crystals and glasses, the use of ion beams to modify such crucial optical parameters as refractive index, reflectivity, colour centre content, and luminescence has now passed this 20-year apprenticeship, and the subject is expanding to include the valuable application phase. Appreciation of possible uses of ion implantation is gaining momentum, in part as a result of the ease with which one can fabricate optical waveguides and waveguide lasers and tailor electro-optic and non-linear properties of the key materials of modern optics. Our own experience with these ion implanted property changes, and potential applications, encompasses a diversity of examples, from lasers to studies of fundamental imperfections in insulators, to fabrication of car rearview mirrors.

Since Sussex has been among the pioneers in the study of work with optical materials, we have written a text which has perhaps presented a disproportionate number of examples using our own data. They do, however, typify many aspects of the subject. The topics cover the basic ion beam interactions in solids, followed by the optical effects of absorption and luminescence. We have then included a chapter on waveguide theory and analysis in order to lead into the very exciting examples of ion implanted lasers, second harmonic generation and non-linear waveguide optics. The final chapter gives a flavour of current or potential applications.

In the 1960s and 1970s many of the publications of optical effects of ion implantation were associated with groups in Sandia, Lyon, Padua, Vanderbilt and Sussex but more recently the usage has become so widespread that a catalogue is no longer possible.

We hope that this book will convey our enthusiasm and confidence in this continuing rapid growth and commercial usage of ion implantation effects for optical modification.

Our particular thanks go to Barry Farmery for his operation of the Sussex 3 MeV accelerator and to David Hole, John Barton and Robert Wood for lower energy implantation. During the last 25 years many colleagues have participated in our attempts to understand the optical effects produced by ion beams and more recently we have benefited from a close collaboration with the Optical Research Centre in Southampton. Finally, and not least, we wish to thank the SERC and Società Italiana Vetro for considerable financial support.

Sussex
March 1993

1
An overview of ion implantation

1.1 Development of ion implantation

The control of surface properties is of paramount importance for a wide range of materials applications, and craftsmen and technologists of all scientific disciplines have battled with problems of corrosion, surface hardness, friction and electrical and optical behaviour for many hundreds of years. Even for the simplest of articles, whether they be knives, bottles or non-stick frying pans, the manufacture of materials which have the desired surface properties is often incompatible with bulk performance, and so there is an emphasis on finding ways to modify surface layers. Processes such as thermal quenching prove effective for hardening steel and glass bottles but lack the finesse which is required for more sophisticated technology. Instead, these use more controllable treatments, including the deposition of surface coatings or diffusion of impurities into the surface layer and, of course, ion implantation.

Historically, ion implantation has generally been the last of the treatments to receive widespread acceptance. The reason for this is that, compared with coating and diffusion treatments, it appeared to require more complex and expensive equipment which was not readily available. Figure 1.1 indicates that implantation systems may come in several levels of complexity. There are those similar to sophisticated laboratory research machines, which have ion sources, pre-acceleration, mass analysis followed by additional acceleration and then the target region. Commercial applications with requirements of uniformity and a large sample throughput may result in sample handling and beam sweeping equipment as complex and expensive as the accelerator. In less demanding situations Figure 1.1(b) demonstrates that it is possible to use a broad area source,

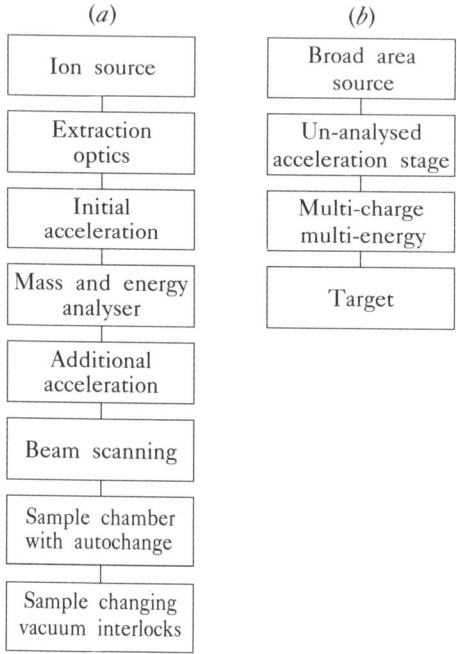

Fig. 1.1. Examples of the layout for ion implantation facilities. (a) is typical of the sophisticated arrangement used for semiconductors and (b) is appropriate for a simple metal implantation.

for example using nitrogen gas, an unanalysed acceleration stage and direct implantation into a large target area.

Although the advantages of surface modifications made by ion implantation were recognised at an early stage, there was not the commercial willpower to develop the subject. This chapter will offer a broad perspective of the advantages and problems of ion implantation in a diverse range of materials. The chapter concludes with a list of general references relating to ion implantation, imperfections and defect properties. Specifically, optical properties will be discussed in more detail in later chapters.

In semiconductor technology the introduction of dopants by thermal diffusion appeared to be both simple and low cost. Hence, even though an ion implanted p–n junction had been made as early as 1950, the ion beam route was not considered to be a high priority. With the development of ever more complex semiconductor circuitry, with many processing steps and higher densities of components, the limitations of

1.1 Development of ion implantation

diffusion technology became apparent. The addition of ion implantation processing offered the possibilities of low temperature introduction of accurate concentrations of different ions in precisely defined regions and a reduced number of thermal cycles. Consequently, modern semiconductor technology has embraced the ion implantation process as a key element in chip production. The electronics industry must therefore be credited with providing the industrial confidence to build reliable high current ion implanters. A non-trivial factor in this acceptance was that the high added value in the very small components justified the capital investment. By contrast, the processing of metals and insulators is often concerned with relatively large areas and, at least initially, the economic advantages are less obvious. For metals, items of small size and high value, such as razor blades which retain their cutting edge, or hip joints and heart valves which do not corrode, wear or fragment, formed a natural set of initial commodities which were suitable for exploitation by ion implantation. As the understanding of surface improvements developed it became apparent that the changes were sufficiently great that many other items were considered viable. The present situation now includes items as large as car engines.

Last, in implantation history, but certainly not least in importance, are insulating materials. These range from items as large as glass windows to exotic and expensive crystals used in solid state tunable lasers and optoelectronics. In the latter case some of the basic steps require the formation of optical waveguides and, naturally, these were first made by diffusion technology. However, the advantages of implantation, the cost of the target materials, and the ability to import all the lithographic techniques of the semiconductor industry, suggest immediately that the time to use ion implantation in insulators has arrived.

The aim of this text is to introduce the basic concepts of ion implantation and demonstrate both the present state of progress and some of its future potential for insulating materials. The implantation literature for insulators is still relatively sparse but all the requisite background theory of ion implantation ranges, control of defect states and types of surface modification which are possible can be found in the established literature. Thus, in principle, the only difficulty is attempting to select, from the vast range of insulator applications, those features which are suitable for improvement. With more initiative, developments which are totally novel may be embarked upon.

4 An overview of ion implantation

Table 1.1. *Properties influenced by surface features*

Mechanical	Chemical	Electrical	Optical
Microhardness	Corrosion	Resistivity	Colour
Friction	Passivation	Photoconductivity	Reflectivity
Adhesion	Diffusion	Electron mobility	Transmission
Wear	Reactivity	Semiconductivity	Optoelectronics

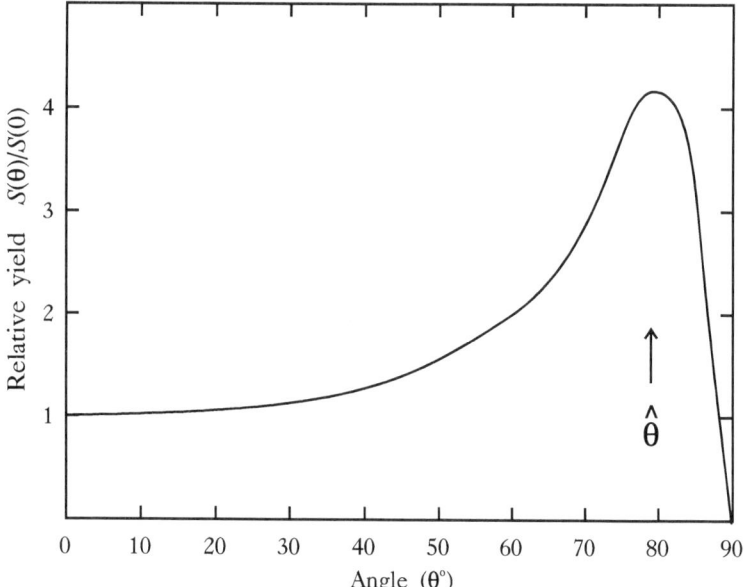

Fig. 1.2. A schematic of the sputtering yield (ions removed per incident ion) as a function of incident angle.

1.2 Properties influenced by ion implantation

Since the penetration depth of ions from most accelerators is limited to a few microns, then the only properties which can be strongly influenced are those governed by the near surface layer. In practice, these comprise a surprisingly wide range which includes not only bulk material but systems with coatings or diffused layers. Table 1.1 summarises many properties which are sensitive to the outer few microns of the material. These broad topics offer clues to the areas where ion implantation may be applied.

1.2 Properties influenced by ion implantation

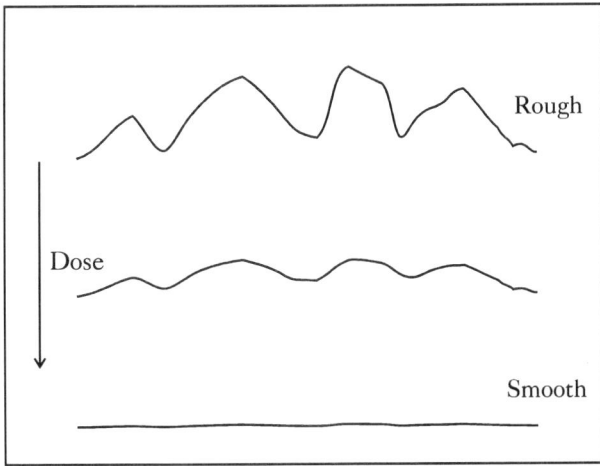

Fig. 1.3. A sketch of how a rough surface may be steadily smoothed by rotating it relative to an ion beam which sputters material from the surface.

1.2.1 Mechanical and chemical properties

Mechanical properties which may be modified include wear, friction, hardness and abrasion. These changes are not specific to insulators and, indeed, much of the initial implant work on surface properties has been directed at metal surfaces. At first sight it may not be obvious why bombarding a surface with energetic ions will improve the quality. However, it should be realised that the ion beam displaces many of the original atoms locally, sputters some material from the surface and, by injecting new material, may induce compressive stresses. The rate of sputtering is strongly direction sensitive, Figure 1.2 (i.e. the yield increases as the secant of the incidence angle), hence surface roughness can in general be smoothed if there is sputter removal. A standard method is to rotate the target with a low energy ion beam at non-normal incidence. Figure 1.3 illustrates how this may lead to a flattening of the surface in favourable cases. Excitation of surface atoms will promote relaxation and diffusion across the surface, which can further flatten the layer. Overall, the presence of improved smoothness will reduce friction and, by reducing the number of asperities, will subsequently reduce the number of fragments which break off and cause wear. Frequently cited examples of success in such improvements in smoothness and wear are the implantation of hip and knee joint replacements which, at least in principle, have a working life far beyond that of a patient.

The introduction of the implant ions and associated stresses can inhibit the motion of dislocations and increase the packing density of glass networks. Hardness and resistance to surface initiated fracture are both increased. Again, the majority of the examples in the literature are for metals, but studies of mechanical properties of ion implanted insulators have demonstrated that identical phenomena occur for the insulators as for the metals. In all these mechanical property modifications the friction, wear and hardness parameters can typically be improved by at least a factor of two. In favourable situations the improvements can be by factors of ten.

A secondary benefit of surface compaction can be to reduce the number of sites which are suitable for chemical attack. Corrosion resistance may therefore ensue. This is not a universal result, and therefore it is difficult to predict whether the immediate effects of implantation will reduce or enhance chemical activity of the surface. For example, if the atomic displacements caused by the passage of the ion beam generate many vacant lattice sites, then the net effect may be the production of a lower density amorphous layer. This will be most obvious if the conversion is from a well structured crystalline surface. The results can be quite spectacular with examples such as ion implanted YAG or $LiNbO_3$, dissolving some 1000 times faster after implantation. Conversely, if the damage effects in the lattice are annealed, then the additional stresses and changes in surface potential caused by the presence of newly added impurities can result in reduced chemical activity. Studies in metals to maintain corrosion resistant steel have taken advantage of the benefits of implanting trace quantities of expensive chemicals, such as yttrium, which are undesirable in the bulk of the steel but beneficial for anticorrosion. One bonus is that the yttrium is retained even while the outer layer is attacked, hence the effectiveness of corrosion protection persists for longer than expected on the basis of the original depth of implantation. No comparable range of studies has yet been published for insulators, and, indeed, the underlying chemistry of surface stabilisation of insulators is in its infancy.

1.2.2 Electrical properties

Inevitably, the examples of ion implanted semiconductors show that for the larger energy gap material it is possible to fully control the electrical conductivity by implantation of small concentrations of dopant ions. Key advantages of implantation in semiconductors are that the electrical

1.2 Properties influenced by ion implantation

Table 1.2. *Advantages/disadvantages of semiconductor doping methods*

ION IMPLANTATION
ADVANTAGES
Accurate dose and depth control
Applicable to most ions
Insensitive to dislocations and impurities
Several dopants may be added
Implantation may be at low or ambient temperature
Doping may be made with masks and/or a passivating layer
Minimal lateral spread of dopants beneath a mask
DISADVANTAGES
High capital costs
Channelling may occur and so distort the depth profile
There is considerable radiation damage – annealing is needed
DIFFUSION DOPING
ADVANTAGES
Relatively low cost
Simple equipment
DISADVANTAGES
Limited choice of ions and dopant profiles
High temperature process
Different dopants disturb one another
Diffusion is sensitive to dislocations and grain boundaries
Lateral spreading of dopants beneath a mask

measurement of ion dose offers extremely accurate and predictable control of the impurity profile within the surface layer. Variations in energy offer the ability to tailor the depth profile of the dopants. Further, several impurities may be added separately without altering their separate depth profiles. Such features are extremely difficult to realise by alternative methods such as thermal diffusion doping. Table 1.2 summarises the major advantages and disadvantages for impurity doping of semiconductors by ion implantation and diffusion. Relevant features will be discussed in detail in later sections.

Doping of semiconductors may not only define and enhance n and p type conductivity within the material, but, additionally, the reverse step of forming regions of electrical isolation can be made by implantation. Routes used to reduce conductivity include the introduction of trapping levels, as in laser diodes, or, more drastically, injection of such high quantities of impurity that a layer is converted to a new material (e.g. from Si to SiO_2 or to Si_3N_4). Control of electrical resistance or the conversion of an insulator into a semiconductor by doping are in principle no different from the semiconductor examples. However, in practice, if the conversion involves the production of a new phase or compound, then the ion doses needed will greatly exceed those typically used for the more subtle control of semiconductor properties. For example, semiconductors operate with doping levels injected by implantation at doses of some 10^{14}ions/cm^2, whereas to form layers of new compounds the implant levels must be some thousand times greater at levels above 10^{17}ions/cm^2. Such doses have been achieved in many examples, for instance in the production of SiO_2 or Si_3N_4, but they cannot be realised if the implant ion is of low energy and high mass, as the resultant sputtering will limit the total number of ions which can be built up in the surface layer. Note that annealing of damage can proceed in several unrelated ways, and particularly for heavily doped material the newly formed chemically altered layer can either disperse by diffusion of ions into the surrounding material, or it may regrow epitaxially on the substrate, or it may develop as a new crystallographic phase. Control of these alternatives is not always possible.

1.2.3 Optical properties

Conventional discussions of optical properties cover the spectral range from the ultra-violet to the infra-red. Thus in terms of suitable materials which are transparent in these regions, optical materials include both narrow energy gap materials which are semiconductors and also the larger gap insulators. In terms of material development, the semiconductors were primarily concerned with electrical properties but the progress into light emitting diodes, lasers and optoelectronic devices has emphasised that the combination of optical and electrical features based on the same substrate is highly desirable. Nevertheless, the processing of optical components on semiconductors often differs little from the routes used for the electrical properties. A discussion of ion implantation into semiconductor substrates will appear as appropriate in this text but the main emphasis will be on the wider bandgap materials. In part this is because less effort has been devoted to the latter, and in part because the interpretation of the mechanisms for change are sometimes simpler to understand.

Glass, which is a complex mixture of oxides, is the oldest and most familiar optical material. Variations in colour, reflectivity, refractive index or chemical stability were originally controlled by the ingredients of the

1.2 Properties influenced by ion implantation

bulk melt. The material is used in large volume applications such as windows, with thicknesses of millimetres, so, quite clearly, ion implantation is unable to influence the bulk properties of the materials for such systems. However, the addition of surface coatings allows the use of a low cost, transparent substrate, which can subsequently be modified. Thin film coating treatments by deposition of surface layers offer the possibility of selective optical absorption and transmission, altered reflectivity to form anti-reflection coatings or mirrors, and chemical passivation of the surface. Coating technology has become highly sophisticated and examples range from the deposition of multi-layer laser mirrors involving in excess of 20 separate layers, to a simple coating to produce heat reflecting window glass. Despite the significant progress which has taken place in coating technology the deposition methods are difficult to control and both the perfection of the layers and their bonding to the glass substrate can suffer failure. Thus, even in this well established field the possibility of introducing ion implantation, either as an additional step to enhance the bonding, or to correct for deposition problems, can be exploited. The more radical step, to move to the sole use of ion implantation as a direct alternative to coatings of large area structures is also possible, and is being explored by several companies. The present situation is that ion implants may provide the requisite changes in optical properties but their usage will be controlled by economic decisions.

Cost effectiveness rises sharply once the insulator device is reduced in size, and specialised materials are necessary. Hence the fields of optoelectronics and non-linear optics should be considered. The main argument is that for many of these applications the optical signals that are being processed are contained within optical waveguides. The size of such guide dimensions are comparable with the wavelength of the light, i.e. on a micron-size scale. This is precisely the order of magnitude which can be influenced by ion implantation. This compatability, coupled with the use of highly expensive crystals to optimise the optical non-linearities or electro-optic properties overrides the initial investment costs of implantation equipment. We are currently reaching the same point for insulators as for semiconductor devices, where, once established, ion implantation has been proved to be a highly flexible, valuable and indispensable processing treatment. Having emphasised the need to justify the economics of ion implantation we will now concentrate on the details of the results which have been realised so far.

1.2.4 Optical properties controlled by surface layers

We will briefly list the types of optical properties which are controllable by surface treatment as this sets the immediate scope of ion implantation experiments which may be worth investigating for insulating materials. For most of these properties, examples of ion beam modifications exist, but few of them have been subject to detailed study. Indeed, the values quoted in the literature for changes in electro-optic coefficient, pyroelectric effects, changes in surface acoustic wave velocity, dispersion or waveguide loss are often in conflict, not only in magnitude, but also in sign! Fortunately, this only emphasises the shortage of detailed studies, rather than any inherent difficulty in controlling ion implanted changes. Some possibilities have only been explored very briefly but the major items considered so far include the following:

Reflectivity and refractive index
    Index changes to act as anti-reflection coatings
    Wavelength selective mirrors and bandpass filters
    Index matching for coupling optical systems
    Modification of coatings
    Corrections, after deposition of multi-layer coatings
    Alterations of properties of coatings and diffused layers
    Enhanced reflectivity for front face mirrors
    Wavelength biased wide band mirrors
    Refractive index profiles to form optical waveguides
    Surface optics as lenses and diffraction gratings
    Blazed grating optics on waveguides
    Multi-layer coupled optical waveguides
    Index confinement for lasers

Lasers
    Optical confinement in semiconductor lasers
    Definition of waveguide laser structures in insulators
    Impurity additions for lasing regions
    Definition of harmonic generators by mode matching
    Production of quasi-phase matched second harmonics
    Production of second harmonics by Cerenkov radiation

Waveguide features
    Optoelectronics
    Enhancement of four-wave mixing
    Waveguide parametric amplifiers
    Production of optically active layers

Information storage
  Photochromic layers
  Electrochromic layers
  Enhancement of photoswitched ferroelectrics
Miscellaneous
  Production of amorphous layers
  Surface phosphor layers
  Electrochemistry
  Chemical reactions and catalysis
  Surface hardness, friction and wear
  Formation of new alloys and compounds
  Modification of metastable phases
  Academic studies of defects

The preceding topics include many optical features which are potentially suitable for industrial applications as well as being scientifically interesting. Similarly, many of the electrical, mechanical and chemical changes, previously mentioned for metals and semiconductors, are equally applicable for insulators. Implantation can play a role in making a rapid survey for determining the choice, or mixture, of dopants or composition, even though the subsequent commercial scale production may use a lower cost route. There are also many aspects of ion implantation in insulators which provide information on the underlying physics and chemistry of materials. From this viewpoint ion implantation may be used to study colour centre formation, kinetics of defect interactions, phase changes, dopants required to control luminescence and thermoluminescence, surface passivation, enhanced bonding of coating layers, or control of superconducting transition temperatures. With such a wealth of opportunities it is remarkable that the present literature is only just beginning to explore these possibilities. This book will primarily discuss implantation effects in inorganic insulators, although many spectacular changes can occur as a result of ion beam interactions with organics. Not least are examples of changes in electrical conductivity by 12 orders of magnitude.

1.3 Processes occurring during ion implantation

Predictions of ion beam induced modifications to surface layers are difficult as a wide variety of processes can occur. We shall briefly attempt to summarise the various possibilities, but leave more detailed discussions of the factors which determine ion ranges, sputtering and details of defect interactions to be presented later. Since the incident ion arrives with an energy of perhaps 500 keV, but only penetrates a micron into the surface, the rate of energy transfer to the target is very high. For light ions at energies of hundreds of keV, or greater, the energy is transferred primarily by electronic excitation. At the lower energies near the end of the ion range there are more nuclear collisions which displace lattice ions and can lead to cascades of damage which influence many tens of lattice sites. Hence these conditions can induce features which are unexpected on purely thermodynamic grounds from the temperature of the target. Figure 1.4(a) indicates the predicted depth distribution of 500 keV carbon ions implanted into silica. The distribution is approximately Gaussian in shape. However, predictions of the damage are more difficult as many of the displaced atoms may migrate or relax back into the original lattice sites. Nevertheless, it is useful to calculate the probability of a target atom being displaced from the lattice site. Figure 1.4(b) shows the results of such a calculation for three different dose levels of 10^{15}, 10^{16} and 10^{17} ions/cm^2. For low ion doses this damage distribution is similar to that of the impurities; but with increasing ion dose, all the target ions within a much greater depth will have been displaced at least once. Note that the calculation overestimates the retained damage, and the efficiency of damage retention is quite complex as it is a function of temperature, lattice structure and prior damage.

1.3.1 Nuclear collisions and high defect densities

At the end of the ion track the collision events are sufficiently energetic that all types of ion, both cation and anion, are displaced. In many cases the secondary collisions similarly destroy the original lattice and hence large cascades of damage are formed. If one assumes a typical displacement energy of 25 eV, then the deposition of 50 keV could imply some 2000 disturbed ions. Even appreciating that this is an overestimate and that many ions will rapidly return to normal sites, a collision cascade may still retain, say, 100 defects. If the objective of the implantation is to amorphise a region of the crystal, as will be described in the formation of optical waveguides or enhanced chemical etching, then this is a valuable feature. The distribution of the damage within a cascade is not uniform and calculations indicate that the core is vacancy rich and interstitials separate to the outer region.

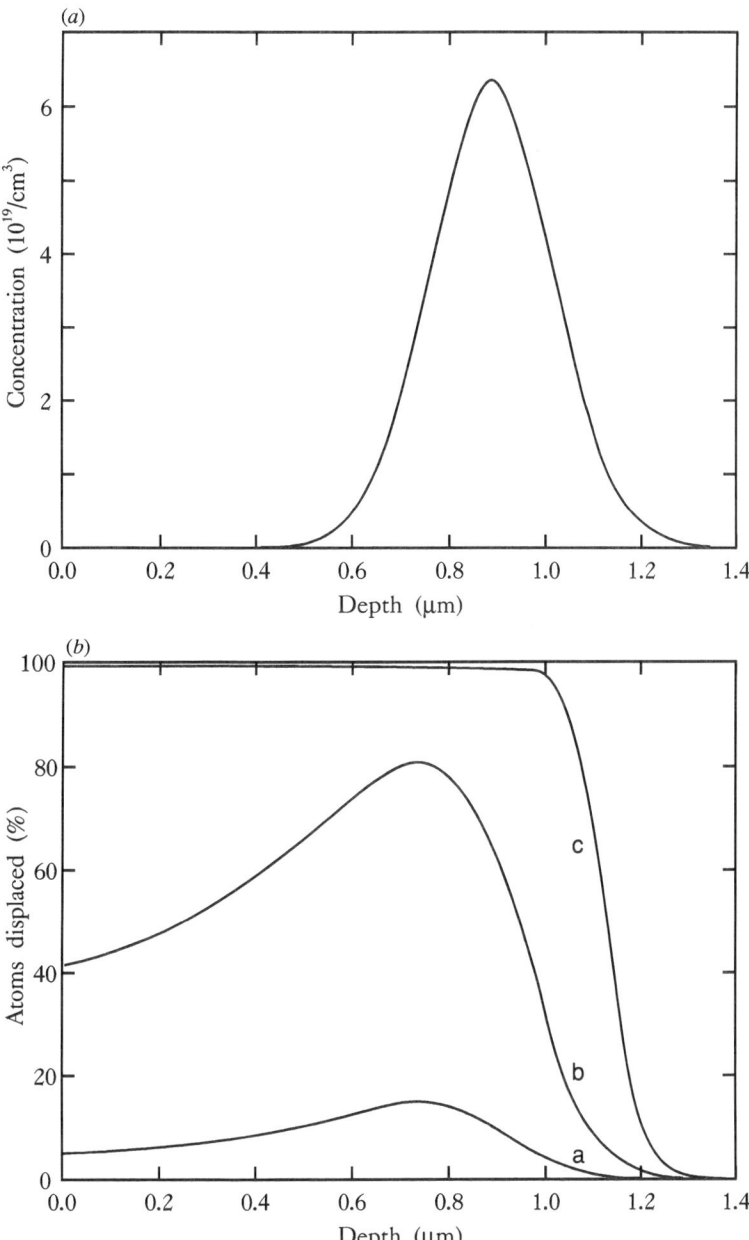

Fig. 1.4. (a) The depth distribution calculated for 500 keV carbon ions implanted into silica. The concentration is for an implant dose of 10^{15} ions/cm^2. (b) The probability of atomic displacements in a silica lattice during implantation of 500 keV carbon at doses of a: 10^{15}, b: 10^{16} and c: 10^{17} ions/cm^2.

1.3.2 Point defects and electronic interactions

For metals, the electronic excitations are unlikely to form displacements in the layer but, particularly for insulators, electronic energy can be efficiently coupled into the lattice and can displace individual atoms. The classic example of purely electronic damage is provided by alkali halides where even exciton energies are sufficient to form defects in the halogen sublattice. In general, electronic processes of defect formation are most evident for the anion sublattice. In addition to the creation of new defects, ionisation will liberate electrons and holes and so colour centres are formed at pre-existing defect sites, such as impurities. Optical absorption bands and charge trapping centres are two immediate consequences of these colour centres. In many cases this can be highly desirable as, for example, in forming laser isolation channels in semiconductor laser structures. Indeed, although we will frequently use the terms 'defects', or 'radiation damage', to imply changes from a perfect lattice structure, the words have emotive overtones which are unfortunate. Many authors and conference organisers now choose to describe such features resulting from implantation damage as 'ion beam modification', since this is both appropriate and marketable. Indeed, it cannot be overemphasised that virtually every modern technology *requires* the presence of imperfections.

1.3.3 Synergistic effects

Although simple theory presents two types of energy transfer, they are not independent. In particular, the presence of ionisation in the region of nuclear collisions can induce further relaxation and diffusion, or the build up of defect complexes within the lattice. Ionisation may additionally alter the displacement thresholds. The net result of such synergistic effects is that defects may be formed in regions which are primarily electronically excited, whereas the material may be stable against ionisation if the electronic energy is provided by a different route, such as an electron beam. Electronically induced annealing of the nuclear collision damage can occur in addition to normal thermally driven recovery of the damage. In some examples, such as for $Bi_4Ge_3O_{12}$ and lithium niobate, the overall effect is to relax the structure into a new or a more perfect phase.

1.3.4 Radiation enhanced diffusion

It is clear that during implantation, the localised energy deposition, the high concentration of vacancies and metastable ionised states can all

1.3 Processes occurring during ion implantation

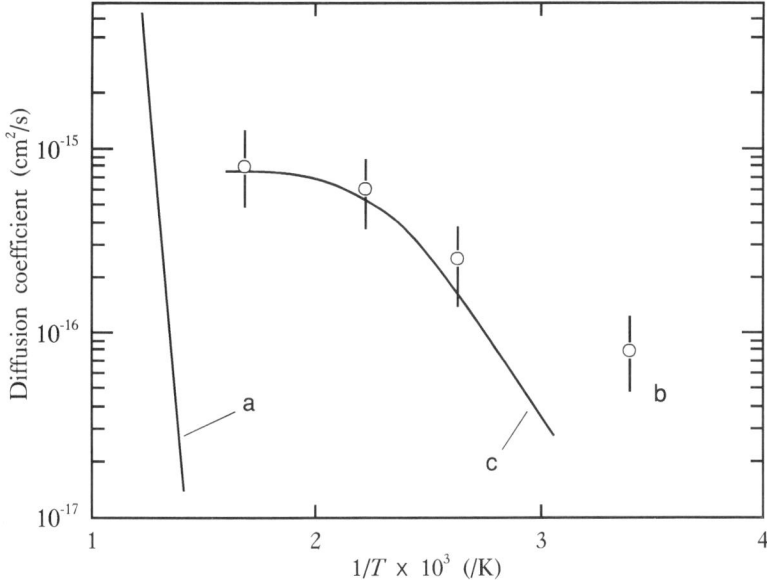

Fig. 1.5. A comparison of a, the thermal diffusion coefficient for nickel in copper with b, the diffusion calculated from self-interstitial motion and c, experimental data for nickel implanted copper. After V. Naundorf, M. P. Macht, H. J. Gudladt and H. Wollenberger (1982) in *Point Defects and Defect Interactions in Metals*, North Holland, Amsterdam, p. 935.

contribute to greater defect movement than would be expected from the bulk temperature of the layer. This is referred to as radiation enhanced diffusion and, since it can be 100 times greater than conventional diffusion, it is important over the scale of the collision cascades and the ion track. Figure 1.5 contrasts the normal thermal diffusion coefficient of nickel in copper with that observed in nickel implanted material. The diffusion enhancement is well modelled by calculations of self-interstitial movement during the irradiation. This non-equilibrium thermodynamic situation also allows the formation of many interacting defect structures. Hence not only will the ion track contain isolated point defects, but there will be regions of considerable disorder and a multiplicity of defect complexes.

Further features are that different defects and impurities will diffuse at different rates. Hence, even if the initial composition of target and impurities is uniform, the modified material may alter in composition as a function of depth. Examples will be given later for alkali motion in ion

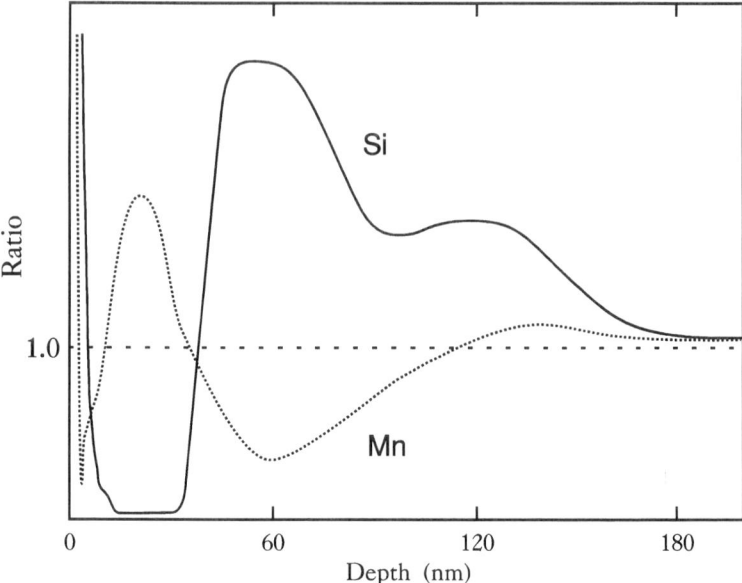

Fig. 1.6. Measured depth profiles of silicon and manganese solutes in a nickel crystal after irradiation with Ni^+ ions at 500 °C. The implant range was 18 nm. After A. D. Marwick and J. E. Hobbs (1984), AERE Rept. AERE-R 11527.

implanted glass and $LiNbO_3$, but a particularly clear demonstration of changes in composition was provided for nickel, uniformly doped with Mn and Si impurities. The material was implanted at 500 °C with 75 keV Ni^+ ions, to give a penetration of some 18 nm. Figure 1.6 shows that the Mn impurities flow into the damage region, whereas the Si moves down the defect concentration gradient.

1.3.5 Thermal effects

The target temperature will influence diffusion to some extent, particularly the migration away from the bombarded track. If the aim is to preserve the induced disorder, for example during amorphisation, then defect retention will be helped by using a low temperature during implantation. Subsequent annealing of the damage will not be influenced by the transient ionisation, and annealing usually follows the expected pattern for damage generated by other routes. Typically, interstitials move at very low temperature (i.e. less than 50 K), vacancies move just above room temperature and the larger the defect cluster the less likely

1.3 Processes occurring during ion implantation

it is to diffuse. One consequence of this is that the point defects formed in the ionisation region of the material tend to be less stable than the defect clusters in the nuclear collision region. Similarly, purely electronic features, i.e. colour centres, disappear at lower temperatures than the structural rearrangements. As with all defect studies where there are high densities of mobile atoms, there can be growth and dissolution of very large defect complexes such as colloids and voids. Since clustering is related to the dynamic rate of defect formation, control of such features will be strongly influenced not only by the temperature during implantation, but also by the rate of defect generation, which in practice means the ion beam current.

It should be noted that defect diffusion may be anisotropic in a crystal lattice because, in some structures, such as quartz or sapphire, there are open channels. Consequently, implantation in directions perpendicular and parallel to these easy flow directions will not produce the same types of damage and defect stability. Additional asymmetry in defect migration will result from thermal gradients across the layer during implantation, from electric fields generated by charge injection and secondary electron emission, and post-implantation annealing may be influenced by stress gradients. The problems of secondary electron generated surface potentials are often poorly considered despite the fact that they pose major problems for ion implantation of insulators and semiconductors. For example, implants into oxide regions of semiconductors can result in electrical destruction of the devices, and for implants into large areas of voltage unstabilised insulators, changes of the surface potentials in excess of 50 keV are commonly reported.

Experience with semiconductors and metals has indicated that very rapid annealing, either by a thermal pulse, or laser heating, can lead to recovery of the original structure sufficiently quickly that the impurities remain close to the depth of implantation. The presence of the impurities can, of course, redefine the perfect structure. For example, new lattice arrangements may include anti-site defects stabilised by the dopant.

1.3.6 Compositional effects

Insulators are more prone to compositional change during implantation than most other materials. There are several reasons for this. The first is that sputtering from the surface can occur via the electronically enhanced mechanisms and when this happens it is generally by loss of anions. As just mentioned, a second problem, for example in a multi-component glass, is that some ionic species move under electric fields and/or radiation enhanced diffusion.

Compositional changes are an inherent feature of implanting impurity ions but in chemical terms the concentrations are often so low as to be negligible. However, this is not the case if the impurities form strong bonds which so distort the lattice as to stabilise defect formation, or trigger a relaxation into a new crystal phase. At high implant doses the formation of new compounds is inevitable. The upper limit to the implant concentration is set by sputtering of the surface. If the surface recedes to the depth of the implant, then an equilibrium situation will develop.

1.4 A summary of the advantages of ion beam processing

Of the various alternative methods of altering surface properties, ion implantation has a number of unique features. Also, although it may be contrasted with the alternatives such as diffusion doping, it is not an exclusive processing route and may be usefully combined with other treatments. The advantages may be listed as follows:

(1) An ion beam carries electrical charge, hence accurate dose control is possible merely by measurement of current. Caveats occur as a result of secondary electron production and beam neutralisation. Secondary electron yields from insulators are significant, and errors of 20–50% in beam current can exist if electron suppression is not included in the system. Similarly, one must be cautious to avoid beam neutralisation at any point in the ion beam path, as this can allow neutrals to reach the target. The neutrals penetrate to different depths as they will include a range of energies, and even if they contribute to the desired effects, they will make current monitoring difficult.

(2) The depths at which the dopants are injected, and lattice disorder is introduced, are directly related to the ion energy and the masses of the ion and target material. As a first approximation the depth profile of the impurities and the structural changes may be tailored by variations of ion dose and energy. In semiconductor work this has been particularly useful in the implantation of several dopant species. These may be introduced individually without significant distortion or interaction. Atomic diffusion will occur in subsequent annealing treatments, but if only one annealing cycle is used the movement of one impurity is not perturbed by heat treatments for each ion type, as would be the case for diffusion doping. In specifically optical examples the control of refractive index profiles has

1.4 A summary of the advantages of ion beam processing

been used to form multiple buried waveguides with controlled coupling between them. This type of structure would normally be extremely difficult to achieve by diffusion methods.

(3) Ion implantation is a very direct and energetic process, hence it is not sensitive to the detailed lattice structure or the presence of impurities or dislocations. In this respect it differs greatly from a diffusion approach, where dopant ions may be pinned by capture at impurity sites, and, more obviously, the diffusion coefficients along dislocations and grain boundaries may greatly exceed impurity movement in the bulk material. Ion implanted systems may therefore avoid the problems of grain boundary or dislocation decoration.

(4) In accurately aligned crystals small ion doses may be directed deeper into the crystal than expected from normal range calculations because of reduced ion-target interactions. This is termed 'ion beam channelling' and can be utilised for low concentrations of dopants in semiconductors, but for the optical examples it is probably of only minor importance because the dose levels tend to be higher.

(5) Many materials for electro-optic applications are chosen for non-linear or ferroelectric properties arising from their crystal symmetry. Such interesting crystal phases may exist over rather limited temperature ranges. For example, $KNbO_3$ has excellent frequency doubling properties but the relevant phase disappears by 225 °C. Formation of a waveguide for frequency doubling that is compatible with waveguide laser structures is excluded by a high temperature diffusion step, but implantation driven changes may be made with the target material at any chosen temperature.

(6) High temperature processing may lead to decomposition of the outer surface, e.g. as for GaAs. Low temperature implants can reduce this difficulty. Equally, implants may be made through a protective overlay.

(7) Ion implanters can be designed for specific applications and so may be chosen to deliver high currents with poor energy resolution or alternatively to provide isotopically pure beams at low current, as contrasted in Figure 1.1. Hence the choice of implanter is a function of the application. Although rarely used at this time, isotopic purity has many advantages for defect studies, for example in EPR or IR vibrational structure measurements. Future applications may include the injection of isotopically pure beams to fabricate waveguide lasers.

(8) Finally, because the impurity injection only depends on the ion energy, it is not constrained by thermodynamic considerations. This means that virtually any type of ion may be implanted into any host, but the

subsequent fate of the implants, in terms of substitution into the lattice, precipitation, or phase separation will be governed by conventional thermodynamics. However, these regions can be generated below the original surface. An advantage, even if the solubility limit is exceeded, is that such implants may act as a buried diffusion source. For example, conventionally, optical waveguides have been formed in $LiNbO_3$ by titanium diffusion at high temperature. However, introduction of the titanium at the surface is problematic as the outer layers of the niobate may decompose and titanium oxides can form which inhibit diffusion. A buried metal source layer can minimise the consequent reproducibility problems.

The preceding advantages were phrased in terms of impurity injection, as for semiconductors, but it is simple to make implants of self-ions to retain the original composition, or in the case of material which grows non-stoichiometrically, to redress the balance of the composition to some new level. The $LiNbO_3$ crystals referred to above are typical of mixed oxide structures and can vary significantly in the ratio of Li to Nb ions. Indeed, from the phase diagram, the congruent growth structure is not stoichiometric. Implants of Li and O could be used to adjust this ratio and hence the bandgap. Sputter deposited or evaporated optical coatings rarely, if ever, match the density and composition of bulk material. Therefore, compositional adjustments may be considered to improve evaporated optical coating layers.

1.5 Pattern definition

Optical waveguides require lateral confinement and so the implantations must be combined with lithographic definitions of the optical pathways. This poses similar problems to those encountered, and overcome, in the semiconductor industry. The only major difference is that most of the semiconductor masks are thinner than those required for optical systems, where the guides may be several microns deep. A fortuitous feature of ion implantation, compared with thermal diffusion doping, is that the lateral spread of the ions is much less than the projected range into the solid. Hence, even for micron-scale penetration, the lateral broadening of the implants will be submicron in scale and the surface mask definition will be preserved.

1.6 Energy and dose requirements

The more abundant literature for semiconductor and metal implantation offers two very different extremes of beam conditions. Semiconductor applications have frequently required isotopically selected beams of ions with energies of some 200 keV and small doses of 10^{13}–10^{15} ions/cm^2. By contrast, the metal effects have been achieved with the formation of new materials and so involve very large concentrations of implanted ions. Economically, this is possible by compromising the mass and energy resolution of the ion beam, for example by operating with unanalysed gas ion beams of 100 keV at dose levels above 10^{17} ions/cm^2. More recent semiconductor work to form buried isolation layers has modified the conventional semiconductor range of conditions, and for SiO_2 or Si_3N_4 formation high current accelerators of several MeV are now in use to give doses up to the 10^{17} level. Many optical effects will typically require structures comparable in dimensions with the wavelength of the light and so depth scales of microns are needed, which in turn implies high beam energies up to the MeV level and above. Dose requirements will depend on the application but experience so far has indicated that for many features the doses are in the range of 5×10^{15} to 5×10^{16} ions/cm^2.

1.7 Summary of implantation effects

Ion implantation into insulating materials can generate a very wide range of property changes. The problems are sufficiently complex that predictions of the effects are extremely difficult, although many general guide lines are beginning to emerge. To emphasise the predictive problems, the published examples include the conversion of crystalline material to an amorphous phase, as well as the converse, in which glasses crystallise. Complex defects may develop, annealing may cause anti-site defects or clusters, and nearly every property of the original insulator can be modified. On the positive side, it is this flexibility which makes ion implantation so attractive. On the negative side, the complexity makes prediction of detailed change very difficult. It is hoped that the subsequent examples will demonstrate that in many cases a recognisable pattern of change is beginning to emerge.

References

The following references include a selection of post-1980 books on ion implantation, review papers which emphasise progress in implant technology of semiconductors, metals and insulators, and references to conference proceedings which include numerous examples of ion implantation effects. Some books which discuss ion implanted defects are listed.

Reviews

Agullo-Lopez, F., Catlow, C.R.A. and Townsend, P.D. (1988). *Point Defects in Materials* (Academic Press, London).
Bahan, R., Kleeman, M. and Poirer, J.P. (eds.) (1981). *Physics of Defects* (North Holland, Amsterdam).
Behrisch, R., Hauffe, W., Hofer, W., Laegreid, N., McClanahan, E.D., Sundqvist, B.U.R., Wittmaack, K. and Yu, M.L. (1991). *Sputtering by Particle Bombardment III* (Springer Verlag, Berlin).
Briggs, D. and Seah, M. (eds.) (1992). *Practical Surface Analysis II – Ion and Neutral Spectroscopy* (John Wiley, Chichester).
Chadwick, A.V. (ed.) (1987). *Defects in Solids* (Plenum, New York).
Chadwick, A.V. and Terenzi, M. (eds.) (1986). *Defects in Solids – Modern Techniques*. NATO ASI series 147 (Plenum, New York).
Dearnaley, G. (1980). *Ion Implantation Metallurgy* (AIME, New York).
Dresselhaus, M.S. and Kalish, R. (1992). *Ion Implantation in Diamond, Graphite and Related Materials* (Springer Verlag, Berlin).
Eckstein, W. (1991). *Computer Simulation of Ion–Solid Interactions* (Springer Verlag, Berlin).
Hayes, W.A. and Stoneham, A.M. (1985). *Defects and Defect Processes in Non-Metallic Solids* (Wiley, New York).
Hirvonen, J.K. (ed.) (1980). *Ion Implantation, Treatise on Materials Science and Technology. vol.18* (Academic, New York).
Hirvonen, J.K. (1984). *Ion Implantation and Ion Beam Processing of Materials* (North Holland, Amsterdam).
Johnson, R.A. and Orlov, A.N. (eds.) (1986). *Physics of Radiation Effects in Crystals* (North Holland, Amsterdam).
Lannoo, M. and Bourgoin, J.C. (1981). *Point Defects in Semiconductors*, Vols I and II (Springer, Berlin).

Mazzoldi, P. and Arnold, G.W. (eds.) (1987). *Ion Beam Modification of Insulators* (Elsevier, Amsterdam).
Picraux, S.T. and Choyke, W.J. (eds.) (1982). *Metastable Materials formed by Ion Implantation* (North Holland, Amsterdam).
Pranevicius, L. (1993). *Ion Beam Activated Processes in Solids* (Maceny, New York).
Prutton, M. (1983). *Surface Physics* (Oxford University Press, Oxford).
Pulker, H.K. (1984). *Coatings on Glass* (Elsevier, Amsterdam).
Smidt, F.A. (ed.) (1983). *Ion Implantation for Materials Processing* (Noyes Data).
Spohr, R. (1990). *Ion Tracks and Microtechnology* (Vieweg, Braunschweig).
Tilley, R.J.D. (1987). *Defect Crystal Chemistry* (Blackie, London).
Townsend, P.D. (1987). Optical effects of ion implantation. In *Rept. Prog. Phys.*, **50**, 501–58.
White, C.W., McHargue, C.J., Sklad, P.S., Boatner, L.A. and Farlow, G.C.(1989). *Mat. Science Repts.* **4**, 41–146.
Ziegler, J.F. (1984). *Ion Implantation Science and Technology* (Academic Press, New York).
Ziegler, J.F., Biersack, J.P. and Littmark, U. (1985). *The Stopping Range of Ions in Solids* (Pergamon, New York).
Ziegler, J.F. (1992). *Ion Implantation Technology* (North Holland, Amsterdam).

Conference proceedings

European Materials Research Society Symposium Proceedings (1992). From Elsevier, Amsterdam, vol.29.
Materials Research Society Symposium Proceedings. From Elsevier, New York (1981) vols. 2, 3; (1983) vol. 14; (1984) vols. 24, 27, 31; (1986) vols. 60, 61; (1992) vols. 235, 244; (1993) vol. 279.
Proceedings of the conference series 'Ion Beam Modification of Materials', in *Radiation Effects* (1980) vols. 47–49; and *Nucl. Inst. Methods* (1981) vols. 182, 183; (1983) vols. 209, 210; (1985) vols. B7, B8; (1987) vols. B19, B20; (1989) vol. B39; (1991) vols. B59, B60; (1993) vols. B80, B81.
Proceedings of the conference series 'Radiation Effects in Insulators', in *Radiation Effects* (1982) vols. 64, 65; (1986) vols. 97, 99; and *Nucl. Inst. Methods* (1984) vol. B1; (1988) vol. B32; (1992) vol. B65.

2
Ion ranges, damage distributions and sputtering

2.1 Predictions of range distributions

An essential first step in the consideration of ion implantation effects is to understand how energy is coupled into the target material. We will first present examples of energy transfer and ion range, and then indicate how these features have been calculated. In practice there has been a continuous interaction between the theoretical and experimental assessments of ion ranges. This has resulted in modifications to the theories so that there are now tabulations and computer codes which predict ion ranges in virtually any ion/target combination. These computations are accurate to within 5–15%. Consequently, although it is useful to know the underlying assumptions of the range theories, and hence their limitations, the majority of the profiles for the distributions of implanted ions are calculated from standard computer simulations. Since knowledge of the ion range, damage distribution or surface sputtering involves many factors in addition to the initial ion range, the existing level of accuracy is perfectly acceptable. Indeed, divergence between measured and computed ranges is frequently not a result of a failure of the computation, but, rather, it results from the fact that such computer codes do not allow for subsequent migration and secondary processes. As has already been mentioned briefly in Chapter 1, there are two main processes which slow down the incoming ion. These are electronic excitations and nuclear collisions. The rate of energy transfer for each process is a function of the nuclear charge and mass of the incoming ion (Z_1, M_1), and the target (Z_2, M_2), as well as the energy. Figure 2.1 shows an idealised example of the rates of energy transfer for the two processes for a light ion entering a target.

Figure 2.2 gives more detail by indicating the depth and calculated

2.1 Predictions of range distributions

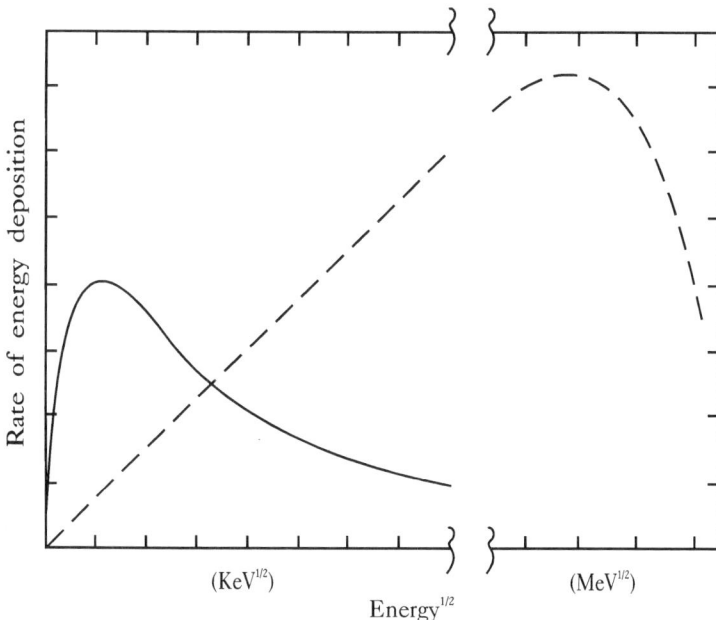

Fig. 2.1. A comparison of the rates of energy transfer for electronic stopping (dashed curve) and nuclear interactions (solid line) of a light ion entering a target.

energy transfer rates which occur in implantation of ions such as He, N, Ar or Xe into an amorphous SiO_2 target. Different types of calculation are discussed at the end of this chapter. For this example a starting energy of 500 keV has been chosen to demonstrate that, for the light ions, two distinct regions are distinguishable. Figure 2.2(a) demonstrates the ion range distributions for these ions whereas Figures 2.2(b) and (c) indicate the depth dependence of the normalised rate of energy deposition from nuclear stopping and electronic interactions. Figure 2.2(a) shows the effect of ion mass on the overall range for a silica target but to demonstrate the effect of ion range effects on average target mass, Figure 2.3 indicates how the integrated ratio of energy deposited into electronic and nuclear collisions varies with Z_1 for two different values of Z_2. Note that in a compound system one can approximate the Z_2 value by a weighted average of the component elements. Within the limitations of the computer simulations of the interactions this is an acceptable simplification.

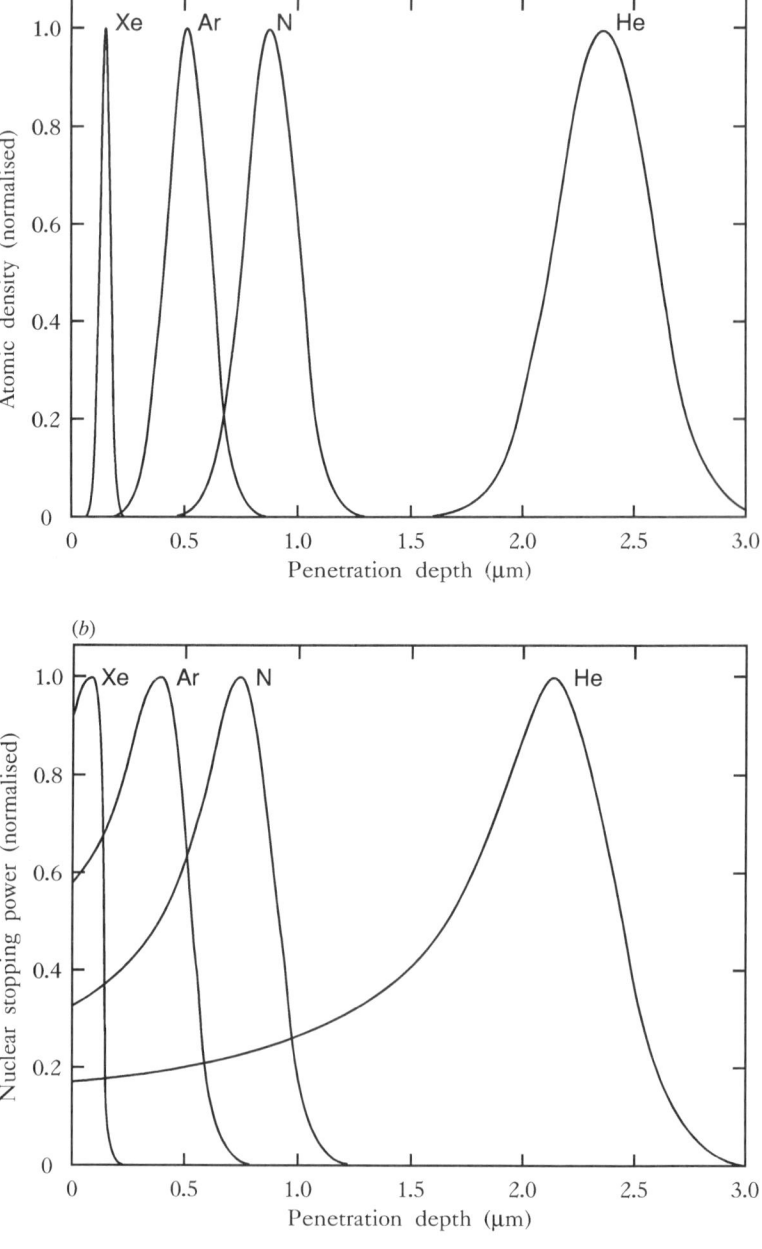

Fig. 2.2. For legend see facing page

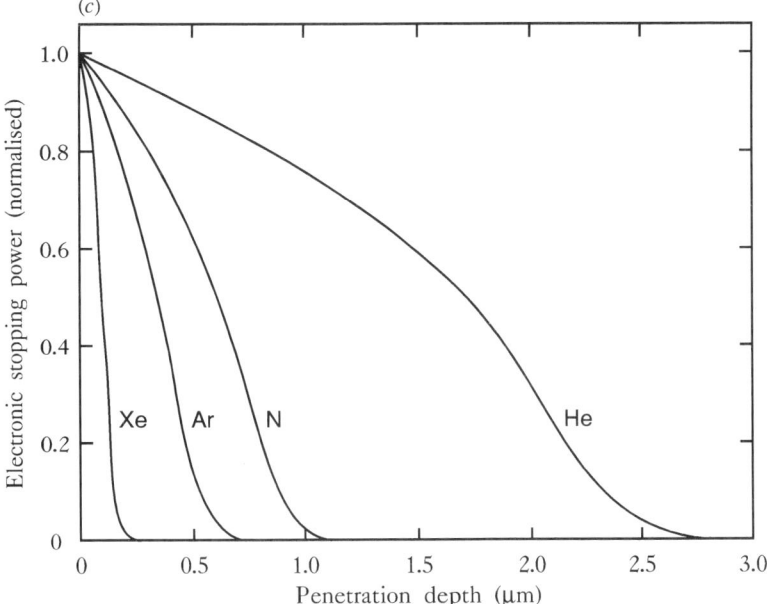

Fig. 2.2. (a) the ion range distributions for 500 keV ions implanted into silica; (b) and (c) show the normalised nuclear collision and electronic excitation rates as a function of depth.

2.1.1 Nuclear collisions

The first assumption in calculating ion beam scattering and energy transfer between the implant and a target ion is to assume that the two ions are isolated from the rest of the material. This enables one to use a simplification of a two-body collision event. Figure 2.4 shows the kinematics of such a collision in the centre of mass (CM) frame. In the laboratory frame the incoming kinetic energy would be $E_1 = 1/2\, M_1 u_1^2$, and the target would be stationary ($u_2 = 0$). The velocities are v_1 and v_2 after the collision. These parameters relate to the CM units to give

$$v_2 = 2V_{CM}\sin(\Phi/2) = 2\mu u_1/M_2 \sin(\Phi/2) \qquad (2.1)$$

where Φ is the collision angle in the CM system, μ is the reduced mass $M_1 M_2/(M_1 + M_2)$, and V_{CM} is the velocity of the centre of mass in the laboratory frame. The kinetic energy transferred to the target atom is

$$T = 1/2\, M_2 v_2^2 = \frac{4M_1 M_2}{(M_1 + M_2)^2} E_1 \sin^2(\Phi/2) \qquad (2.2)$$

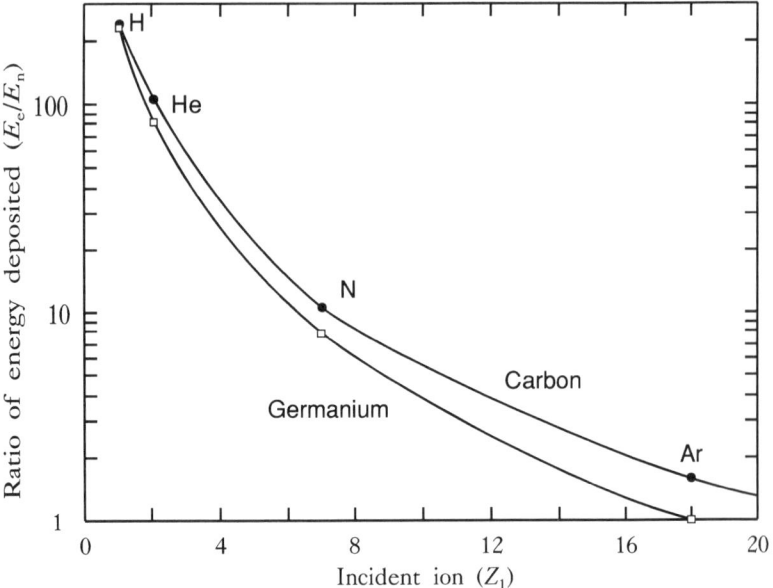

Fig. 2.3. A comparison of the integrated electronic and nuclear stopping powers for a variety of 500 keV ions implanted into carbon and germanium.

From this we note the maximum energy transferred, T_m, is in a head-on collision when $\Phi = \pi$, so

$$T = T_m \sin^2(\Phi/2) \tag{2.3}$$

Reference to Figure 2.4 shows that the important kinematic parameter is the distance b of closest approach, which is given by

$$b = 2Z_1 Z_2 e^2/(\mu u_1^2) \tag{2.4}$$

where $Z_1 e$ and $Z_2 e$ are the net charges on the ions and b relates to the impact parameter, p, via Figure 2.4, by $p = b\cos(\Phi/2)$.

The relationship between T and the impact parameter p depends on the interaction potential. The integration of the differential energy transfer cross-section over all impact parameters gives the total energy loss cross-section and hence the nuclear stopping power, i.e. the energy lost by the ion per unit path length, $S_n = -(dE/dx)_n$, which is

$$S_n(E_0) = \int_0^\infty T(E_0, p) 2\pi p \, dp$$

2.1 Predictions of range distributions

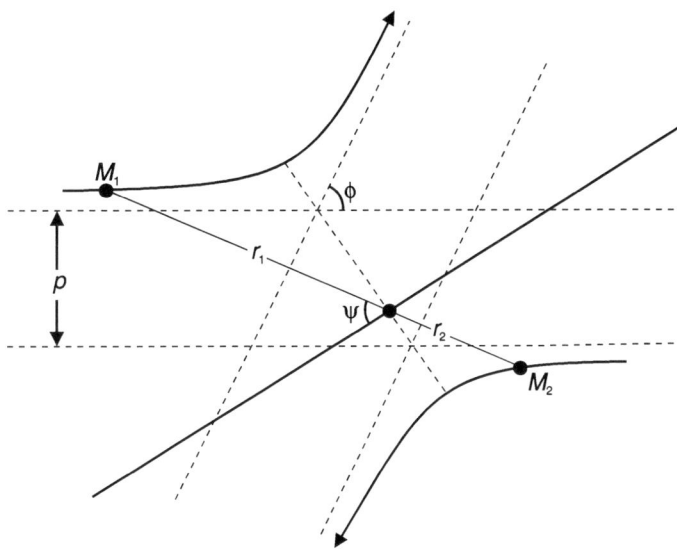

Fig. 2.4. The collision path for impact in centre of mass co-ordinates.

$$= 2\pi \frac{4M_1 M_2}{(M_1 + M_2)^2} E_0 \int_0^{p_{max}} \left(\sin \frac{\Phi}{2}\right)^2 p\,dp \qquad (2.5)$$

where E_0 is the initial energy of the ion and Φ is the scattering angle of deflection. This is a function of the interaction potential, $V(r)$. The first unified approach to nuclear stopping (and range theory, which will be discussed later) was made by Lindhard, Scharff and Schiøtt (1963), and their approach is commonly called the LSS theory. An excellent extended review of the earlier theory was provided by Sigmund (1972).

2.1.2 Differential cross-section

To consider the differential cross-section for energy transfer during the collision it is necessary to specify an interaction potential. Historically, a wide range of potential functions have been used. In the earlier years they were chosen as analytical functions, but with increased computing power more complicated or non-analytical functions were introduced which have resulted in closer agreement between theory and experiment. Examples of the initial choice of functions included a screened Coulomb potential $V(r) = Z_1 Z_2 r^{-1} \exp(-r/a)$, or variations on this, with the exponential replaced by a function of $\Phi(r/a)$. Such variations include those of Brinkman, Firsov, Lenz-Jensen and Lindhard (e.g. Biersack, 1987; Ziegler et al., 1977–80,1985; Townsend et al., 1976). The length, or screening parameter, a, is related to the Bohr radius, a_0, (0.529 Å) by functions such as

$$a_u = \frac{0.8854 a_0}{(Z_1^{0.23} + Z_2^{0.23})}$$
$$a_l = \frac{0.8854 a_0}{(Z_1^{2/3} + Z_2^{2/3})^{1/2}} \qquad (2.6)$$

The exponents are semi-empirical and are chosen to match experimental results. In order to simplify the calculation for nuclear collisions it is useful to express lengths in reduced units that are normalised by the screening length, so $b = p/a$. Similarly, energies are written in units of $Z_1 Z_2 e^2 a^{-1}$. This gives the energy in the centre of mass frame in the collision as

$$\epsilon = E_{CM} / (Z_1 Z_2 e^2 a^{-1}) \qquad (2.7)$$

In the original formulation by Lindhard et al. (1963) a slightly different version of a was used, so they wrote the 'universal' energy parameters and ion ranges as

$$\epsilon = \frac{a_0 M_2 E_0}{Z_1 Z_2 e^2 (M_1 + M_2)} \qquad (2.8)$$

and

$$\rho = \frac{4\pi (0.8853 a_0)^2 M_1 M_2 N R}{(Z_1^{2/3} + Z_2^{2/3})(M_1 + M_2)^2} \qquad (2.9)$$

where N is the atomic density and R the path length. A rewriting of Equation (2.2) in this reduced energy scheme gives

$$S_n(\epsilon) = \int_0^\infty T(\epsilon, b) \pi a^2 d(b^2) = \int_0^\infty T(\epsilon, b) d\sigma \qquad (2.10)$$

where $d\sigma$ is the differential cross-section. In more practical units:

$$S_n(E_0) = \frac{8.462 \times 10^{-15} Z_1 Z_2 M_1 S_n(\epsilon)}{(M_1 + M_2)(Z_1^{0.23} + Z_2^{0.23})} \quad \text{eV cm}^2/\text{atom} \qquad (2.11)$$

The reduced nuclear stopping power is calculated as

$$S_n(\epsilon) = \frac{\ln(1 + 1.138\, 3\epsilon)}{2[\epsilon + 0.013\, 21 \epsilon^{0.212\, 26} + 0.195\, 93 \epsilon^{0.5}]} \quad (\epsilon \leq 30)$$

$$S_n(\epsilon) = \frac{\ln(\epsilon)}{2\epsilon} \quad (\epsilon > 30) \qquad (2.12)$$

2.1 Predictions of range distributions

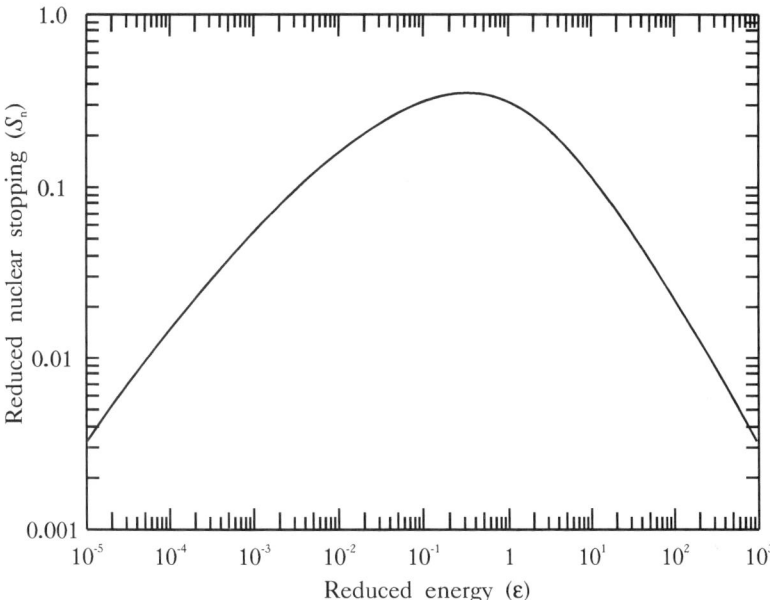

Fig. 2.5. The 'universal' description of nuclear stopping versus ion energy as given in the 'reduced' units.

The upper equation corresponds to heavy projectile and target atoms, and low projectile energy; the lower equation corresponds to light projectile and target atoms, and high projectile energy.

In the more modern semi-empirical formulations, where the potential has been adjusted to fit experimental data, the expression for the reduced nuclear stopping versus the reduced energy parameter provides an excellent fit for all practical applications, and the form of the curve is shown in Figure 2.5.

2.1.3 Electronic stopping

The nature of the electronic energy loss is complicated by there being several possible origins such as:

(a) direct kinetic energy transfer to target electrons, mainly due to electron–electron interactions;
(b) excitation of band-electrons and conduction-electrons, i.e. weakly bound or unlocalized target electrons;

(c) excitation or ionization of target atoms, i.e. promotion of localized (strongly bound) target electrons;
(d) excitation, ionization or electron-capture of the projectile itself.

Electronic energy loss is not only a simple case of inelastic energy loss, as the atomistic nature of the electronic energy loss processes indicate a strongly varying behaviour over the different species of projectile and target. In practice, the calculation of the electronic energy loss of an ion can be divided into two categories: one is for light ions, the other for heavy ions. A brief example of the stopping for light ions, such as He$^+$, is given here.

At low velocities (in the regime $E \leq 1$ MeV for He$^+$) the electronic stopping power may be determined from the Lindhard–Scharff treatment (1961), which suggested that the electronic stopping power is proportional to velocity. It may be expressed as

$$S_e(\epsilon) = K_L \epsilon^{1/2} \qquad (2.13)$$

where S_e is the reduced (dimensionless) electronic stopping power corresponding to Equation (2.11) and

$$K_L = \frac{0.0793 Z_1^{2/3} Z_2^{1/2} (M_1 + M_2)^{3/2}}{(Z_1^{2/3} + Z_2^{2/3})^{3/4} M_1^{3/2} M_2^{1/2}} \qquad (2.14)$$

However, experimental data have indicated that the exponent in Equation (2.13) may differ from 1/2. The electronic stopping power will decrease in the high velocity regime, when the ion velocity exceeds the orbital velocities. It is then treated by Bethe–Bloch theory (see Ahlen, 1980). A convenient expression is the velocity-dependent stopping power

$$S_e(\epsilon) = \frac{4\pi e^4 Z_1^2 Z_2}{m_e v^2} L(v) \qquad (2.15)$$

in which Bethe's stopping number essentially is expressed as

$$L(v) = \ln\left(\frac{2 m_e v^2}{I}\right) - \ln\left(1 - \frac{v^2}{c^2}\right) - \frac{v^2}{c^2} - \frac{c}{Z_2} - \frac{\delta}{2} \qquad (2.16)$$

where I is the target mean excitation potential, c/Z_2 the shell correction and δ an ultra-relativistic density correction which can be neglected (Sternheimer and Peierls, 1977). The mean excitation potential is defined as

$$\ln I = \sum f_n \ln E_n \qquad (2.17)$$

where E_n are all possible energy transitions of the target atom and f_n the corresponding dipole oscillator strengths. The calculations of I are usually based on a statistical model of the target atoms. The shell correction accounts for the deviations from the requirement in Bethe's derivation that the projectile velocity should be much larger than that of the bound electrons. Bethe's calculation is based on a first-order quantal perturbation treatment. So for ions of higher atomic number or of low velocities Bloch's treatment is more correct than that of Bethe (Andersen et al., 1977). The substantial change is that Z_1 enters into the stopping number, and, thus, that the position of the maximum depends on the atomic number of the projectile ion.

A problem that is closely related to the stopping power is the charge state of the penetrating ions. The projectile electrons with orbital electron speeds exceeding the projectile velocity will stick to the projectile while the slower ones, belonging to the outer shells, will be stripped. According to Bohr's estimate the average ion charge increases with the ion velocity. However, regardless of the initial charge state, which may be far from the average state, an ion beam will reach the equilibrium charge distribution after passing a very short distance.

2.1.4 Summary of nuclear and electronic stopping

The nuclear and electronic stopping (energy deposition) is summarised in Figure 2.1. At the lowest velocities ($\epsilon \leq 1$) nuclear stopping dominates for heavy and medium mass ions, and is competitive for light ions. It reaches a maximum value in the kilo-electronvolt energy range and approaches the classical Rutherford distribution at higher energies. At higher velocities ($v \geq Z_1^{2/3} e^2/\hbar$), electronic stopping takes over according to Equation (2.13). In this regime, the ion is predominantly neutral, and the rate of electronic energy deposition into the target is proportional to ion velocity for energy below about 1 MeV. Beyond the stopping power maximum, the Bethe regime, Equation (2.15), is approached where the ion is practically stripped of all electrons; the stopping power decreases with increasing energy and the target becomes more 'transparent'. The relative importance of these two mechanisms depends on the choice of the ion mass and energy.

2.1.5 Ion range distributions

Energy loss of a particle moving through a random medium has been defined as

$$\frac{dE}{dx} = -S(E) \qquad (2.18)$$

where $S(E)$ is the stopping power, the sum total of nuclear, electronic, and other interaction stopping powers, i.e.

$$-S(E) = \sum_i \left(\frac{dE}{dx}\right)_i = \left(\frac{dE}{dx}\right)_{nucl.} + \left(\frac{dE}{dx}\right)_{elec.} + \left(\frac{dE}{dx}\right)_{other}$$

Each stopping power is a function of the mass and the velocity of the ion and the target. Therefore, the mean penetrated path length $R(E)$ of an ion before coming to rest can then be estimated by integration:

$$R(E) = \int_0^E \frac{dE'}{N\,S(E')} = \int_0^E \frac{dE'}{\sum_i (-dE/dx)_i} \qquad (2.19)$$

In principle, from predicted functions of $(dE/dx)_i$, it is possible to calculate the total path length from the given ion and target. Because of ion scattering, the ion range normal to the surface, termed the mean projected range $R_p(E)$, is smaller than $R(E)$ by a factor which depends on the mean path, scattering angles, and, consequently, on the specific path of an individual ion, e.g. as shown in Figure 2.6. Furthermore, as a result of multiple collisions the ions will be deviated from their original direction and there will be a lateral spreading, $R_\perp$ (straggling effect) of the ion beam in the target. For a single element target, the ratio of R_p/R is $\ll 1$ for $M_1 \ll M_2$ (light ion, e.g. $\epsilon < 1$), and close to 1 for $M_1 \gg M_2$ (heavy ion, e.g. $\epsilon > 1$). In a compound target material case, the calculation of the ion range is more complicated due to the need to consider the combination of interactions between the ion and different atoms of the compound. In practice, statistical fluctuations in the actual stopping mechanism will give a spread to the range R and so will contribute to the spread ΔR_p. The degree of spreading will increase with increasing depth into the target and it is a function of the mass ratio M_2/M_1. There is only a small deviation for heavy ions incident into a light atom target; for $M_2/M_1 \simeq 0.1$, $\Delta R_p \simeq 0.2 R_p$. The spreading increases with increasing M_2/M_1 ratio and is about $0.5 R_p$ for $M_2 = M_1$. The straggling effect results in an ion distribution which is approximately Gaussian in form. For low energy implants, the concentration of ions at a depth x for an incident ion dose φ (ions/cm^2) in a target of atomic

2.2 Damage distributions

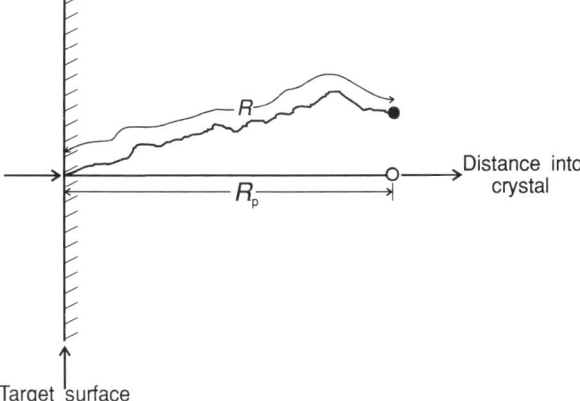

Fig. 2.6. The relationship between the projected range R_p and the total ion path R as the ion is scattered from the forward direction.

density N is characterized by R_p and ΔR_p (Agullo-Lopez et al., 1988), e.g.

$$C(x) = \frac{\varphi}{(2\pi)^{1/2} N \, \Delta R_p} \exp\left[\frac{-(x - R_p)^2}{2\pi \, \Delta R_p{}^2}\right] \quad (2.20)$$

Early calculations of such range spreading, both along the projected range direction, and laterally, were discussed by Schiøtt (1970) and are typified by Figure 2.7. One notes that for light target ions, relative to the incoming ion mass, the lateral and projected range straggling is similar, but for heavy targets there is a greater rate of increase in the lateral spreading. An approximate analytical expression for the ratio is $R/R_p = 1 + M_2/3M_1$. Similarly, a useful simplified estimate of the average impurity concentration within the implant region for a Gaussian type distribution is

$$< C(x) > \, = \, (Total \; dose)/(2.5 \, \Delta R_p)$$

2.2 Damage distributions

To a first approximation the damage distributions are in a similar form to the distribution profile of the implanted ion. However, the peak will be closer to the surface, as the damage (atomic displacements) occurs not only from primary ion–atom collisions but also from the secondary collisions between atoms of the target. The energy required to displace an atom from a lattice site during a dynamic event, in which

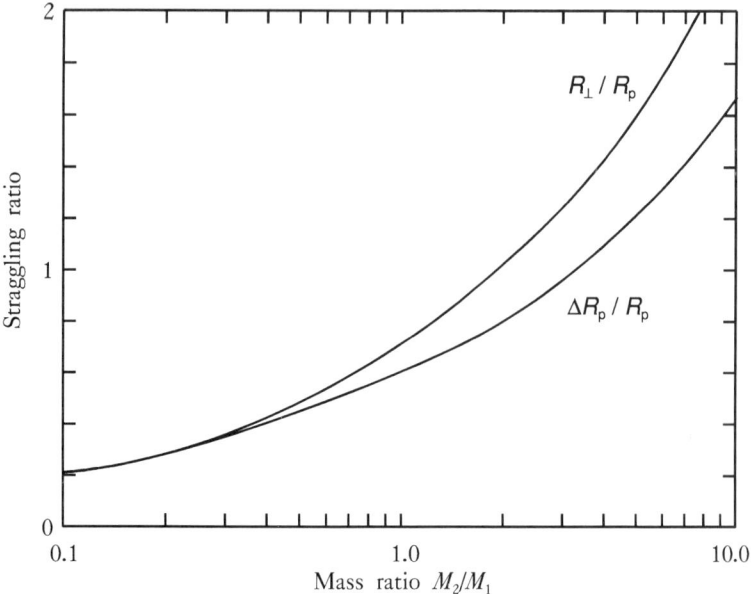

Fig. 2.7. The effects of input ion and target masses (M_1 and M_2) on the range straggling (ΔR_p) and lateral spread ($R_\perp$) relative to the projected range (R_p).

the displaced ions are forced through the unrelaxed lattice, such as an ion–atom collision, is typically quoted to be of the order of 25 eV. For many situations this is a convenient approximation, but classical damage studies indicate that the range can be from, say, 5 eV in some semiconductors up to 75 eV in strongly bonded oxides. Further, in a crystalline target, this displacement energy, E_d, will be direction sensitive, and it will probably decrease with increasing disorder in the lattice, i.e. it will decrease with lower atomic packing density. Energy is transferred to target atoms throughout the ion range, hence displacements will occur even near the surface, whereas the distribution of implanted ions will be peaked at greater depths. Secondary collisions from the displaced target ions will generate more damage, and in the case of heavy ions the collision cascade may involve many hundreds of lattice ions. The passage of the ion will leave a trail of disorder which can subsequently be seen by precipitation of impurities, or chemical etching. Such particle tracks are clearly visible and form familiar features of some minerals, such as mica, which have tracks resulting from uranium impurities. They are referred to as fission tracks.

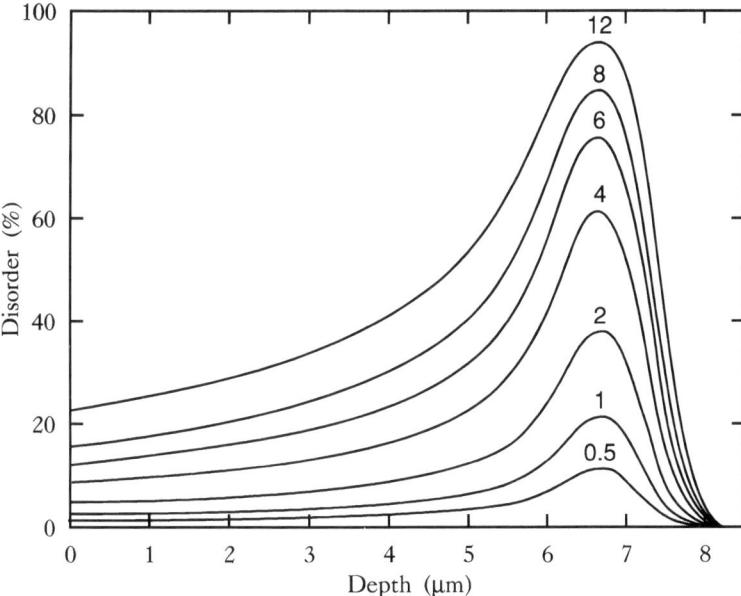

Fig. 2.8. Calculations of lattice disorder generated by 2.5 MeV He ions implanted into quartz. Dose units are for 10^{16} ions/cm^2

There are various estimates of the total numbers of displaced ions (e.g. as summarised by Agullo-Lopez *et al.*, 1988) and these may be related to the maximum energy in a head-on collision and the displacement energy by

$$Total\ displacements\ =\ 1/2\ \ln(T_m/E_d)$$

Examples of calculated damage distributions for different implant dose levels are given in Figure 2.8 for helium ion implants into quartz. The ion energy is 2.5 MeV and the dose units are 10^{16} ions/cm^2. For large ion doses target atoms may be displaced on more than one occasion and in this situation it is useful to quote 'displacements per atom', as was first used for nuclear reactor damage. It is obvious that this estimate only refers to the initial damage situation during the passage of the ion into the material, i.e. in principle there could be complete lattice recovery after the passage of each ion. The components of a collision cascade will include vacancies and interstitials, together with more complex arrangements of the defects. It appears from computer simulations that the core of the cascade will be vacancy rich whereas interstitials will predominantly appear at the outer parts of the track. Nevertheless, this separation of vacancies and interstitials is insufficient to stabilise them and a very large percentage of the damage recovers on the timescale of nanoseconds. Overall, one might assume that implantation of a 250 keV ion could generate some 10^4 displacements, but probably only 10^2 are retained. In the optical work for waveguide formation (Chapter 5) it is desirable to retain the disorder at the end of the ion track, and to do this several strategies have been used. The first is to make the implantations at low temperature. This reduces the diffusion and hence the annealing of defects. Although the damage cascade is not in thermodynamic equilibrium, nor are bulk diffusion coefficients appropriate, cooling the samples does nevertheless inhibit loss of the defects. A second route is to select the orientation of the crystal relative to the incident ion beam, as diffusion coefficients for defects are sensitive to the crystallography. A third alternative is to use impurities which induce a relaxation of the lattice structure and, by producing distortions, stabilise defect structures. An example of waveguide formation in sapphire by carbon implants utilised all three possibilities (Townsend *et al.*, 1990).

2.2.1 Electronic defect formation

A major difference between ion implantation in insulators and metals or semiconductors results from the need to consider the role of energy transferred to the lattice by electronic excitation. In insulators, electronic excitation can be used extremely efficiently to produce point defects and also to induce sputtering. Purely electronic mechanisms can operate, the classic example being that of illumination of alkali halides with exciton light, since this results in halogen displacements and even halogen ejection. The effects have been reasonably well understood since the late 1960s and are summarised in detail in many reviews (e.g. Agullo-Lopez *et al.*, 1988; a special issue of *J. Phys. Chem. Solids* 1990; and *Proc. DIET Conferences*).

Atomic displacement requires both energy and momentum, and for the purely electronic route in alkali halides the energy is derived from the formation of an exciton. This has a relatively long lived excited state. Hence, if the exciton is trapped at a specific lattice site, there is sufficient time for lattice relaxation, and this allows the phonons of the structure to provide the momentum needed for defect migration. In the detailed model for the alkali halides the vacancy and interstitial are separated by a replacement collision cascade of halogens along a $<110>$ crystal

2.2 Damage distributions

direction. Effectively, this can be viewed as the transport of an interstitial which is temporarily packaged within a halogen molecular ion. This specific excitonic route to damage formation is not universal and colour centre formation by purely ionising radiation is negligible in many other insulators, e.g. pure oxides, despite it being highly efficient in many alkali halides.

2.2.2 Displacement threshold effects

One should note that although the displacement energy may be 25 eV for a dynamic displacement in which an atom is forced through the neighbouring sites of an unperturbed lattice, electronic processes, particularly if they involve a long excited state lifetime, allow lattice relaxation and hence a greatly reduced barrier to atomic motion. An effective displacement threshold may thus fall to 5 or 10 eV.

The main alternative model to the exciton route has been one of multiple ionisation, but its efficiency has not been clearly demonstrated. Although ionisation provides adequate energy for a displacement it does not necessarily cause a sufficiently long lived excitation to allow a relaxation or addition of momentum to an excited atom. From the viewpoint of implantation these purely electronic models are interesting as they indicate that there are mechanisms for displacing anions which do not need nuclear collisions. In addition, during ion bombardment there is transfer of both energy and momentum to the lattice from the ion beam. Hence, both the key components are already available in all insulators, and so defect production may not be limited to those materials in which there is a long excited state coupled to a lattice relaxation. Consequently, in insulators the synergistic features of electronic and nuclear collisions may result in defect formation in a wide range of materials. The damage is most probably in the form of isolated point defects on the anion sublattice.

2.2.3 Diffusion, relaxation and amorphisation

The point defects and lattice disorder formed during implantation are not in thermal equilibrium and so there is a high probability of migration within the region of implantation. Attempts to assign an 'equivalent temperature' to the cascades resulting from heavy ion implantation leads to initial values near 10^4 K. This is unhelpful as it merely suggests that plasma-like conditions briefly exist and the critical features

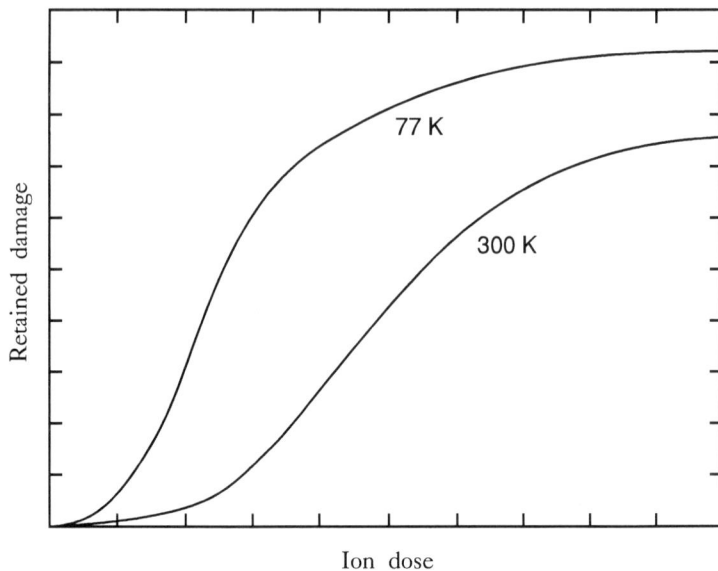

Fig. 2.9. Schematics of how the retained damage may depend on ion dose and target temperature.

of the mechanism of energy dissipation are less predictable. However, if pseudo-thermodynamic situations exist, then defects which require an activation energy, E_f, can be formed with a probability proportional to $\exp(-E_f/kT)$, where T is the effective temperature. This includes defect clusters with E_f values of perhaps 10 eV, i.e. these would have been totally inhibited by any normal thermal treatment. The net effect of this is that in insulators implanted with energetic light ions there will be regions of point defects, primarily on the anion sublattice, near the surface, and more complex defect structures near the end of the ion range. If the damage is retained within the target, then simple production of local disorder will develop into an amorphous region. As damage tracks overlap, the induced disorder in the lattice may help to stabilise the amorphised region.

Damage production is a function of implant temperature and the pre-existing state of lattice disorder. More defects are retained at low temperature, but at all temperatures one typically observes dose dependent changes of the form sketched in Figure 2.9 as a crystalline material becomes amorphous. At the lower doses, mainly point defects exist, but once sufficient disorder is formed by damage, impurities or stress, then

2.2 Damage distributions

the rate of defect production increases rapidly. In part this may reflect a lowering of the displacement threshold energy. At the highest dose levels an 'amorphous' layer develops and so the curves saturate. Further, observations indicate that the structure of this amorphous layer will frequently depend on the implant temperature and the ion species being used. Many properties show this overall dose dependence but it is essential to examine the form of the curve in detail before associating property changes to the same state of amorphisation, as the curves may be displaced along the dose axis by factors of up to 10. Clearly, some features merely require high defect densities to reach saturation, whereas other properties require complete amorphisation. Note also that further attempts to increase the damage levels may result in such high stresses as to induce plastic flow and defect relaxation. Figure 2.10 (Chakoumakos *et al.*, 1991) shows related, but not identical, results for lattice parameters, hardness, elasticity, density, etc., of zircon amorphised by alpha particles. For natural zircon such amorphisation may occur naturally from radioactive inclusions within the zircon crystals. Note that in Figure 2.10(*b*) 'amorphisation', as monitored by the lattice parameter change, has occurred after only 30% of the ions are likely to have been displaced, whereas the density variation requires at least a 100% probability of atomic displacements.

A separate compilation of dose dependent properties with $LiNbO_3$ crystals is shown in Figure 2.11. The $LiNbO_3$ was implanted with 150 keV nitrogen at 300 K. (Götz and Karge, 1983). In Figure 2.11 the volume changes, chemical etch rate, refractive index and lattice disorder (as assessed by RBS) are contrasted. The normalisation of the ordinates disguises the scale of the changes, the most spectacular being the enhancement of hydrofluoric attack by a factor of 1000 compared with the original crystal. The obvious feature is that despite similarity in the response to dose, the curves are distinctly separate. Even more surprising is that both the damage and refractive index data are clearly sensitive to the crystal face which is bombarded. However, the $LiNbO_3$ is particularly instructive as, by contrast with zircon or quartz, the crystal does not readily form a truly amorphous structure. There is lattice disorder, and many defects, but, overall, the presence of the underlying crystal forces some epitaxial regrowth or directionality on the damage layer. Consequently, parameters such as refractive index, or RBS, monitor rather different changes according to the face implanted. Note that these RBS damage studies primarily sensed off-axis changes of Nb ions. Therefore, if there is an axial restructuring along the z axis the X, Y and Z cut crystals will show different responses. As will be

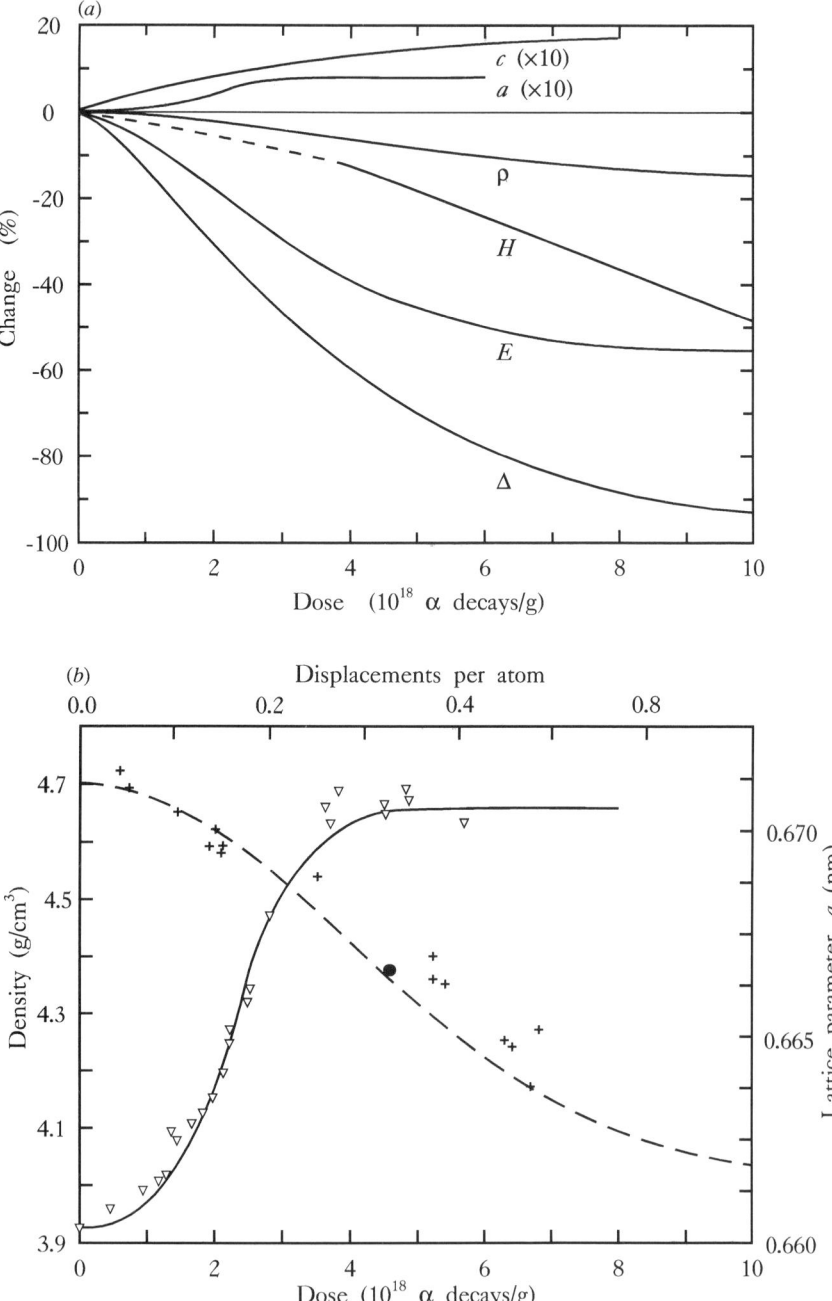

Fig. 2.10. (a) Comparisons of properties of zircon during He irradiation. The properties monitored are lattice parameter (a,c), density (ρ), hardness (H), Young's modulus (E) and birefringence (Δ). (b) An expanded view of the comparison between He damage induced decrease in density (crosses) and expansion of lattice parameter (triangles) with ion dose. The estimated displacements per atom are included.

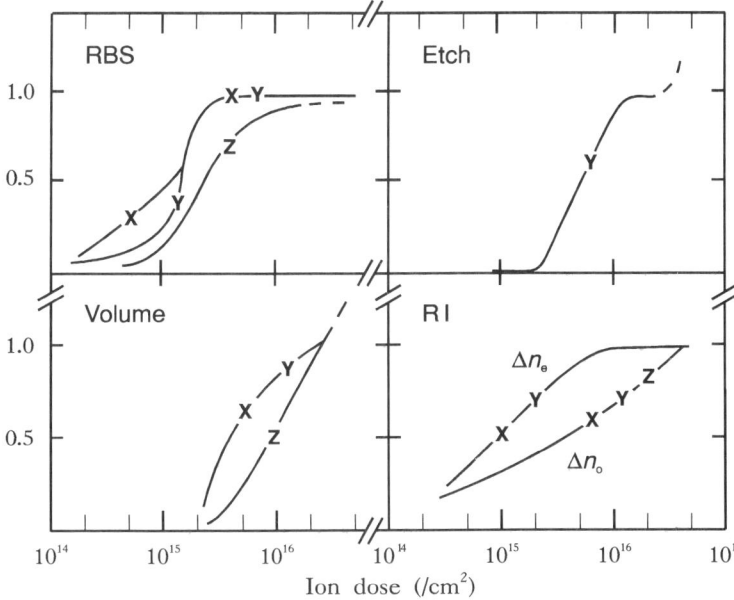

Fig. 2.11. Property changes in crystallinity (RBS), volume expansion, etch rate enhancement and refractive index decrease in LiNbO$_3$ during N$^+$ ion implantation. Note the effects are sensitive to the crystal face being implanted.

mentioned in Chapter 5, there is even greater complexity in the depth profiles of the refractive indices, but an isotropic value that is consistent with amorphisation is never reached.

Destefanis et al. (1979) attempted to describe the latter part of this pseudo-amorphisation process in terms of the energy deposited by nuclear collisions and a single displacement energy, E_d. Figure 2.12 shows that at room temperature, and for low ion doses, the data fit a 'universal' curve for all the ions used if the abscissa is plotted in units of energy derived from nuclear collisions. The curve can be fitted with a function of the form

$$\Delta A = A_{sat.} \left[1 - \exp\left(\frac{-0.4 \, v(E) \, D}{E_d \, s \, e} \right) \right] \quad (2.21)$$

in which ΔA is the fraction of lattice amorphised, $A_{sat.}$ is the equilibrium value, $v(E)$ is the energy in nuclear collision events, D is the ion dose and s and e are atomic density and width of the damage layer. The fit gives a displacement energy average of 32 eV, which is of the order expected for oxides. Saturation occurs by some 10^{22} keV/cm^3 which similarly

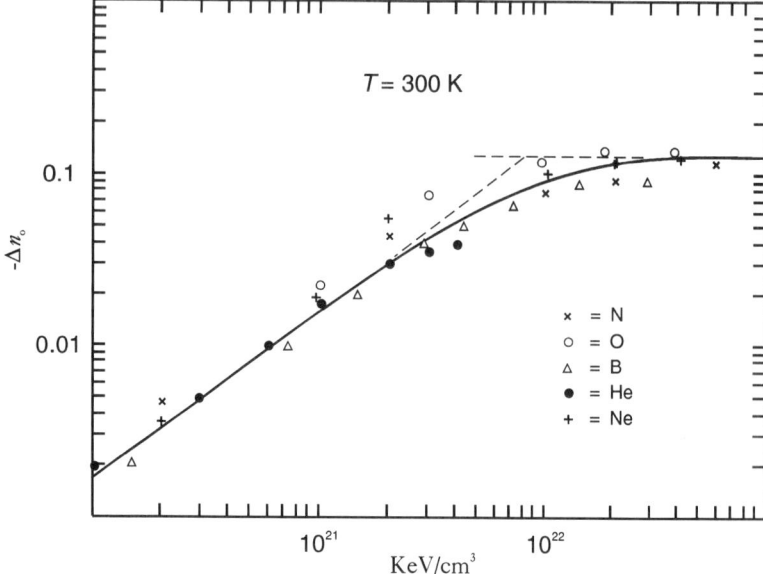

Fig. 2.12. Evidence that the reduction in refractive index of $LiNbO_3$ is primarily linked to the energy deposited in nuclear collisions as described by Equation (2.21).

compares favourably with other quoted measurements of 10^{20} keV/cm^3 for silica (Norris and EerNisse, 1974), 10^{21} keV/cm^3 for Si_3N_4 (EerNisse, 1974) or 5×10^{21} keV/cm^3 for MgO and Al_2O_3 (Krefft et al., 1975; Krefft and EerNisse, 1978; and Arnold et al., 1974).

Figure 2.12 also indicates that the saturation value depends on the dopant concentration. Chandler et al. (1986) indicate that the degree of amorphisation is modified by the depth of the damage layer beneath the surface.

The second factor which influences the diffusion, relaxation and amorphisation, is the development of stress. Typically this builds up to a critical level and is followed by a collapse with consequent plastic flow, precipitation of defects and relaxations. Figure 2.13 shows Arnold's (1988) data for the stress which develops during ion implantation of Na—Mg—Al—Ca—silicate glass. For the silicate example the stress produces a compaction of the glass network and there are contributions from both electronic and collision processes. However, similar examples exist for many crystals; one such, for colour centre production, is shown later in Chapter 3.

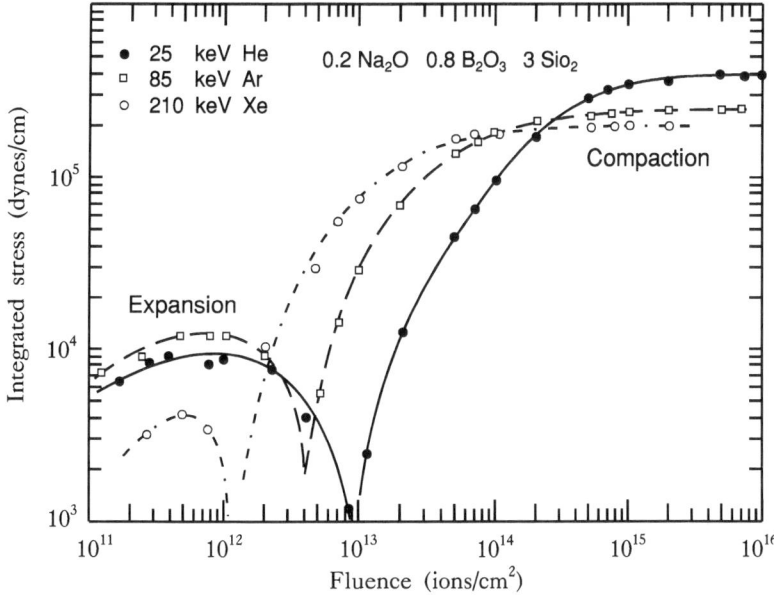

Fig. 2.13. The effect of nuclear collision damage on the build up of stress in ion bombarded glass. Note that both expansion and compaction are possible consequences.

As indicated above, the energy deposited to amorphise a lattice is dependent on the type of bonding and the retention of displacements (Naguib and Kelly, 1975). Burnett and Page (1986) sketched a curve which relates these parameters, Figure 2.14. This offers a useful first guide to the ion dose required but does not include temperature, directional or impurity effects. There are also some differences between the Burnett and Page numbers and those quoted elsewhere. It is obvious that since amorphisation is associated with large lattice parameter changes, and therefore stress, the ion dose for amorphisation will vary with the depth of the amorphisation beneath the surface. The extensive review of amorphisation of surfaces of oxides by White *et al.* (1989) provides some scale for the ion doses required to amorphise the crystals, rather than in 'universal' units of energy/cm^3. For a particularly stable material, sapphire, the room temperature amorphisation dose is some 2×10^{17} ions/cm^2 for 150 keV Cr^+ ions, but this falls to as low as 2×10^{15} ions/cm^2 when the implant is made at 77 K. This is probably an extreme example, in part because chromium has the same valence as the aluminium, and chromium ions only introduce a small degree of

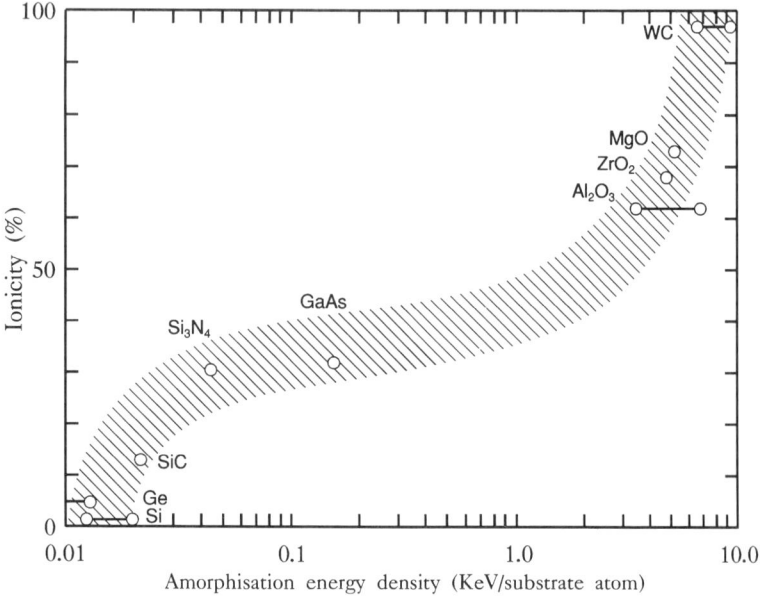

Fig. 2.14. A schematic of how the nuclear collision energy required to amorphise a crystal lattice depends on the ionicity of the lattice.

lattice mismatch if they move to lattice sites (i.e. as in ruby). Chemically stabilised disorder eases the process of amorphisation and, for sapphire, implants with group IV elements (C or Si, and also Zr) require a factor of 10 less dose than comparable mass ions of different valency. One suggestion for this mechanism is that the carbon forms carbide bonds by substitution on to the oxide sublattice. These so distort the rhombohedral sapphire lattice that further damage is readily retained. In summary, the energy needed for amorphisation is primarily governed by retention of defects, rather than difficulty in displacing lattice atoms.

Light ions such as helium should in principle require far greater ion doses as they deliver less nuclear collision energy than heavy ions. Hence, to amorphise sapphire the calculated doses might be estimated to be some 10^{18} He$^+$ions/cm^2. However, at these dose levels the composition has so changed that the target is no longer sapphire. Further, with an inert gas, there will be bubble formation which will greatly distort the lattice. In attempts to amorphise sapphire with helium at high doses a problem occurs whereby the gas bubbles form a layer which causes the surface to peel or eject entire surface blocks (this is termed blistering or exfoliation).

2.2 Damage distributions

In other crystals, such as LiNbO$_3$, light ion implantation never produces amorphisation, as indicated by the fact that the two refractive indices do not reach a common value. This suggests that the damage only generates a highly disordered LiNbO$_3$/He layer. One should emphasise that with many materials helium ion doses of, say, 4×10^{16} ions/cm^2 readily induce amorphisation. Such examples include quartz and zircon (paratellurite). Not only are these materials less ionic but, at least for the first two, they exist naturally in an amorphous phase. One therefore suspects that such factors must be considered in attempting to amorphise crystalline lattices.

2.2.4 Stability of amorphised layers

The stability of the damage is revealed during annealing treatments. In multi-component crystals the various sublattices may recover damage or amorphisation at different rates. In general, isolated point defects anneal at lower temperatures than more gross damage regions, and for the extreme examples of amorphous phases (e.g. silica in quartz) stability may exist well above 1000°C. Evidence for these statements is often provided by the RBS analyses of the damage depth distributions. The annealing curves are sensitive to the total damage, the presence of impurities and neighbouring layers which act as epitaxial growth planes. Figure 2.15, from Destefanis *et al.* (1979), gives details of the recovery of refractive index changes generated in LiNbO$_3$ by a variety of ion species and ion doses. The key point to recognise is that none of the curves have the same form, and, at least at first sight, they scarcely seem to involve the same annealing stages. Related problems are mentioned in Chapter 3, where they are linked to preferential diffusion, phase precipitation and changes in charge state.

2.2.5 Amorphisation of semiconductors

Amorphisation may be advantageous or a problem, depending on the application. Diamond and silicon are covalently bonded so are easily amorphised. In the case of silicon this is a useful step in the formation of semiconductor devices, since amorphisation with a silicon beam allows subsequent doping to be made into an amorphous layer. This avoids problems of channelling and lack of control on the dopant profile. For silicon, the amorphous layer can be thermally recrystallised. By contrast, implants in diamond readily form amorphous carbon, or possibly

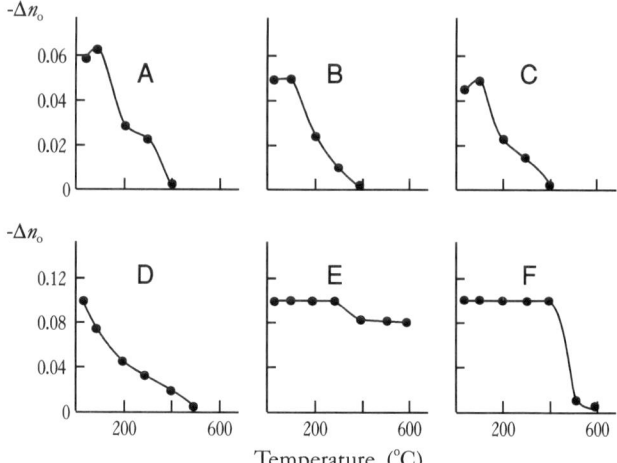

Fig. 2.15. Examples of thermal stability of refractive index changes produced by different ion species implanted into LiNbO$_3$.

graphite, and recrystallisation is not possible. There have been numerous attempts to avoid this difficulty, both to grow large diamonds by carbon implants into seeds, and to exploit semiconducting properties of diamond. Partial success has been reported for both objectives by implanting at high temperatures with careful control of the ion energy and dose rate. Rusbridge and Nelson (1980) used carbon implants between 600 and 900 °C, whereas Braunstein and Kalish (1981, 1983) went to higher implant temperatures followed by annealing to 1400 °C. For their semiconducting diamond they used dopants of Sb, P and Li. Optically, the samples became darker if a graphite layer developed.

The earlier discussion on amorphisation concentrated on ionic insulators in which in all cases the amorphisation of a crystalline lattice resulted in an amorphous layer of reduced density. Potentially, densification as a result of amorphisation should not be ignored. Several examples exist including damage of silica and silicate glasses which causes relaxation into a more compact glass. At first sight, a similar effect should be considered for silicon and germanium as ion beam damage of these covalent crystals increases the refractive index of both these semiconductors as they become amorphous. However, it should be noted that there are many factors which control the refractive index and an increase in index in a semiconductor need not imply an increase in density. Indeed, the Naval Research Laboratory studies (Hubler *et al.*, 1979a,b, 1984; Fredrickson

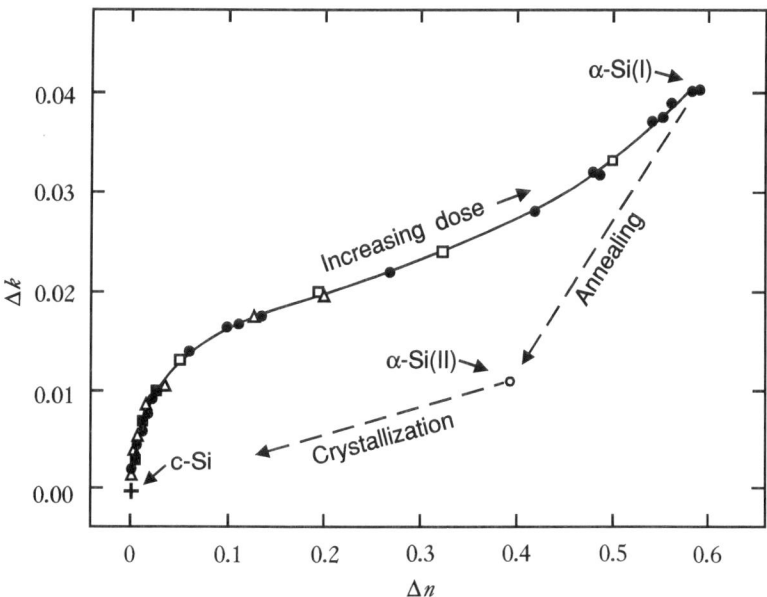

Fig. 2.16. Growth and annealing of index and absorption changes in silicon caused by Ge ion implants. Note at the highest doses the silicon is amorphised.

et al., 1982; Waddell et al., 1982) showed that the density fell by 2% in each case whilst the index increased by 8% for Ge and 11% for Si. Annealing revealed two stable amorphous forms and so the increase was attributed to terms due to a less stable feature from broken bonds, a more stable feature from an extended band edge absorption tail, and, finally, a 20% enhancement in bond polarisability.

These changes in index, optical absorption and types of amorphisation were seen by Heidemann (1981). He noted that the index variations followed the nuclear collision damage depth profile. On separating the real (n) and imaginary (k) parts of the refractive index, $N = n + ik$, for silicon damaged with 4 MeV Ge$^+$ ions, he recorded the changes during annealing. Data shown in Figure 2.16 clearly demonstrate both a knee in the growth phase and a two-part annealing curve.

2.2.6 Stability of point and cluster defects

The difference between the point defects from electronic damage and the larger defect structures and/or point defects on both anion and

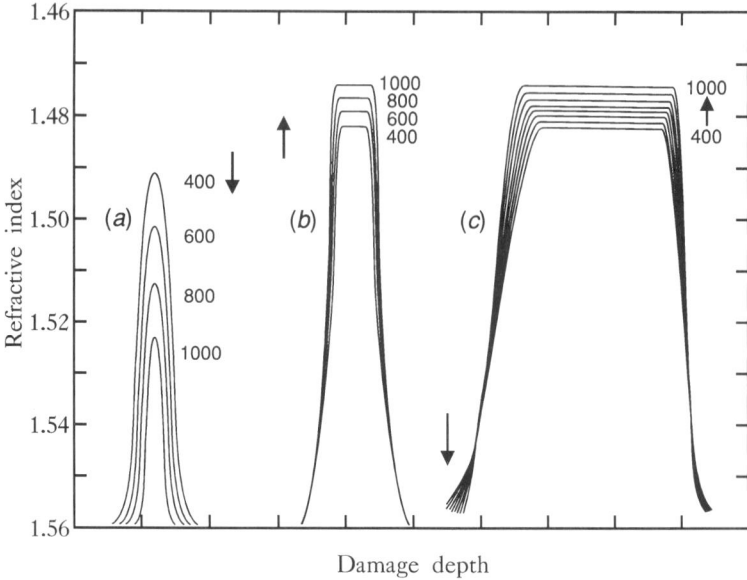

Fig. 2.17. Stability of damage in He implanted quartz. In (a) the quartz is primarily crystalline so there is recovery of the lattice perfection during annealing to 1000 °C. (b) Amorphised silica relaxes into a lower density version. (c) For a broad region with long tails, then both effects occur simultaneously.

cation sublattices will result in different thermal stability. Since the general rule is that point defects are more mobile than larger entities, they will anneal at lower temperatures. There are thus many examples in which annealing of defects is non-uniform throughout the implanted layer. Early measurements of refractive index changes in crystalline quartz (Presby and Brown, 1974) recorded a faster recovery of electronic damage. Subsequent data of Chandler et al., 1990 provide more detail, and, as shown in Figure 2.17, they report that the 'purely' electronic damage near the surface anneals by 200 °C, whereas the amorphised material (i.e. silica glass) does not reconvert into crystalline quartz even at temperatures as high as 1200 °C. Figure 2.17 indicates that annealing behaviour is also strongly related to the degree of damage. The data show that if the damage, as assessed by the index decrease, is less than the amorphisation level, as in section (a) of Figure 2.17, then all the lattice recovers as the temperature is raised towards 1000 °C. This is as expected if the glassy silica reverts to crystalline quartz. A narrow region of amorphised material decreases in index as it relaxes into the annealed variant of silica, curves (b). More surprising is that we see in section (c) of Figure 2.17 that for a broad damage region, which includes both tails of lower damage concentration and a central amorphised region, the tails recover with annealing, but the amorphised core relaxes into the lower density glass.

A related example of the difference in annealing behaviour between electronic and collision damage regions is given by index changes which take place during ion implantation of silica glass. As implied in the quartz data, silica glass exists over a range of densities, and hence, refractive indices. When manufactured it is in a lower density version. Addition of energy by either electronic or collision processes induces a structural relaxation into the higher density versions, presumably aided by the defects which are formed. Figure 2.18, curve A, shows that after nitrogen ion implantation (with 10^{16} ions/cm^2 at 0.18 MeV) the density, and refractive index, increase throughout the range and saturate at the value of compacted silica (Faik *et al.*, 1986). However, after annealing at 450 °C, the density of the layers closest to the surface is reduced to give the index profile labelled B. The calculated damage profile, curve C, is in close agreement with this and, clearly, the distribution of the nitrogen, curve D, does not determine the index profile.

2.2.7 Defect diffusion and crystallography

Diffusion of defects is not isotropic in crystalline materials and so implants into different crystal faces will not be identical in terms of the defects retained, since in some cases the directions of rapid diffusion will lie in the plane of the damage production, whereas in others the rapid diffusion can separate the defects from the region of the implant. As mentioned above, Götz and Karge (1983) compared the effects of nitrogen ion implantation into the three non-equivalent faces, X,Y and Z of LiNbO$_3$. Their results, Figure 2.11, can be described in broad terms as evidence for an accumulation of defects in the plane of the projected range, which at a sufficiently high concentration leads to a pseudo-amorphous layer. However, the retention of defects is governed in part by diffusion of displaced atoms along crystal channels, hence the dependence on crystal cut.

Crystal axis effects have also been documented in studies of silver colloid formation during implantation in LiNbO$_3$, Al$_2$O$_3$ and quartz (Rahmani *et al.*, 1988, 1989). Again, the rate of colloidal growth is sensitive to the face implanted and is interpreted in terms of fast diffusion

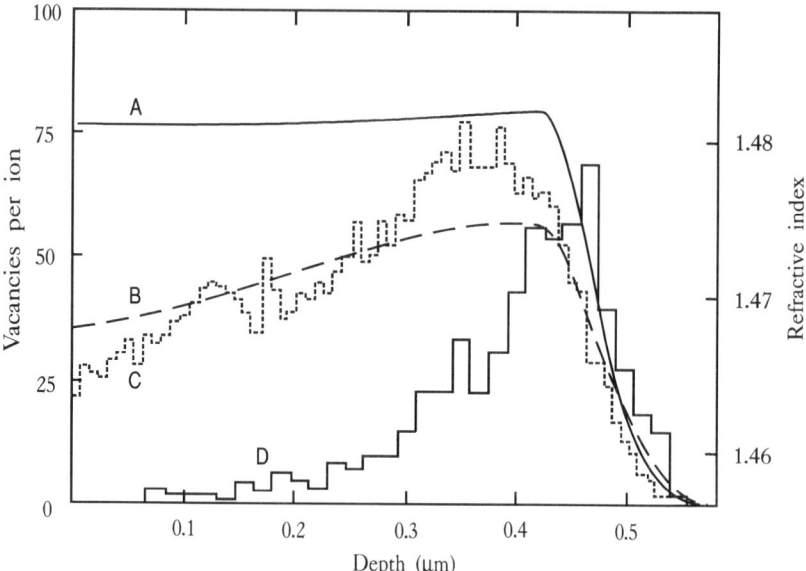

Fig. 2.18. Densification of silica glass by a combination of electronic and nuclear collision damage using N^+ ions. Curve A is the index profile after implantation, B is after annealing of the electronically induced damage, C is a computer prediction of the nuclear collision effects. The implant distribution is distinctly different, as seen in curve D.

along the z axis of the materials. Not only are anisotropic effects apparent during the implant, but also the subsequent stability of the defects depends on the face implanted, as shown by Figure 2.19.

Directional effects which influence intrinsic colour centre formation during ion implantation are reported in the sapphire lattice by Dalal et al. (1988) for optical absorption, and by Al Ghamdi et al. (1990) for luminescence. A combination of directional damage sensitivity, defect retention with low temperature implants and the use of chemical defect stabilisation were effective in making optical waveguides in sapphire (Townsend et al., 1990).

2.2.8 Structural and compositional changes

Intuitively, it is not unexpected that the deposition of energy, and the formation of defects, can induce relaxation during implantation and hence lead to glass structures of different density. Less obviously, it may also lead to greater perfection in crystalline structures. Haycock (1988, un-

2.2 Damage distributions

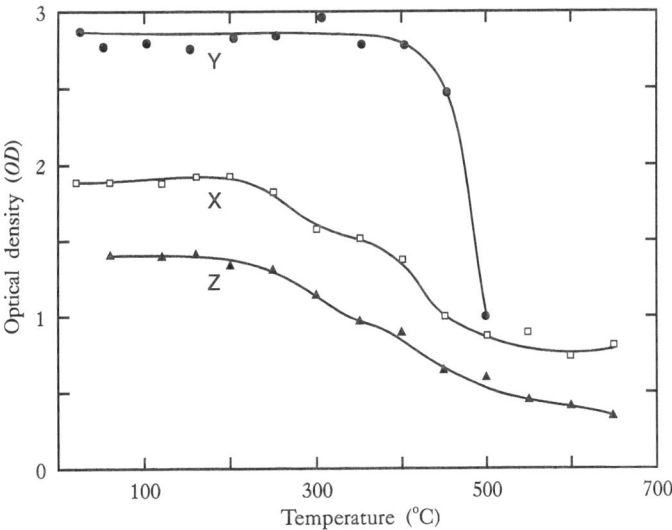

Fig. 2.19. The stability of optical density changes caused by Ag$^+$ implants in LiNbO$_3$ are a function of the face implanted. All samples received the same ion dose and annealing treatments.

published) has noted that the LiNbO$_3$ lattice shows less microcrystallinity after helium ion implantation. This is a non-stoichiometric crystal and so contains growth defects and stresses. The passage of the ions produces excited states so that local relaxations can occur which lead to an improvement in crystallinity. Refractive index measurements of this material (see Chapter 6) show that the implantation damage produces quite different index profiles for the ordinary and extraordinary indices. The reasons for this are interpreted (Zhang et al., 1991,a,b) in part as arising from preferential diffusion of lithium ions within the non-stoichiometric structure.

A frequently cited example of compositional readjustments resulting from ion implantation is that of alkali migration within a multi-element glass. The effects occur not only as a result of ion implantation, but are equally apparent during electron irradiation of glasses. Consequently, the process is not related to collision cascades or the more dramatic damage production caused by the ion beams. Nevertheless, radiation damage can greatly enhance diffusion rates such that chemical rearrangements may be more obvious after ion implantation of glass. The driving force for the preferential alkali diffusion is thought to be the development of electric fields originating from secondary electron ejection from the surface and/or injection of charged species. The problem of such field production has been considered by Cazaux (1986) but, although the development of electric fields is almost inevitable during implantation of insulators it is extremely difficult to predict their magnitude and even their sign. For a surface which is not grounded, and with incident beam currents of microamps, the development of field strengths of 10^3–10^5 volts/cm may occur. Direct evidence for surface charging can be observed by sparking across a surface. Alternatively, charging effects are apparent during RBS analysis of insulators in which the energy of the reflected ions can be shifted by as much as 50 keV from the expected energy value. To a large extent the problem can be overcome by surface potential stabilisation during implantation using secondary electron suppression or a grounding contact layer.

Nevertheless, there are many examples of compositional changes caused by implantation. Numerous cases have been noted by the Italian group (e.g. Chinellato *et al.*, 1982; Arnold, 1982; Mazzoldi, 1983), and by Bach and Hallwig (1984); and have been reviewed by Mazzoldi and Arnold (1987). Typically, the alkali ions separate from the glass matrix and diffuse towards the surface. They may even then leave the structure by preferential sputtering. If heavier alkaline earth elements are included in the glass, e.g. Ca, these may move into the ion damaged volume. The problem is exacerbated if the implant ion is likely to take part in such ion exchange reactions (e.g. protons).

Transport of impurities or defects, or energy in the form of excitons which in turn produce defects, can thus arise for a variety of reasons. Further examples are cited in the introductory chapter and later chapters on optical absorption (Chapter 3), luminescence (Chapter 4), refractive index measurements (Chapter 5) and from chemical etch rates (Webb *et al.*, 1976) and depth profiling.

2.2.9 Conclusions on damage distributions

The overall conclusion is that although a detailed assessment may be made of how energy is transferred to a target material, it would be unreasonable to expect any computer modelling of the process to be able to predict the final damage distribution. This point is often ignored and it is not uncommon for authors to cite the computer calculation of the damage distribution as though this is the situation which is retained by

2.3 Channelling

the target, whereas in fact it is only an indication of the initial starting point during the first few picoseconds.

2.3 Channelling

In the discussions of stopping power we have tacitly assumed that crystallographic factors could be ignored, and in practice this is broadly correct if the dose is high or the implant ion is heavy. However, it has been realised for a very long time (Stark, 1912) that the interactions between an incoming ion and a lattice structure will be minimised if the ions are directed down a crystallographic axis or plane. Not only will the ions travel in a potential minimum, but also, small deviations from the channel direction will be corrected by the potential wall of the channel. Figure 2.20 contrasts three trajectories, A, B, and C in which there are random collision scattering events, partial channelling until the oscillations in the channel reach a critical angle, and finally a long range path resulting from channel guidance. The topic has received considerable attention (e.g. Morgan, 1973; Gemmell, 1974; Townsend et al., 1976; Feldman et al., 1982; Feldman and Poate, 1982) as the net result of careful alignment between the beam direction and a channel is to give an enhanced penetration into the solid which can be as much as $5R_p$. Positive features of the channelling are that for a very small ion dose, before radiation damage has caused channel distortions and de-channelling, one could implant impurities deep into the crystal with reduced energy. A second more valuable feature is that in channelling studies of ion backscattering (i.e. RBS) the clarity of the channels gives a measure of imperfection in the crystal. If there are impurities in the lattice, then channelling data from several axes can frequently offer positive identification of the impurity location. Standard texts are Chu et al. (1978), or Feldman and collaborators (1982, 1987).

On the negative side, it is clearly undesirable to allow channelling to occur if one wishes to have an accurately tailored impurity profile in the crystal. This can be a major problem in semiconductor production as the ion doses tend to be small and the materials, e.g. Si or GaAs, are extremely high quality crystals. The unwanted doping at depths beyond R_p degrades the dopant profile. The two standard approaches to overcome channelling are to implant at, say, 7° off the main crystal axes, or to use a large dose of self-ions, e.g. Si^+ into Si, to amorphise the target. After the impurity implants have been made the amorphous material is rapidly annealed to regenerate the crystalline structure. Very

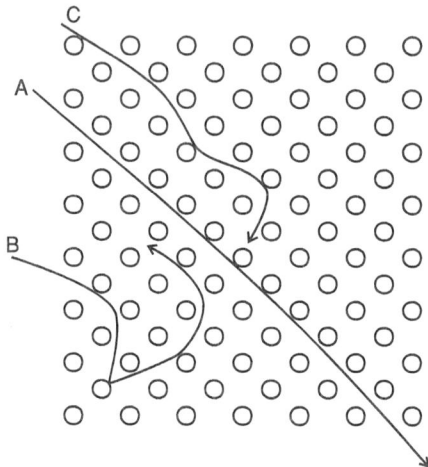

Fig. 2.20. Examples of A a long range channelled ion implanted path, B a random short path and C a path in which the ion is initially channelled.

rapid annealing is successful as the epitaxial regrowth of the crystal proceeds faster than the broadening of the impurity profile by normal thermal diffusion.

Although channelling poses a number of problems it should be realised that the onset of de-channelling occurs by a misalignment of about 1° for axial channels and 0.25° for planar ones. Consequently, large ion doses of light ions, such as 10^{16} ions/cm^2, or low doses, 10^{15} ions/cm^2, of heavier ions, will introduce sufficient lattice disorder to destroy channelling possibilities. Nevertheless, for crystalline targets the possibility of channelling should always be considered. Experimentally, channelling effects may be difficult to distinguish from the features of rapid diffusion of defects or impurities along open crystallographic directions.

In Figure 2.11 various parameters were compared as a function of ion dose for a set of LiNbO$_3$ crystals, including an assessment of displaced Nb by noting changes in the channelling signal. Ideally, the channelling signal can be used to distinguish the disorder on each sublattice, either by the energy of the backscattered ions, or by detection of characteristic X-ray emission or by a nuclear reaction. In an example by Newton and Hay (1981) a NaCl crystal was damaged with 1 MeV He$^+$ ions and simultaneous measurements were made for the optical absorption to note the halogen vacancy F centres, RBS to detect overall lattice disorder, and production of chlorine X-rays from off-axis halogens. Figure 2.21 shows

2.4 Sputtering

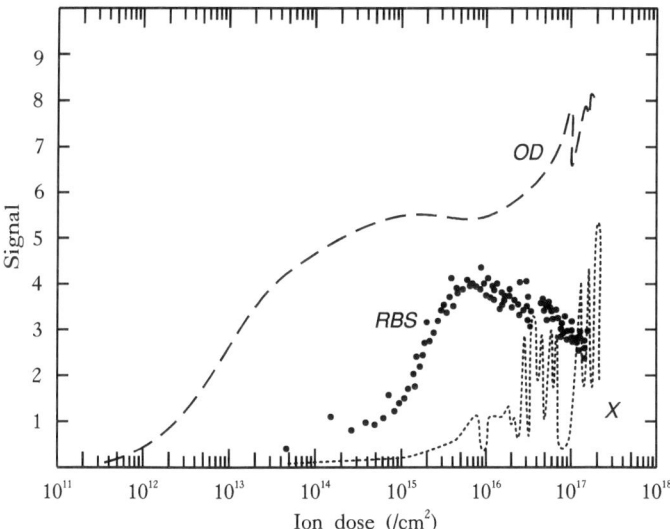

Fig. 2.21. A comparison of optical density (*OD*), halogen X-ray production (*X*) and lattice disorder (*RBS*) as a function of He$^+$ damage in NaCl.

the dose dependencies of these three signals. The most obvious feature is that the three signals do not change identically with dose. Instead, one may suggest that there is a rapid saturation of the basic colour centres; when the dose is increased by a further factor of 20 the lattice damage similarly saturates, and then relaxes as a result of the stress; a further factor of 10 in ion dose is needed to generate a significant interstitial chlorine signal. The interstitial chlorine signal arises from precipitation of halogens into clusters or interplanar sheets. The broad features of this interpretation are consistent with earlier comments in this chapter, and further examples will be presented in later chapters.

2.4 Sputtering

Deposition of energy into the surface of a material not only causes atomic displacements, but also leads to ejection of ions and neutrals from the surface. Inevitably, the sputtering yield, that is, the number of ejected atoms per incident ion, is a function of the incoming ion energy E_0 and the masses of the ion and target atom M_1, M_2. Further, it is a function of the depth of the energy deposition so the yield varies strongly with angle of incidence, as shown in Figure 2.22. This demonstrates how this yield

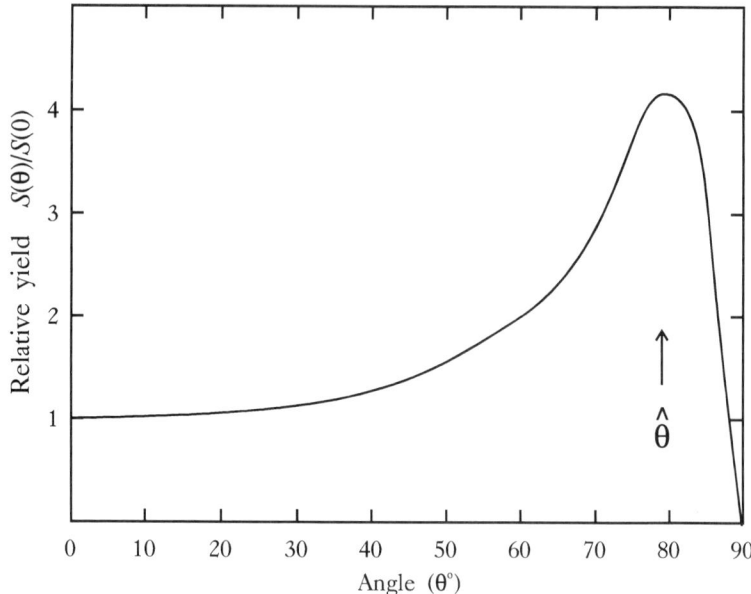

Fig. 2.22. The relative sputtering yield as a function of incidence angle.

varies with incidence angle. At the lower angle values, i.e. close to normal incidence, the curve shape approximately follows the $\sec\theta$ function.

The original theoretical descriptions such as those of Thompson (1968) and Sigmund (1969, 1972) were made in terms of single element targets. The predictions were for idealised systems and describe the initial sputtering yields from perfect amorphous solids caused by the deposition of energy via nuclear collision processes. This lead to some apparent conflict with experimental data as, for experimental reasons, it is usually simpler to measure the steady state or large ion dose sputtering yields. These are really a measure of the yield from a new compound formed by additions of the implants into the target, not the idealised situation. However, for most applications this is probably a more useful yield value. Figures 2.23 and 2.24 illustrate typical predicted sputtering yields for M_1 and M_2 combinations, as a function of energy E_1. In Figure 2.23 the calculation is for a variety of ions incident on an aluminium target. Figure 2.24 considers how changes in the incident M_1 influence the sputtering of copper, relative to Cu self-sputtering.

One consequence of the sputtering is that there is an upper limit to the number of impurity ions which can be implanted, and retained, within a

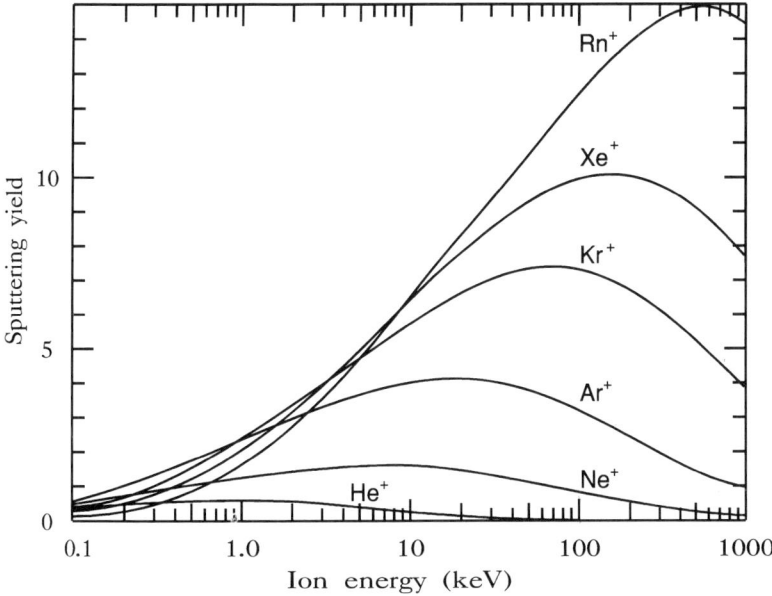

Fig. 2.23. Calculated sputtering yields for ions incident on aluminium.

target. Figure 2.25 shows how such impurity profiles approach saturation as the surface recedes during implantation. In the case of semiconductor implants with ion dose levels of only 10^{12}–10^{13} ions/cm^2 and energies such as 100 keV, then the sputtering will only remove a few monolayers of the surface. In practice this may act as a cleaning procedure to remove stray surface contaminants. At the other extreme, implants in metals often require doses of 10^{17}–10^{18} ions/cm^2 to form surface alloys. Such alloy concentrations may be self-limited by an erosion which keeps pace with the formation of new material.

Insulator studies, so far, generally lie within these extreme limits, and so must be viewed carefully to decide if sputtering is a problem, or just a convenient method of maintaining a clean surface during implantation. Factors which need to be noted, and were not relevant for the initial theory, are that in multi-element compounds the sputtering yield is not the same for all elements. If one assumes that the sputtering cascade is in thermal equilibrium, then the lighter ions are travelling faster and so are more likely to escape from the surface. Hence, multi-element systems are most likely to be depleted by the lighter elements. The situation is slightly

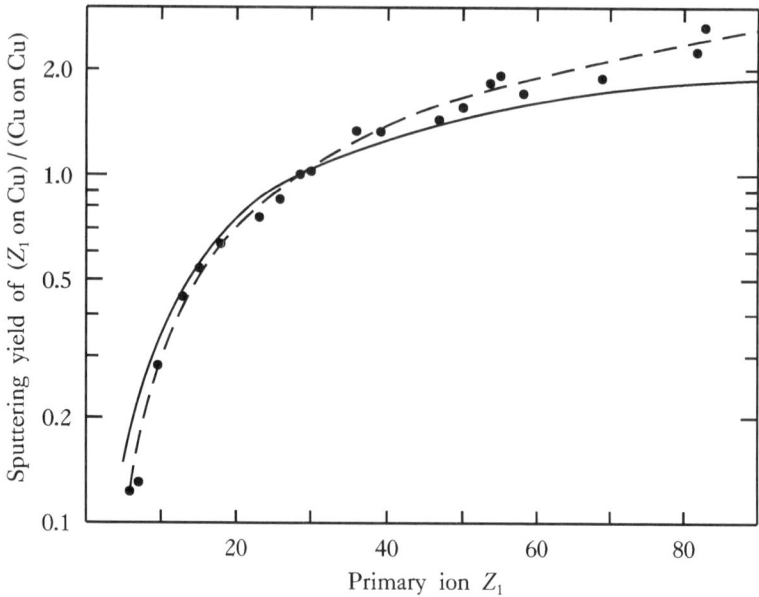

Fig. 2.24. Dependence of the copper sputtering yield, relative to self-sputtering, as a function of Z_1. The data are compared with the theoretical predictions of Sigmund (1972).

more complicated as gaseous elements, such as oxygen or nitrogen, tend to be lost more readily than metallic elements of comparable mass.

The majority of theories concerned with ion beam sputtering are concerned solely with the energy deposited by nuclear collisions. They are thus well suited to metal sputtering. However, for insulators the deposition of electronic excitation energy is a significant factor both in the production of point defects and stimulated desorption (i.e. sputtering) of species from the surface. Since electronic damage mechanisms invariably only lead to displacements of anions, and surface loss increases for gaseous ions, there may be additional mechanisms which sputter oxygen, halides or other volatile anions. In extreme cases the surface layers may be totally depleted of one element. Ion implantation in weakly bonded Van der Waals systems, such as the ices of inert gases, destroys the material primarily by electronic mechanisms and the sputtering yields are quoted in hundreds of ejected atoms per incident ion. The range of yields in systems which are predominantly sputtered by nuclear energy deposition can be much lower than for similar compounds which include an electronic mechanism. A fairly extreme example attributed to these

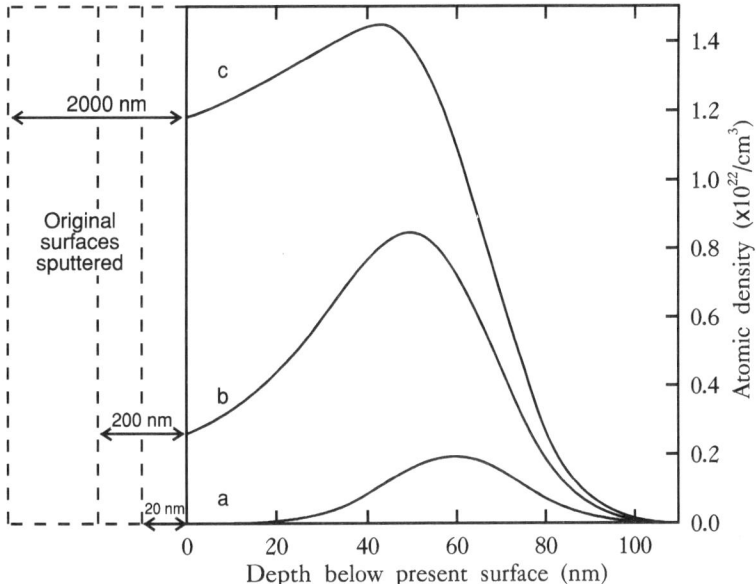

Fig. 2.25. Ion implanted impurity distribution as a function of dose in a case where there is a significant sputtering yield. The calculation is for Kr$^+$ doses of 1.5×10^{16}, 10^{17} and 10^{18} ions/cm^2 at an energy of 100 keV into SiO$_2$. The dashed lines indicate the extent of removal of the original surface.

differences occurs in the comparison of UO$_2$ and UF$_4$ sputtering. Their yields differ from 0.06 to 200 for the same ion beam conditions (Lama, 1986; Tombrello, 1984).

Fortunately, for many of the optical applications the depth of penetration of the ions is of the order of microns and in order to achieve this light ions of very high energy are used, e.g. MeV He$^+$. Consequently, the nuclear collision route to sputtering is suppressed and only the more subtle electronic processes determine the sputtering.

Low energy ion beam sputtering is a valuable method of shaping surfaces of insulating materials and the details of how the topography evolves during beam incidence from different angles, or with non-flat surfaces, has been reviewed in several books (e.g. Townsend *et al.*, 1976; Auciello and Kelly, 1984; Mazzoldi and Arnold, 1987).

2.5 Computer simulations

Ion range and damage distribution predictions are essential for ion implantation and a wide variety of computer simulations have been developed. An inherent problem is that the generation of a realistic simulation of all the events that occur as an energetic ion travels into a target requires a detailed knowledge of the types of interaction between the ion and target electrons and atoms. Ideally, the simulation should take account of the simultaneous motion of all the displaced atoms as well as the incoming ion and follow their motion until the material relaxes down to thermal energies. It should also include the possibility of annealing of the damage, sputtering from the surface, stabilisation of defect complexes and formation of new compounds. To include all these possibilities is beyond our present understanding of defect and implantation processes so the simulations normally only address the problem of range distributions and initial damage creation. Problems can be minimised for amorphous targets but the difficulties increase with a crystal lattice, since not only will the ion trajectories be steered by the planes and channels of the lattice, but, additionally, the separation of vacancies and interstitials will be sensitive to the diffusion rates along crystalline directions. Even if one could specify the full details of the problem the computational time would be prohibitively long. Consequently, all the computer algorithms used are aimed at making acceptable predictions within realistic, and affordable, timescales. Many of these simulations are well refined and widely used, to the point that they are quoted as being totally reliable. However, they are only a guide and this fact should not be forgotten as there is a temptation to assume that computer simulations are infallible. Some of the more widely used programs, have been upgraded continuously over the years from their original versions, e.g. the TRIM (TRansport of Ions in Matter) of Biersack and Haggmark (1976), has been modified both by Biersack and his co-workers, and by owners of copies of the program. Unfortunately, many users of the code just cite a 'TRIM calculation', without specifiying the version used. This has sometimes led to apparent anomalies in range or lateral spread of the implanted ions which far exceed the normal 10% accuracy expected from a modern version of TRIM.

Computer simulations of ion ranges have been reviewed at regular intervals and include excellent items by Biersack and Ziegler (1982), Ziegler (1984) and Webb (1992). The problems of sputtering and computer simulation in general include other books or reviews such as those of Behrisch and Wittmaack (1991), Harrison (1983, 1988), Robinson (1981), Andersen (1987), Frenkel (1989) and Eckstein (1991).

2.5.1 Simulation approaches

Broadly, there are two classes of approach. These are to either simulate the trajectories of the ions and displaced lattice ions or to use analytical expressions to encompass overall transport properties. Trajectory simulation is conceptually simple in that one specifies an ion of a given energy spatial position and direction, defines an interaction potential and computes the result of the collision on the primary and struck ion. This is a simple binary collision which, for an isolated pair of ions, is soluble. For a solid target the reality is rather different. One begins by defining the matrix of an amorphous or crystalline target and the relative point of impact with respect to the target. Even if the first collision is clearly defined, the trajectories of both ions will cause further collisions and in turn develop a cascade of moving ions. The predicted outcome of the range and disorder will obviously depend on the sequence in which each binary pair is considered. Simultaneous collisions with several ions will similarly present difficulties and may be excluded in some algorithms. Algorithms thus vary in how they treat the sequence of collisions, for example some first compute the secondary displacements from the fastest moving ions and then follow the paths of the slower moving ones into the modified lattice. The sequence in which each binary collision should be considered is a matter of opinion and the various algorithms vary in their 'event store codes'. For a localised event it will clearly matter if one first computes the movement of the incoming ion and then the target ions, or follows the fastest moving ion, or the slowest moving ions, or initially ignores ions scattered back towards the surface, but it is assumed that for a large number of randomly generated events the errors will not seriously bias the form of the overall distributions. However, this assumption cannot be fully justified and consequently the various algorithms will lead to variations in predictions of the range and damage distributions, and accuracy may be lost in order to run more rapid calculations.

Programs based on the TRIM code are widely used for amorphous targets and low ion doses; they include interaction potentials based on a very large experimental data base and hence offer reliable predictions of the impurity distributions. The damage profiles are more difficult to model as they require knowledge of the survival rates of the different types of defect.

TRIM, as with other simulation programs, follows a large number of individual ion trajectories in a target. Each trajectory begins with a given position, direction and energy of the ion. It is then followed through a sequence of collisions with the target atoms, making the assumption of straight free-flight paths between collisions. The particle's energy is reduced for each free-flight path by an amount due to electronic energy loss, and then (after the collision) by the nuclear energy loss which is the result of transferring momentum to the target atom in the collision. It has been shown that the program calculations are consistent with the experimental results in most common cases. However, considerable error will be produced in two instances; one is for low doses in crystalline targets in which case the channelling effect may become significant. Hence, the assumption is no longer true in the TRIM program that the target can be considered amorphous with atoms at random locations, which means that any directional properties of the crystal lattice are ignored. The other is the low energy case, in which the ion may interact with more than one target atom at the same time. This breaks down the assumption in TRIM that the ion's history is determined by a series of consecutive unrelated binary encounters with the target atoms.

Examples of the electronic and nuclear damage distributions and the ion range profile which are obtained from TRIM are shown in Figure 2.26 for the case of He^+ with energy 2 MeV implanted into $LiNbO_3$. For He^+ implantation to produce optical waveguides, its energy needs to be in the MeV region in order to affect the crystal at the required depth of several wavelengths (e.g. 2–3 μm). Nuclear damage (collisional atomic displacements) causes extreme disorder or even amorphisation associated with about a 10% decrease in density and hence refractive index. From the TRIM calculation depicted in Figure 2.26 one can clearly see that most of the energy of the ion is lost in electronic processes (E). The total number of target vacancies created by nuclear collisions is distributed in a peak near the end of ion track (N). The centre of the distribution is at about 4.2 μm. Very few nuclear collision events happen in the region where the ion is travelling fast, which is dominated by ionization. The ion range is also distributed in a peak, the centre of which is at about 4.4 μm (A), that is, slightly deeper than the damage distribution centre. This example demonstrates that the simulations can be successfully used at quite high energies, even though they were initially designed for low energy semiconductor calculations. However, as will be seen in Chapter

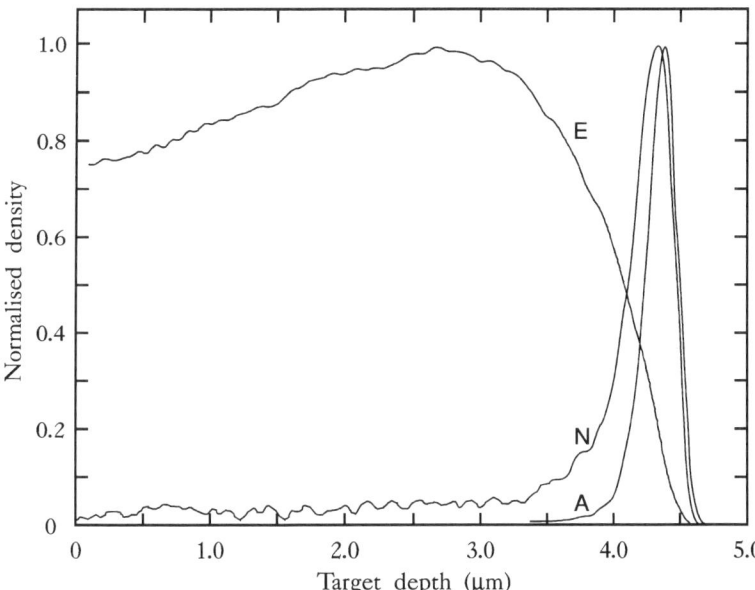

Fig. 2.26. An example of TRIM predictions for the normalised depth distributions of electronic (E), and nuclear (N) collision energy deposition, and the implanted atoms (A). The example is for an energetic light ion (2 MeV He$^+$) incident on LiNbO$_3$.

6, the simulations predict the depth of the damage and, for LiNbO$_3$, the ordinary refractive index profile. They fail totally to predict that of the extraordinary index because the simulation is unable to consider such factors as preferential diffusion of lithium within the damaged layer. The example thus demonstrates both the strengths and the weaknesses of computer simulations.

Variants of TRIM include codes for sputtering, high dose damage estimates and formation of new compounds. Crystalline target versions of the code are still under development. A different binary-type code which is designed for crystalline targets is named MARLOWE (e.g. see Webb, 1991).

2.5.2 Molecular dynamics

The previously mentioned binary event calculations followed individual trajectories in turn but in principle it should be possible to solve simultaneously the equations of motion for all ions in the region of interest.

The choice of interaction potential is critical and may require attractive as well as repulsive components In the region of a collision cascade the molecular dynamics approach is preferable to the simple binary sequences. However, it is computationally much slower and requires more computing power (Diaz de la Rubia *et al.*, 1987; Averback *et al.*, 1991). To economise on the computational time there have been attempts to use the simple binary collisions for low interaction rates and the molecular dynamics within the cascade core (Webb *et al.*, 1986).

2.5.3 Boltzmann transport equation

Rather than discuss individual events one can use analytical methods. The most common of these is the Boltzmann transport equation in which the energy, energy transfer and depth into the solid are linked by a single equation via a differential cross-section for energy transfer. This approach was initially described for ion beam sputtering by Sigmund (1969). In that application atoms are lost from the surface but the transport equation becomes more complex when back scattered ions remain within the solid. A fast simulation program, termed PRAL (Projected Range ALgorithm), based on the Boltzmann transport theory was developed by Biersack (1981, 1982). More recently a faster version of this program has been developed (SUSPRE, Surrey University Sputter Profile from Energy deposition).

2.5.4 State of simulation programs

Existing simulation programs are widely used and offer moderately accurate predictions of implanted ion ranges and initial damage distributions for amorphous targets, but crystalline targets require the inclusion of channelling, which is not always successfully predicted. In all cases there is a compromise between computer speed and accuracy and although running more events in a Monte Carlo set of initial conditions may improve the smoothness of a predicted range profile, it does not avoid the inherent shortcomings of the algorithms used. The present level of prediction by simulation programs is valuable but it is unreasonable to expect such approaches to offer more accuracy or more detailed information.

References

Agullo-Lopez, F., Catlow, C.R.A. and Townsend, P.D. (1988). *Point Defects in Materials* (Academic Press, London).
Ahlen, S.P. (1980). *Rev. Mod. Phys.,* **52**, 121.
Al Ghamdi, A. and Townsend, P.D. (1990). *Nucl. Inst. Methods,* **B46**, 156.
Andersen, H.H. (1987). *Nucl. Inst. Methods,* **B18**, 321.
Andersen, H.H., Bak, J.F., Knudsen, H. and Nielsen, B.R. (1977). *Phys. Rev.,* **A16**, 1929.
Arnold, G.W. (1982). *Rad. Effects,* **65**, 17.
Arnold, G.W. (1988). *Nucl. Inst. Methods,* **B32**, 504.
Arnold, G.W., Krefft, G.B. and Norris, C.B. (1974). *Appl. Phys. Letts.,* **25**, 540.
Auciello, O. and Kelly, R. (1984). *Ion Bombardment Modification of Surfaces: Fundamentals and Applications* (Elsevier, Amsterdam).
Averback, R.S., Diaz de la Rubia, T., Hsieh, H and Benedek, R. (1991). *Nucl. Inst. Methods,* **B59/60**, 709.
Bach, H. and Hallwig, D.J. (1984). *Rad. Effects,* **81**, 129.
Behrisch, R. and Wittmaack, K. (eds.) (1991). *Sputtering by Particle Bombardment III* Springer Topics in Appl. Phys. **64**.
Biersack, J.P. (1981). *Nucl. Inst. Methods,* **182/183**, 199.
Biersack, J.P., (1982). *Z. Phys. A Atoms and Nuclei,* **305**, 95.
Biersack, J.P. (1987). In *Ion Beam Modification of Insulators,* Chap. 1, P. Mazzoldi and G.W. Arnold (eds.) (Elsevier, Amsterdam).
Biersack, J.P. and Haggmark, L.G. (1976). *Nucl. Inst. Methods,* **132**, 647.
Biersack, J.P. and Ziegler, J.F. (1982). *Springer series in electro-physics*, **10**, 122.
Braunstein, G. and Kalish, R. (1981). *Nucl. Inst. Methods,* **182/183**, 691. ibid (1983). **209/210**, 387.
Burnett, P.J. and Page, T.F. (1983). *Proc. Brit. Ceram. Soc.,* **34**, 65.
Burnett, P.J. and Page, T.F. (1986). *Rad. Effects,* **97**, 283.
Cazaux, J. (1986). *J. Appl. Phys.,* **59**, 1418.
Chakoumakos, B.C., Oliver, W.C., Lumpkin, G.R. and Ewing, R.C. (1991). *Rad. Effects and Defects in Solids,* **118**, 393.
Chandler, P.J., Glavas, E., Lama, F., Lax, S.E. and Townsend, P.D. (1986). *Rad. Effects,* **98**, 211.
Chandler, P.J., Zhang, L. and Townsend, P.D. (1990). *Nucl. Inst. Methods,* **B46**, 69.
Chinellato, V., Gottardi, V., LoRusso, S., Mazzoldi, P., Nicoletti, F. and Polato, P. (1982). *Rad. Effects,* **65**, 31.

Chu, W.K., Mayer, J.W. and Nicolet, M.A. (1978). *Backscattering Spectrometry* (Academic Press, New York).
Dalal, M.L., Rahmani, M and Townsend, P.D. (1988). *Nucl. Inst. Methods,* **B32**, 61.
Destefanis, G.L., Gailliard, J.P., Ligeon, E.L., Valette, S., Farmery, B.W., Townsend, P.D. and Perez, A. (1979). *J. Appl. Phys.,* **50**, 7898.
Diaz de la Rubia, T., Averback, R.S., Benedek, R. and King, W.E. (1987). *Phys. Rev. Lett.,* **59**, 1930.
DIET (Desorption Induced by Electronic Transitions, Conference Series) (1983), vol. I; (1985), vol. II; (1988), vol. III; (1990), vol. IV (Springer Verlag).
Eckstein, W. *Computer Simulation of Ion–Solid Interactions* Series in Materials Science, **10** (1991) (Springer Verlag).
EerNisse, E.P. (1974). *J. Appl. Phys.,* **45**, 167.
Faik, A.B., Chandler, P.J., Townsend, P.D. and Webb, R. (1986). *Rad. Effects,* **98**, 233.
Feldman, L.C. and Mayer, J.M. (1987). *Fundamentals of Surface Thin Film Analysis* (North Holland, Amsterdam).
Feldman, L.C., Mayer, J.M. and Picraux, S.T. (1982). *Materials Analysis by Ion Channelling* (Academic Press, New York).
Feldman, L.C. and Poate, J.M. (1982). *Ann. Rev. Mat. Sci.,* **12**, 149.
Fredrickson, J.E., Waddell, C.N., Spitzer, W.G. and Hubler, G.K. (1982). *Appl. Phys. Lett.,* **40**, 172.
Frenkel, D. (1989). In *Simple Molecular Systems at Very High Density*, pp. 411. A. Polian, P. Loubeyre and N. Boccara (eds.) (Plenum Press).
Gemmel, D.S. (1974). *Rev. Mod. Phys,* **46**, 129.
Götz, G. and Karge, O. (1983). *Nucl. Inst. Methods,* **209/210**, 1079.
Harries, D.R. and Marwick, A.D. (1980). *Phil. Trans. R. Soc. London,* **A295**, 197.
Harrison, D.E., (1983). *Rad. Effects,* **70**, 1.
Harrison, D.E., (1988). *Critical Revs in Solid State and Material Science,* **14(1)**, S1.
Haycock, P.W. (1988). Private communication.
Heidemann, K.F. (1981). *Phil. Mag.,* **B44**, 465.
Hubler, G.K., Holland, O.W., Clayton, C.R. and White, C.W. (eds.) (1984). *MRS Symp.,* **27** (North Holland, Amsterdam).
Hubler, G.K., Malmberg, P.R., Carosella, C.A., Smith, T.P. Spitzer, W.G., Waddell,C.N. and Phillippi, C.N. (1979a). *Rad. Effects,* **48**, 81.
Hubler,G.K., Waddell,C.N., Spitzer, W.G., Fredrickson, J.E., Prussin, S. and Wilson, R.G. (1979b). *J. Appl. Phys.,* **50**, 3294.
J. Phys. Chem. Solids (1990). Special issue on defects and impurity centres in ionic crystals, **51**, (7).
Krefft, G.B., Beezhold, W. and EerNisse, E.P. (1975). *IEEE Trans. Nucl. Sci.,* **NS-22**, 2247.
Krefft, G.B., and EerNisse, E.P. (1978). *J. Appl. Phys.,* **49**, 2725.
Lama, F.L. (1986). Sputtering Processes in UO_2 and UF_4. D. Phil. Thesis, Sussex University.
Lindhard, J and Scharff, M. (1961). *Phys. Rev.,* **124**, 128.
Lindhard, J., Scharff, M. and Schiøtt, H.E. (1963). *Mat. Phys. Medd. Dan. Vid. Selsk.,* **33**, no 14.
Mazzoldi, P. (1983). *Nucl. Inst. Methods,* **209/210**, 1089.
Mazzoldi, P. and Arnold, G.W. (1987). *Ion Beam Modification of Insulators* (Elsevier, Amsterdam).

Morgan, D.V. (1973). *Channelling* (Wiley).
Naguib, H.M. and Kelly, R. (1975). *Rad. Effects,* **25**, 1.
Newton, C.S. and Hay, H.J. (1981). *Rad. Effects Lett.,* **43**, 211
Norris, C.B. and EerNisse, E.P. (1974). *J. Appl. Phys.,* **45**, 3876.
Piller, R.C. and Marwick, A.D. (1978). *J. Nucl. Mater.,* **71**, 309.
Presby, H.M. and Brown, W.L. (1974). *Appl. Phys. Lett.,* **24**, 511.
Rahmani, M., Abu-Hassan, L.H., Townsend, P.D., Wilson, I.H. and Destefanis, G.L. (1988). *Nucl. Inst. Methods,* **B32**, 56.
Rahmani, M. and Townsend, P.D. (1989). *Vacuum,* **39**, 1157.
Robinson, M.T. (1981). In *Sputtering by Particle Bombardment,* chp. 3, pp. 73–144. Topics in Applied Physics, R. Behrisch (ed.), vol. 47. (Springer Verlag, Berlin).
Rusbridge, K.L. and Nelson, R.S. (1980). Paper B1 presented at the 1980 IBMM Albany Conference, unpublished.
Schiøtt, H. (1970). *Rad. Effects,* **6**, 107.
Sigmund, P. (1969). *Phys. Rev.,* **184**, 383.
Sigmund, P. (1972). *Rev. Roum. Phys.,* **17**, 823, 969.
Stark, J. (1912). *Phys. Z.,* **13**, 973.
Sternheimer, R.M. and Peierls, R.F. (1977). *Phys. Lett.,* **A63**, 359.
Thompson, M.W. (1968). *Phil. Mag.,* **18**, 377.
Tombrello, T.A. (1984). *Nucl. Inst. Methods,* **B2**, 555.
Townsend, P.D., Chandler, P.J., Wood, R.A., Zhang, L., McCallum, J.C. and McHargue, C.W. (1990). *Electr. Lett.,* **26**, 1193.
Townsend, P.D., Kelly, J.C. and Hartley, N.E.W. (1976). *Ion Implantation, Sputtering, and their Applications* (Academic Press, London).
Waddell, C.N., Spitzer, W.G., Hubler, G.K. and Fredrickson, J.E. (1982). *J. Appl. Phys.,* **53**, 5851.
Webb, A.P., Houghton, A.J. and Townsend, P.D. (1976). *Rad. Effects,* **30**, 177.
Webb, R.P. (1992). *Practical Surface Analysis II, Ion and Neutral Spectroscopy,* (Appendix 3) D. Briggs and M. Seah (eds.) (John Wiley, Chichester).
Webb, R.P., Harrison, D.E. and Jakas, M.M. (1986). *Nucl. Inst. Methods,* **B15**, 1.
Weeks, R.A., Hosono, H., Zuhr,R., Magruder, R.H. and Mogul,H. (1989). *Mater. Res. Soc. Proc.,* **152**, 115.
White, C.W., McHargue, C.J., Sklad, P.S., Boatner, L.A. and Farlow, G.C. (1989). *Mat. Sci. Repts.,* **4**, 41.
Zhang, L., Chandler, P.J. and Townsend, P.D. (1991a). *J. Appl. Phys.,* **70**, 1185.
Zhang, L., Chandler, P.J. and Townsend, P.D. (1991b). *Nucl. Inst. Methods,* **B59/60**, 1147.
Ziegler, J.F. (1984). *Ion Implantation Science and Technology* pp. 51 (Academic Press).
Ziegler, J.F. (1977–80). *Stopping Power and Ranges of Ions in Matter,* vols. 2-6 (Pergamon Press, New York).
Ziegler, J.F, Biersack, J.P. and Littmark, U. (1985). *Stopping Power and Ranges of Ions in Matter,* vol. 1 (Pergamon Press, New York).

3
Optical absorption

3.1 Analysis methods using absorption, ESR and RBS

Optical methods of studying defects have the advantage that if each defect has characteristic energy levels which lie within the forbidden energy gap, then they show separable optical absorption and luminescence bands. Higher photon energy absorption generally monitors electronic transitions, whereas infra-red absorption records vibrational spectra. Many of the optical transitions which result from the presence of impurities have energies in the visible part of the spectrum and consequently the defects are referred to as colour centres. Examples of colour centres are widespread and include the impurities which give colouration to ruby and sapphire or stained glass. They are the basis for photographic and photochromic materials, and frequently involve a mixture of impurity and intrinsic defects. Whilst analysis of absorption bands may determine defect symmetry and inter-relationships of different colour centres, it is unusual to be able to confirm precise models of defect sites solely from the absorption data. In this respect the processes which involve hyperfine interactions such as Electron Spin Resonance (ESR), Electron Nuclear Double Resonance (ENDOR), Mossbauer spectroscopy or spin precession techniques provide more specific answers if they can be applied. In the ion implantation literature there are frequent presentations of data from Rutherford Back-scattering Spectrometry (RBS) to give the depth distributions of impurities or damage in the target material. In part, RBS appears to be popular for implantation analysis because it requires a high energy ion accelerator, which is normally a feature of an implantation laboratory. The information is useful but, like electron microscopy, it rarely gives precise details of individual defect arrangements. All these techniques are quite standard, and are reviewed in many defect books such as that by Agullo-Lopez et al. (1988). Reviews of optical absorption features from ion implantation include those of Townsend (1987), Arnold and Mazzoldi (1987) and Weeks (1991).

3.2 *In situ* optical absorption

Ion implantation in insulators generates damage which leads to the formation of new defects, and ionisation which results in defects in new charge states. The defects are apparent as optical absorption bands. Since the ion track generates a very high defect density the potential range of the types of colour centres is much greater than is normally encountered using the simpler methods of production of imperfections such as impurity diffusion doping, or ionising irradiation. The methods in which the sample is close to thermodynamic equilibrium tend to make point defects involving fewer atomic sites than will occur with ion beam implantation. The potential range of defect sites may include localised point defects of impurities, vacancies and interstitials, as well as complex aggregates, colloids and phase precipitates. The stability of the defects, and their rate of production, will be influenced by the implantation temperature and the crystallography of the target and its orientation relative to the ion beam. Consequently, a study of optical absorption of ion implanted material will reveal details of the defect structures, and, at least in principle, dynamic measurements may reflect the ion beam damage mechanisms and lattice rearrangements.

Experimentally, there are advantages and disadvantages to recording absorption within an ion implanted layer. The positive aspects are that because one is effectively using a very thin sample, say, 1 micron thick, then the growth of defects may be followed to extremely high levels of optical absorption. Early examples included the growth of vacancy centres by Pooley (1966) and interstitial defects (Saidoh and Itoh, 1973). In a normal bulk sample with identical reflectivity at each face (i.e. not precisely the situation for the ion implanted layer), then the transmitted light, I_t, reflectivity, R, absorption coefficient, μ, and sample thickness, t, are related by

$$I_t = I_0(1-R)^2 \exp(-\mu t)\,(1 - R^2 \exp(-2\mu t))^{-1} \qquad (3.1)$$

or, ignoring the reflectivity, $I_t = I_0 \exp(-\mu t)$. The reflectivity of the surface is related to the refractive index n by

$$R = ((n-1)/(n+1))^2$$

Modern sensitive spectrophotometers normally monitor the transmission in terms of an optical density given by

$$D = \log_{10}(I_o/I_t) = \mu t/2.3 \qquad (3.2)$$

and operate from D values of under 0.01 to about 6. Hence they can measure transmission values from 98% down to some 0.0001%, giving an overall μt product range from 0.023 to nearly 14. Depending on the shape of the optical absorption band, i.e. Gaussian or Lorentzian, the defect concentration, N/cm^3, the refractive index, n, the oscillator strength, f (i.e. the transition efficiency), the peak absorption strength, $\mu_{\max}$ cm^{-1}, and band width, W eV are related by Smakula's equation

$$Nf = A\ 10^{17}\ n(n^2 + 2)^{-2}\ \mu_{\max}\ W \qquad (3.3)$$

where A has the numerical value 1.29 for a Lorentzian band shape and 0.87 for a Gaussian one. Assuming values of $t = 10\,\mu$, $W = 1$ eV, $n = 1.5$, $f = 1.0$, then the range of defect concentrations that one might ideally detect in the implanted layer is from 10^{17} to some 10^{20} centres/cm^3. This detection threshold is relatively high, but such concentrations are typically less than 1% of intrinsic defect saturation levels, and, more usefully, one may record signals up to saturation, which would normally represent too high an optical density for a standard (e.g. millimetre) thickness sample.

For a beam current of one microamp per cm^2, and a defect formation energy of 10 eV, then, even for a 1 MeV ion beam, the dynamic defect production rate is only 6×10^{17} defects per second per cm^2. The majority of these defects anneal rapidly but, if the defects have lifetimes of seconds, then these newly formed defects may be detectable experimentally by optical absorption. Additionally, transient features may be detectable if the transitions within existing defects are electronically excited. Measurements have normally been made by post-irradiation absorption in a spectrophotometer; however to search for transient effects it may be preferable to build a high sensitivity optical spectrometer *in situ*, i.e. for measurement during ion implantation. Even for stable defect absorption bands, removal of the sample to an external spectrometer is not possible with the same precision as in an *in situ* system. Removal would also be highly disruptive, particularly if the target temperature were not ambient, and it would not allow continuous monitoring of changes in spectra with ion beam dose. Further, there are already sufficient examples of changes which occur on the timescale of minutes so that a true *in situ* spectrometer is essential. Note, however, that for good experimental practice when

3.2 In situ optical absorption

studying colour centres the sample should only be illuminated with low power monochromatic light, as secondary bleaching effects and charge transfer can occur. In some cases it may be necessary to add a second monochromator to avoid the problem of luminescence which can saturate the photomultiplier. *In situ* spectrometer systems may simply record optical absorption, e.g. as described by Wood and Townsend (1991), or be variations on highly sensitive ellipsometers, e.g. Andrews *et al.* (1989), or record differential reflectance. Measurements may be made of the optical absorption bands directly by transmission data; however the presence of strong absorption influences the reflectivity. Therefore, the presence of absorption features can be mirrored in the reflectivity data, and for *in situ* measurement this may be simpler to record than transmission data.

Experimentally, the ellipsometry is the most complex of the three, although it yields much the same information as differential reflectance. Presentation of ellipsometric information is normally as ϵ_1 and ϵ_2 for the real and imaginary parts of the dielectric function. These relate to the real and imaginary parts of the refractive index by

$$\epsilon_1 = n^2 - k^2 \quad \text{and} \quad \epsilon_2 = 2nk \tag{3.4}$$

where

$$\mathbf{n} = n - ik \quad \text{and} \quad \epsilon = \epsilon_1 - i\epsilon_2 \tag{3.5}$$

In a comparison of differential reflectance and spectroscopic ellismometry Burns *et al.* (1991) studied ion beam damage in silicon surface layers. They emphasise that, because of experimental inaccuracy, there may not be significant advantage in using the ellipsometry in most cases.

Overall there is value in using *in situ* measurements and this is emphasised by examples in which the apparent growth curve of an absorption band differs in the continuous recording compared with off-line measurements. This is variously attributed to a mixture of short term fading and features which result from re-irradiation of a previously implanted sample. Figure 3.1(*a*) shows an example of the reflectivity changes which occur in a float glass target during implantation with silver ions (Nistor *et al.*, 1993). In general, implantation of monovalent metal ions into insulators frequently leads to the production of metal colloids. Colloids are distinctive in that the peak position of the optical absorption, and optical reflectivity, move to longer wavelengths as the colloids increase in size. Figure 3.1(*a*) shows this effect but indicates additionally that there is an apparent shoulder developing on the blue end of the reflectivity curve. Figure 3.1(*b*) presents data at different stages during the implant.

74 Optical absorption

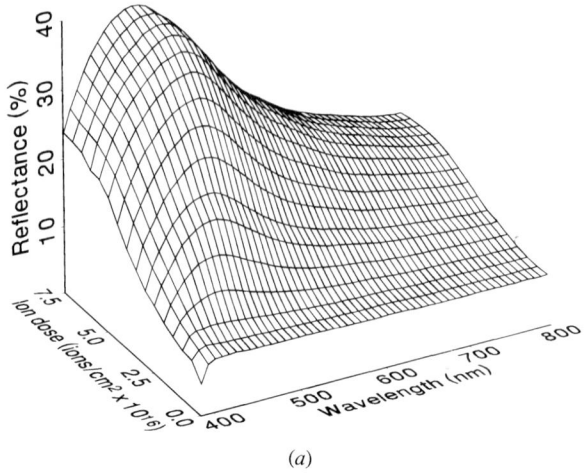

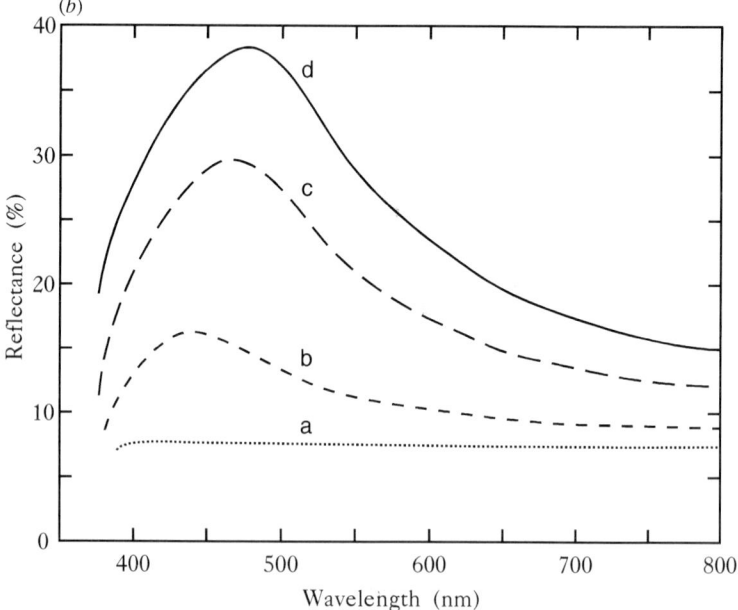

Fig. 3.1. (a) Reflectivity changes caused by 60 keV Ag implantation into float glass. (b) Examples of reflectivity data at fixed implant doses taken from (a). The curves are for a: 0, b: 2, c: 4 and d: 6×10^{16} ions/cm^2.

Whilst the Ag example shows only steadily varying features, this is not always the case. For example, Figure 3.2(a) gives an isometric plot of the reflectivity versus wavelength data during implantation of a 'white optical glass' with argon ions. It is apparent that there is a very sensitive dose dependent peak at the blue end of the reflectivity curve. Such argon implants have been considered for anti-reflective coatings but, as emphasised by Figures 3.2(a) and (b), the dose needs to be accurately chosen and the dose selection may differ depending on the wavelength range of interest. Further details of the dose dependence of this type of reflectivity curve will be given in Section 3.11.2.

3.3 Crystallographic effects on stress and defect motion

Optical absorption bands monitor transitions between ground and excited states of a defect site. They are therefore sensitive to both the local defect symmetry and the crystal field. Consequently, many defect absorption bands are anisotropic. This fact is used routinely to assess models of defect sites by consideration of the relationship of the orientation of the sample and the observed absorption bands, as detected with polarised light. In ion implanted samples there may be additional crystal field terms caused by stress, which is normal to the implanted face. Such stresses will manifest themselves in changes of anisotropy of the absorption bands. Similar features are discussed for the emission of polarised luminescence (see Chapter 4). As for the luminescence, the stresses may change the ratio of related defect signals, or indeed totally suppress some.

In addition to radiation enhanced diffusion caused by the presence of a high vacancy concentration during the implantation process, there are crystallographic factors which will influence the mobility of defects such as interstitials and vacancies. For example, implantation with an ion beam down the c axis of sapphire or lithium niobate will allow rapid separation of point defects by diffusion of interstitials along the channels away from the peak in the damage distribution, whereas implants in the perpendicular plane will lead to a lower rate of defect retention as the vacancies and interstitials move in the plane of the projected range and hence can recombine. Such crystallographic factors are noted in many defect studies.

In discussions of colour centres the convention for describing the defects is to define the anion-type centres with respect to the charge state of the normal lattice. The two main examples selected are for sapphire and LiF as these are probably representative of oxides and halide

76 *Optical absorption*

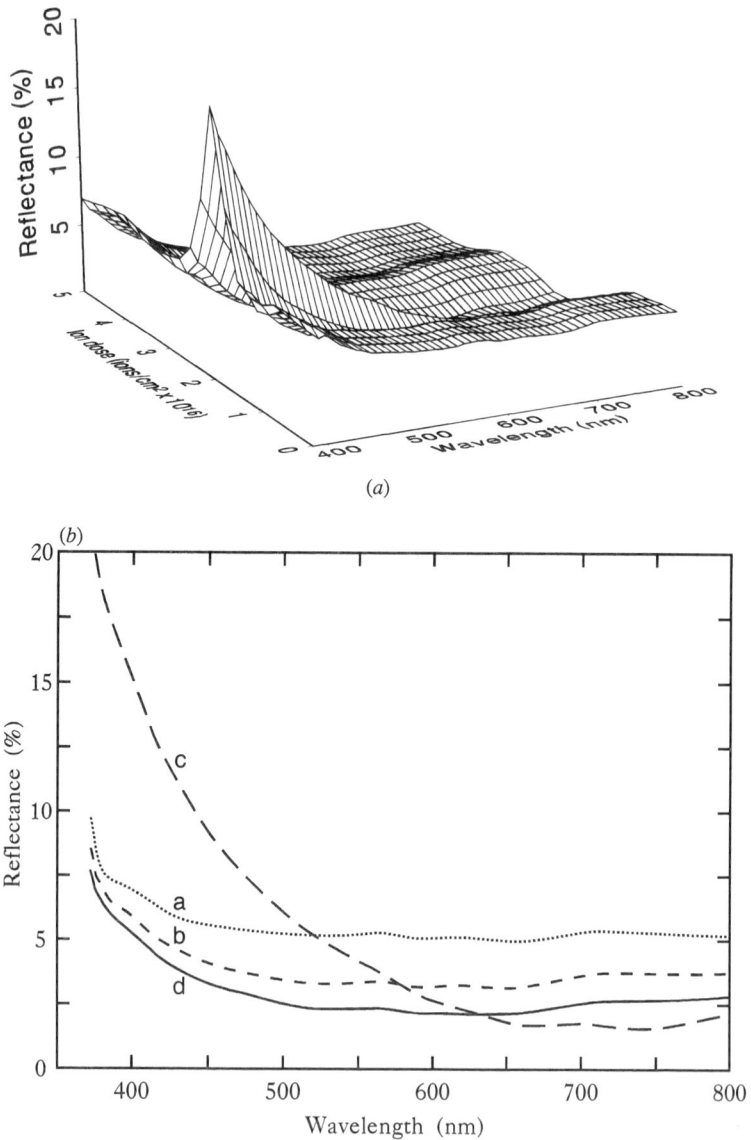

Fig. 3.2. (a) Reflectivity changes in a 'white optical glass' during ion implantation with 80 keV argon. (b) Examples from (a) of the reflectivity at Ar doses of a: 0, b: 1.25, c: 2.0 and d: 3.0×10^{16} ions/cm^2.

Table 3.1. *Anion defects in sapphire and alkali halides*

Sapphire	
F centre	An oxygen vacancy which has trapped two electrons
F^+ centre	An oxygen vacancy with one trapped electron

Alkali Halides	
F centre	A halogen vacancy with a trapped electron
F_2 centre	A pair of adjacent F centres
F_3 centre	Three associated F centres
F' centre	An F centre with an extra electron
F_2^+ centre	An F centre which has lost one electron
V_K centre	A hole trapped on a relaxed pair of halogens
H and I centres	Interstitial halogen-type defects
V centres	Defects on the cation lattice with trapped holes
U centres	Vacancies or interstitials involving H atoms

materials. They are also materials which have a detailed literature on their defect structures and optical absorption bands. Table 3.1 summarises some of the more common defect models and indicates the labelling scheme.

3.4 Sapphire

Both the effects of stress and radiation enhanced diffusion are apparent in colour centre studies of ion implanted sapphire. In early data, Krefft (1977) varied the type of energy transfer from electronic to nuclear collision damage by using ions of different mass, and recorded the ratio of F/F^+ bands. The F/F^+ ratio increased with higher fractions of energy deposited in collisions. In part this may be that ionisation of pre-existing defects in the electronic stopping region overshadow those formed in the collision region. Indeed, the first assumption is that ionisation or stress on the F centres may remove the weakly bonded second electron, and so increase the concentration of F^+ centres. The ion implanted defect situation is more complicated than corresponding studies of bulk crystals with electron or neutron damage. For example, Figures 3.3 and 3.4 show the optical absorption bands of sapphire for He^+ and Ne^+ implants at 77 and 300 K for beams directed down the two major axes (Dalal *et al.*, 1988). It is apparent that the spectra differ in each case. Factors already discussed which influence these differences are that at low temperature more defects are retained in the sapphire and at low temperature the lack of atomic movements may favour the retention of simpler defects, since for the room temperature implants, nearby vacancies and interstitials may annihilate and other defect motion can allow growth of more complex defects. As seen from both Figures 3.3 and 3.4 the implants made at 77 K accentuate the F band at 6.03 eV relative to the lower energy features. A lattice structure such as that of sapphire has differing diffusion rates for defects along and perpendicular to the c axis. Hence both the defect separation and damage induced stress terms are direction sensitive and encourage defect migration along the c axis channels. The familiar point defects of sapphire are the F centre (an oxygen vacancy which traps two electrons) giving the isotropic 6.03 eV absorption band, and the F^+ centre (an oxygen vacancy with one trapped electron), which gives three polarised absorption bands at 6.3, 5.41 and 4.84 eV. In addition to these bands, other absorption bands were resolved over limited dose ranges, occurring near 4.9 and 6.8 eV. The spectrophotometer was only partially polarised but nevertheless major differences are obvious in the ratio of the bands for the two crystal cuts. Again, these features can be interpreted in terms of rapid defect separation along the c axis. Effects of stress, leading to electron loss from the F centres, is most obvious at 77 K, where more disorder is retained in the lattice. In agreement with the earlier data, Dalal *et al.* found that the apparent F/F^+ ratio increased on moving to higher mass ions, but the simple model of F to F^+ conversion is not the only factor.

Additional measurements further into the UV revealed the growth of bands beyond 6 eV, dominated by new absorption bands at 6.8 and 6.97 eV as seen in Figure 3.5(*a*). Defect models were tentatively made in terms of F-type centres imbedded in amorphised sapphire. Alternative interpretations include the possibility of F-type defects in the stress field associated with heavy ion implants.

Relatively few studies have been made of vacuum UV optical absorption bands produced by ion implantation. However, to emphasise that such features are often attributable to familiar colour centres, Figure 3.5(*b*) shows the result of implanting 46.5 MeV Ni^{6+} ions into silica. The data (Antonini *et al.*, 1982) can be analysed as 'standard' oxygen vacancy centres which, for silica, are variously labelled as E, E' or B bands.

Implantation with heavier ions such as In, Zn and Pb have been monitored by absorption and RBS by Mouritz *et al.* (1987) and the spectra show poorly resolved F and F^+ bands, although at high doses a strong band near 3.1 eV emerges (this is often ascribed to Al intersti-

3.4 Sapphire

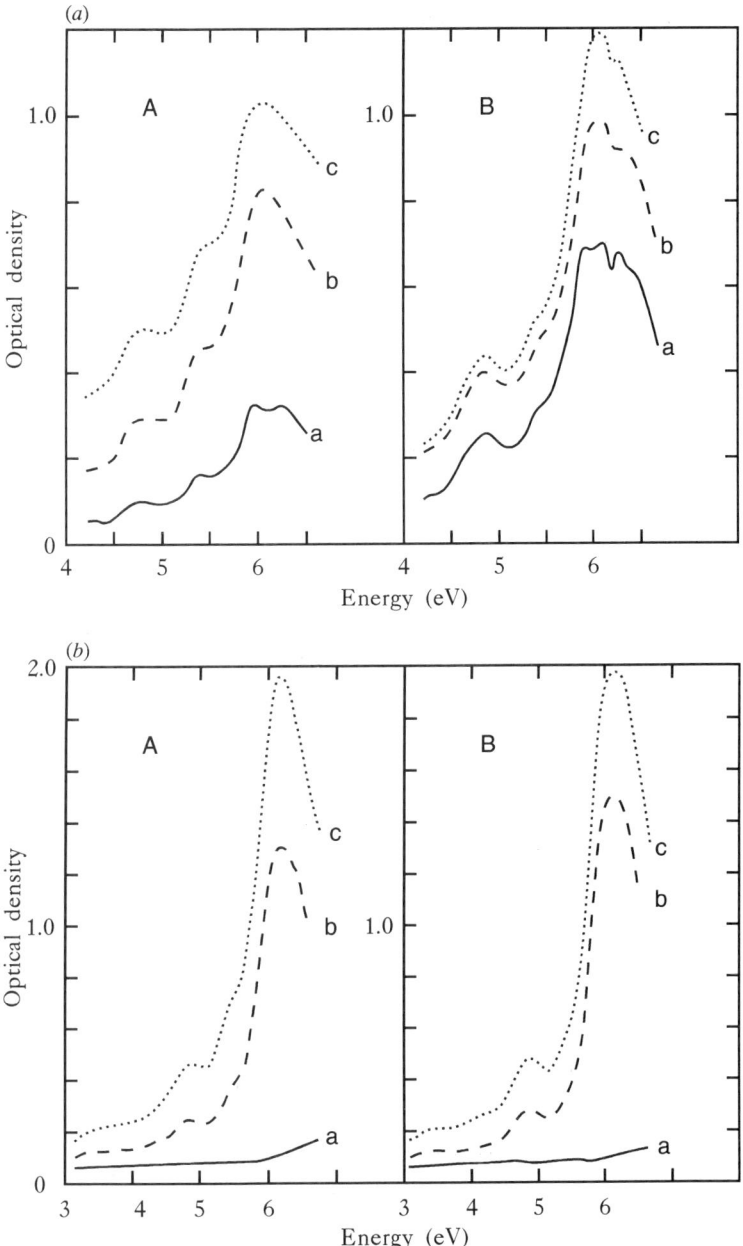

Fig. 3.3. Optical absorption bands induced in Al_2O_3 by 2 MeV He^+ implants. The data in (a) are for implants at 300 K into different crystal faces with set A for $c_\perp$ and B for $c_\parallel$ cut crystals. The doses for set A were a: 1, b: 3 and c: 7×10^{16} ions/cm² and for set B a: 3, b: 6 and c: 8×10^{16} ions/cm². In (b) the data resulted from 77 K implants with the optical measurements taken at 300 K. The doses for set A were a: 0, b: 1 and c: 5×10^{16} ions/cm² and for set B a: 0, b: 2 and c: 5×10^{16} ions/cm².

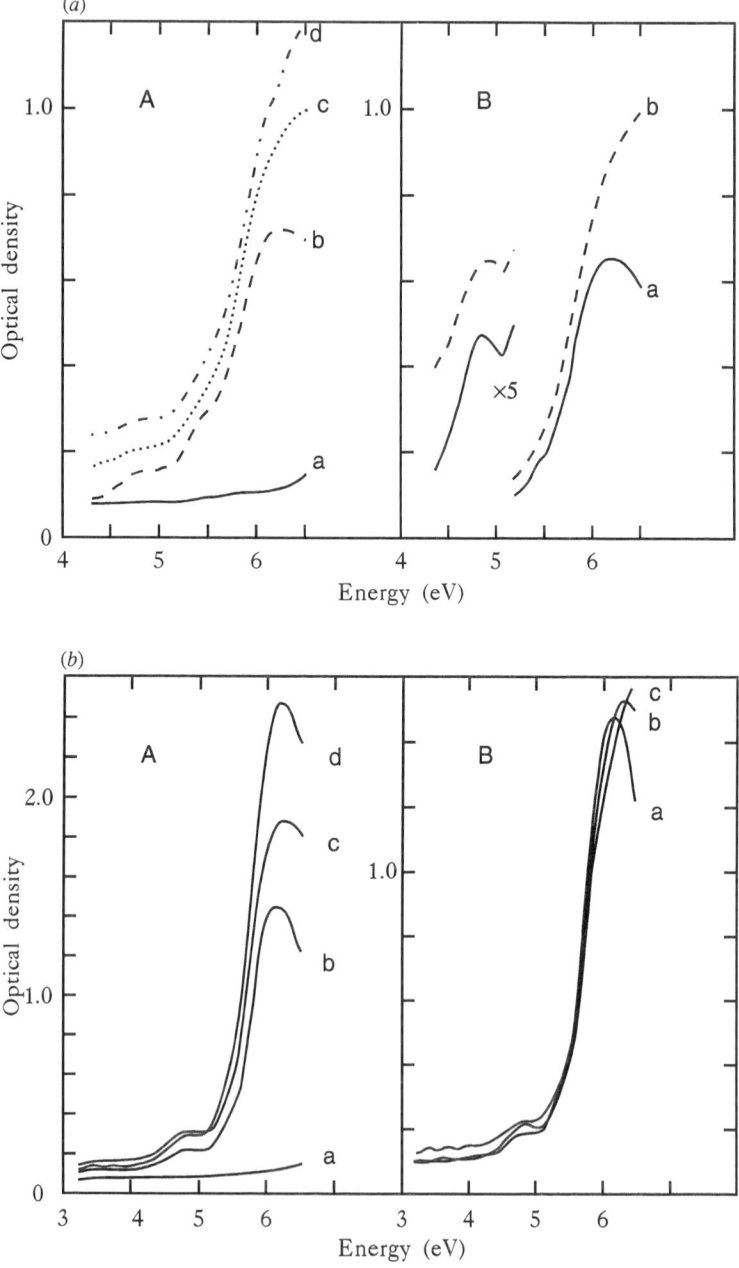

Fig. 3.4. Optical absorption bands induced in Al_2O_3 by 2 MeV Ne^+ implants. The data in (a) are for implants at 300 K into different crystal faces with set A for $c_\perp$ and B for $c_\parallel$ cut crystals. The doses for set A were a: 0, b: 1, c: 3 and d: 5×10^{16} ions/cm² and for set B a: 1 and b: 4×10^{16} ions/cm². In (b) the data resulted from 77 K implants with the optical measurements taken at 300 K. The doses for set A were a: 0, b: 0.4, c: 5.4 and d: 1.4×10^{16} ions/cm² (i.e. higher doses lead to reduced absorption). For set B doses were a: 0, b: 1.4 and c: 2.4×10^{16} ions/cm².

3.4 Sapphire

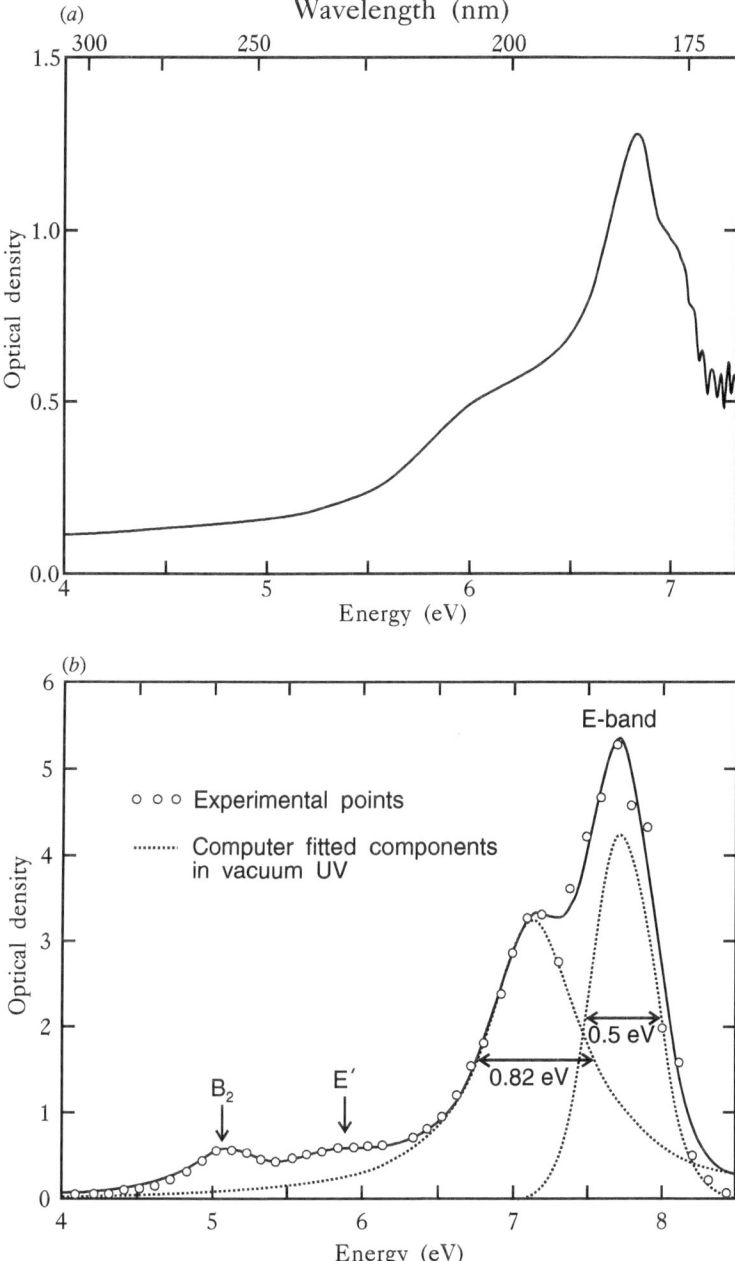

Fig. 3.5. (a) Optical absorption formed by 2 MeV argon ion implants into sapphire at 77 K, but measurements were at 300 K. The implants were for a $c_\perp$ cut crystal. (b) Absorption spectrum produced by 46.5 MeV Ni^{6+} implants to a dose of 10^{14} ions/cm^2 in silica glass (after Antonini et al., 1982).

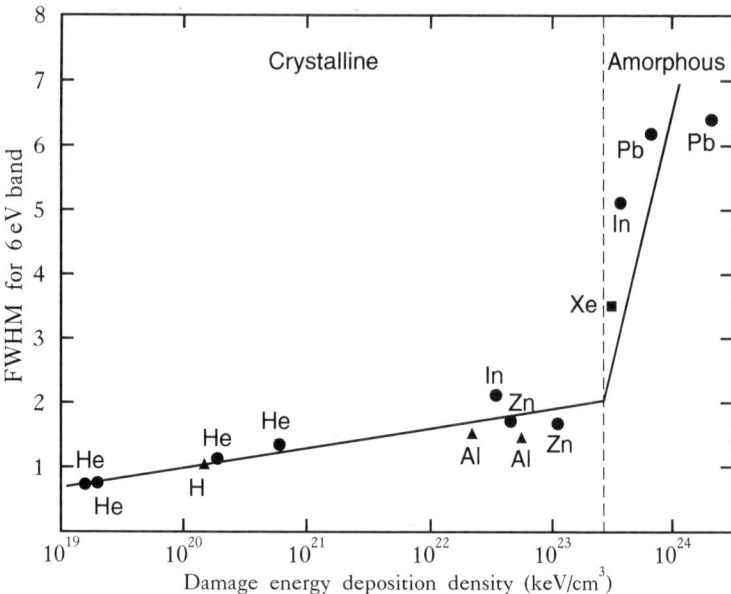

Fig. 3.6. Data of the width of the 6 eV absorption band envelope indicate that a critical energy deposition is required to amorphise the sapphire lattice.

tials). Assuming that the 6 eV absorption monitors point defects, then their data record a clear change in halfwidth of the band at the dose levels associated with amorphisation of the sapphire lattice, as shown by Figure 3.6. The nuclear energy deposition level of some 10^{23} keV/cm^3 is consistent with the Burnett and Page (1984, 1986) estimate of room temperature amorphisation in sapphire. It was seen that annealing of the ion implanted amorphous material differed according to the ion beam used for the destruction of the lattice. This was not surprising, since the peak concentration of the impurities approached some 4% of the atomic sites.

The rate of point defect production and the relative importance of different colour centres are not only a function of the rate of energy deposition into the lattice, but they are also sensitive to the chemistry of the implant. In studies of oxygen vacancy colour centre production in sapphire, Agnew (1992) shows that the low temperature (77 K) defect production rate varies with ion species. In particular, boron and carbon implants are more effective by up to a factor of 3 than ions of mass 4 or 20.

Earlier work with surface amorphisation of sapphire had indicated a rather different chemical influence with greater damage efficiency, using group IV elements (White et al., 1989). Note, however, that in the one case the emphasis is on colour centre determination, and in the other it is on amorphisation.

Other optical absorption work with ion implanted sapphire includes a study by Evans and Hendricks (1977) to implant self-ions of O^+ or Al^+ to search for defects associated with Al interstitials. A study of both optical absorption, luminescence and ESR of defects associated with Fe^{2+} ions implanted at 3.8 MeV by Chen et al. (1991) showed that the only obvious absorption bands were from the F and F^+ centres. However, luminescence data confirmed the presence of F_2, F_2' and F_2^{2+} centres.

Saito et al. (1985, 1991) have used metal ion implants in sapphire to form coloured crystals which offer the possibility of forming novel gem stones, as well as in principle, dopants for scientific purposes. Ions used included Nb^+, Fe^+, Cu^+, Co^+, Ti^+, and Cr^+ to dose levels of 3×10^{17} ions/cm^2. A combination of implants and annealing produced a wide range of colours from light blue to yellowish-brown.

Optical absorption is only one of many techniques for studying defects, and, in the case of ^{57}Fe implants, Mossbauer spectroscopy is a valuable tool. For example, McHargue et al. (1987) recorded the iron in three distinct co-existing states of Fe^{2+}, Fe^{3+} and metallic Fe. The relative signals varied with total iron concentration, as shown in Figure 3.7, and also as a result of annealing. Heating in an oxidizing atmosphere led to a conversion of the iron into Fe^{3+} or Fe^0, together with precipitations of small oxide or metallic iron particles. In order to accomodate such changes there were re-arrangements of the host lattice. The annealing atmosphere is not a passive feature of the lattice recovery and in this study the iron distribution, as assessed by RBS, was significantly different in samples annealed in oxygen or hydrogen (Figure 3.8). The RBS depth profiles were partially confirmed by electron microscopy. The main conclusion is that where there are chemical interactions between the implant, target and atmosphere it becomes difficult to predict the depth distribution and the manner in which the ions are retained within the target.

3.5 Alkali halides

Alkali halides have been a popular test system for colour centre studies and there is a good understanding of the defect structures and production

84   *Optical absorption*

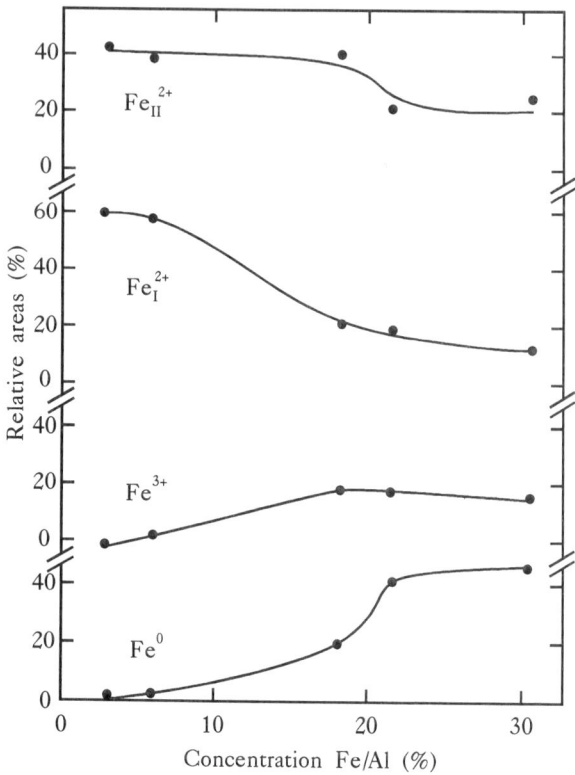

Fig. 3.7. The variation in charge state of iron implanted into sapphire as assessed by Mossbauer spectroscopy.

mechanisms associated with many of the absorption bands. In this section we will concentrate on data taken for ion implantation studies of LiF. The examples will emphasise the wide range of processes which are revealed by the optical absorption. The range of bands and mechanisms may be greater than in more stable systems, e.g. oxides such as sapphire, because the halides damage by electronic processes. Hence, there is a high colour centre density throughout the ion range, and beyond. Note that some 98% of the energy from a 1 MeV H^+ ion is transferred by electronic routes, thus it is possible to vary the efficiency of colour centre formation by changing ion energy or ion mass as these factors alter the ratio of electronic to nuclear collision damage, and, less obviously, the diameter of the excitation track. Perez and his collaborators (1974, 1976) pointed out that, not only does the diameter of the track increase rapidly near

3.5 Alkali halides

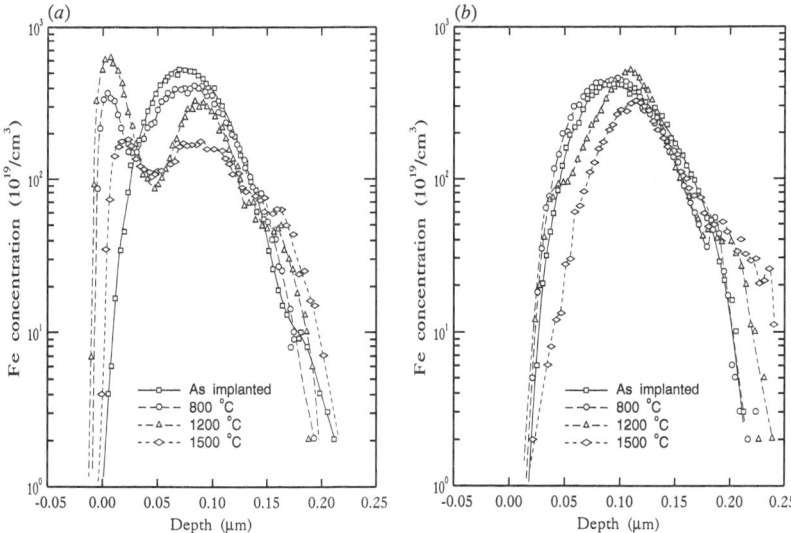

Fig. 3.8. Iron depth profiles as recorded by RBS for Fe^+ implanted at 160 keV into Al_2O_3 at room temperature to a dose of 4×10^{16} ions/cm^2. The depth distribution is altered by annealing and the change is sensitive to the atmosphere used. In (a) the data are for an oxygen atmosphere and in (b) they are for hydrogen.

the end of the ion range for light ions, because of multiple scattering, but also the production of energetic secondary electrons, and thus the track diameter, increase with electronic stopping power. Estimates of the track radius vary from 0.02 to 0.1 to 10 μm for protons of 1, 2 and 14 MeV respectively. For these values the track radius is proportional to (energy)$^{2.5}$.

3.5.1 F and F_2 centres

Such variations in excitation density will influence both the total number and the type of defect formed because the close proximity of defects formed in a track should favour the formation of paired or clustered defects. So one might expect a greater concentration of F_2 or larger defects, compared with isolated F centres, than would be predicted on a purely statistical basis of two randomly produced F centres being close enough to form an F_2 centre. This simple assumption will be perturbed because, as mentioned earlier, many higher order defects such as F_3 centres and those with extra charges, e.g. F_2', are suppressed.

Abu Hassan and Townsend (1986a,b) ion implanted LiF and recorded absorption values at the peak wavelengths of the F and F_2 bands. At first sight their data favoured the production of F_2 centres. The results suggested a high F_2 production rate as they had a relationship of the concentrations as

$$[F_2] = k[F]^n \quad \text{with} \quad n = 1.4 \tag{3.6}$$

However, the direct measurement of the band intensities is misleading in LiF as not only is there an overlap of the tail of the F band with the F_2 band (some 1.5%, Hodgson et al., 1979), but, more seriously, there is a second, higher, excited state of the F_2 centre, the F_2^*, which occurs precisely at the position of the F band. Making corrections for these overlap terms, and assigning oscillator strengths, f, as 0.56, 0.28 and 0.39 for the F, F_2 and F_2^* bands led them to the classic square law power dependence between F and F_2 production:

$$[F_2] = K[F]^2 \tag{3.7}$$

Examples of data are shown in Figure 3.9 which indicate the scale of the correction factors. It is assumed that the oscillator strength is unchanged by the high defect concentrations and the stresses. This appears to be a valid assumption in this case but it might not be correct with all types of implant conditions. Note also that there is not a simple dose dependence leading to a saturation situation; instead there are several linear stages followed by a relaxation. These were tentatively explained in terms of stress relief and plastic flow of the highly disordered lattice (i.e as also suggested for the glass data shown in Chapter 2).

The uncorrected exponent value of 1.4 in Equation (3.6) has also been reported by other authors using gamma irradiation of doped LiF (Sonder and Templeton, 1972) and for H^+ and H_2^+ ion implants in KBr (Saidoh and Townsend, 1977), suggesting that this type of difficulty in resolving overlapping absorption bands from different defects may be quite common in alkali halides, although it is rarely discussed. For the ion beam examples the k value is obviously only an effective value which subsumes a variety of effects including the changes in ionisation density along the ion track.

The electronic stopping power estimated for protons entering a carbon target, which is similar to that in LiF, shows the same energy dependence as the measured k values of Equation (3.6) for protons in LiF. This is shown in Figure 3.10. In addition to the normal energy dependence of the electronic stopping, one can vary the overall track density by using

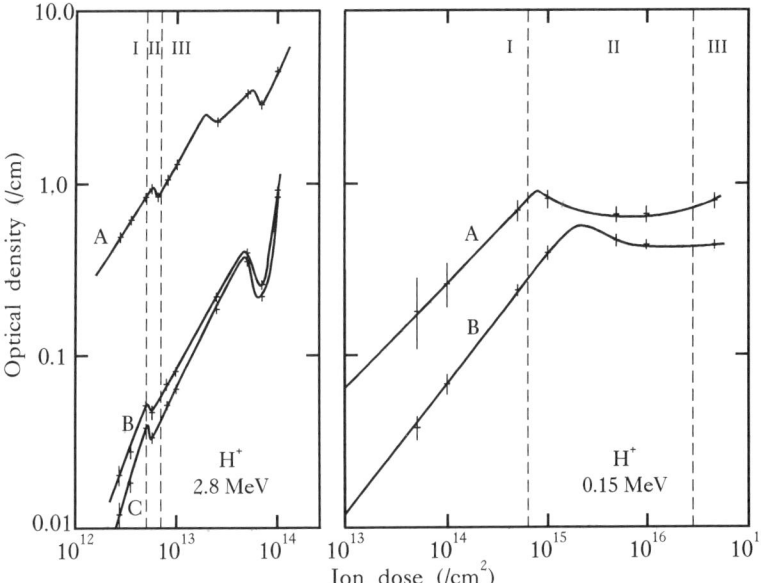

Fig. 3.9. The optical density in the region of the F and F_2 bands (245 and 445 nm) for proton implanted LiF. Note that at certain dose levels the stress is sufficient to cause a relaxation of the damage. The tails of the bands overlap and there is an excited state of the F_2 band, the F_2^*, at the same wavelength as the F band. The dose dependence of the signal is given by curves A at 245 nm, B at 445 nm and C after subtraction of the tail of the F band at 445 nm.

heavier ions or molecular beams of H_2^+ or N_2^+. The data of Figure 3.10 demonstrate that this alteration has a significant effect on both the measured k value and, by implication, the K values of Equation (3.7). In Figure 3.10 the circled data were made for the same projected range of 1 micron. The power dependence of K in $[F_2] = K[F]^2$ goes as $E^{-2/3}$, as does the electronic stopping power for protons, but calculating in terms of the total F-type defects gives a different power dependence of $[F]+2[F_2] \propto E^{-1/2}$. At a constant energy the effective F band production increases as $(dose)^{0.67}$. Previous discussion of defect production involving vacancy defect production rates controlled by interstitial trapping at impurity sites has been modelled by Guillot and Nouailhat (1976, 1979) for alkali halides and by Chadderton (1971) for silicon, and in each case the equivalent exponents lie between 0.5 and 1.0. Thus the actual defect production rate in the LiF is presumably influenced in the same way by the background impurities. However, since the diffusion of interstitials to impurities will be a function of their relative concentrations, one assumes

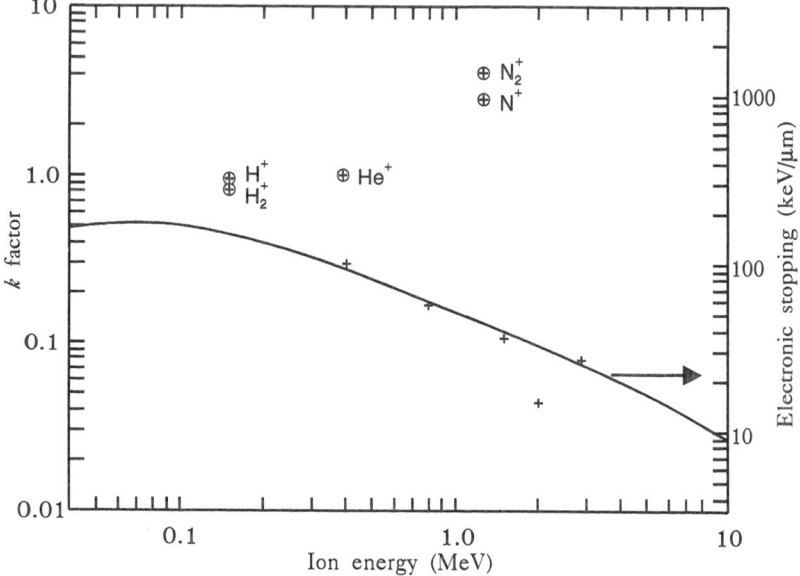

Fig. 3.10. A comparison between the electronic stopping power of protons in carbon (which is similar to that in LiF) with the factor k of Equation (3.6) determined for proton implants in LiF. The circled data are for ions with the same projected range of 1 micron but with different defect production rates.

that both the power dependence of the growth and the saturation level are functions of the initial ion beam energy since the observations sum the variations in defect concentrations along the ion range. This is the observed result, but it is difficult to separate the contributing factors such as the rate of energy transfer, stress, radiation enhanced diffusion and directional effects.

3.5.2 F_3, F_2' and F_3' bands

A particularly obvious example of stress effects occurs in the alkali halides. The simplest defects are the F, F_2, F_3 and F_4 centres which form from electrons trapped at one or more halogen vacancies. In ion implanted alkali halides the F, F_2 and F_4 absorption bands are seen, but the F_3 band is absent (Afonso et al., 1985; Wood and Townsend, 1991). There is also loss of the F_2' and F_3' bands which correspond to the trapping of an additional electron. Although the first supposition is that the F_3 defect does not form, or a weakly bonded electron is lost from the

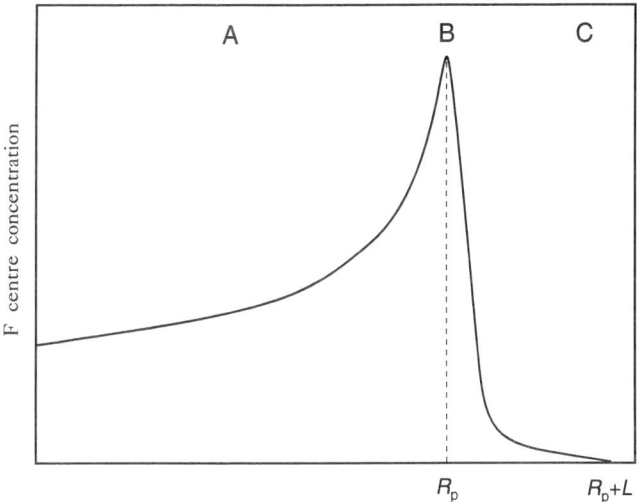

Fig. 3.11. A schematic of the F centre production with depth in ion implanted LiF. Note the damage production extends to a depth of $R_p + L$, which is greater than the expected range spread.

defect, one should also consider the possibility that the stress perturbs the energy states sufficiently that the new optical transition is displaced so far from the original photon energy that it is not immediately recognised as being associated with a distorted version of the same defect.

3.5.3 Other features

The LiF data indicate several other features which are probably quite general in character. For example, Perez (1974) showed that the F centre concentration produced with very high energy ions is highly non-uniform with depth. Not only is there a maximum near the end of the ion range, but in addition many defects are formed at much greater depths (Figure 3.11) than the projected range of the incident ions. In alkali halides damage production beyond the expected ion range is not entirely unexpected since the halides damage by photolytic processes of ionisation or even with photons having less than the bandgap energy, i.e. by exciton production. Factors which can contribute to these alternative mechanisms include diffusion of excitons, vacancies or penetration of energetic secondary electrons. Ion channelling might also be a possibility in some situations.

3.6 Defect complexes

In addition to the basic F- and F_2-type defects, one may observe complexes with impurities. Abu Hassan et al. (1988) and Wood and Townsend (1991,1992) have seen optical absorption bands near 258 nm and 414 nm which they ascribe to impurity versions of the F and F_2 centres. These interconvert with annealing or long term storage. The following set of absorption spectra for proton implanted LiF demonstrates that the colour centres exist in a wider variety of related types than is normally considered in LiF damaged by ionising radiation. Similar complexity probably exists in other materials but the alkali halides have sufficiently resolved features that the detailed problems can be observed. Very similar features have been reported for implants in other alkali halides (Comins et al., 1988), in NaF (Kristianpoller et al., 1992) and also in the related materials such as MgF_2 (Davidson et al., 1993). For LiF, Figure 3.12(a, b) emphasises that in the region of the F band implantation with protons introduces many new absorption features. There is a movement of the overall peak position from 235 to 255 nm resulting from the growth of new defect centres. Deconvolution of the components is not simple as there are obviously impurity factors since different 'pure' samples do not show identical behaviour.

On moving from nominally pure material to the case of LiF doped with Mg and Ti, as used for thermoluminescence dosimeter material, the presence of impurity levels shows yet more complex defect centres. The impurities result in lower limits for the saturation value of the F band. Figures 3.13 and 3.14 indicate examples of spectra taken for different dopant levels. The data for Figure 3.13 were taken for a sample containing 20 ppm of Mg and between 0.3 and 0.8 ppm of Ti. With increasing dose at 77 K new bands develop and the peak moves towards 260 nm. After implantation the defects are not stable and even at 77 K some components disappear. On increasing the Mg content to 150 ppm there are significant changes in the details of the component absorption bands, and their rate of production, although the broad envelope of colouration remains in the same spectral region.

The implantation introduces impurity ions, e.g. protons in this example, and so may add new bands such as the U centres where the hydrogen enters halogen vacancies, or bonds in interstitial sites. These are unstable at room temperature. Consequently, Wood and Townsend (1992) considered the effects of annealing low temperature implants up to 150 °C, and then made further implants to see which of the colour

3.6 Defect complexes

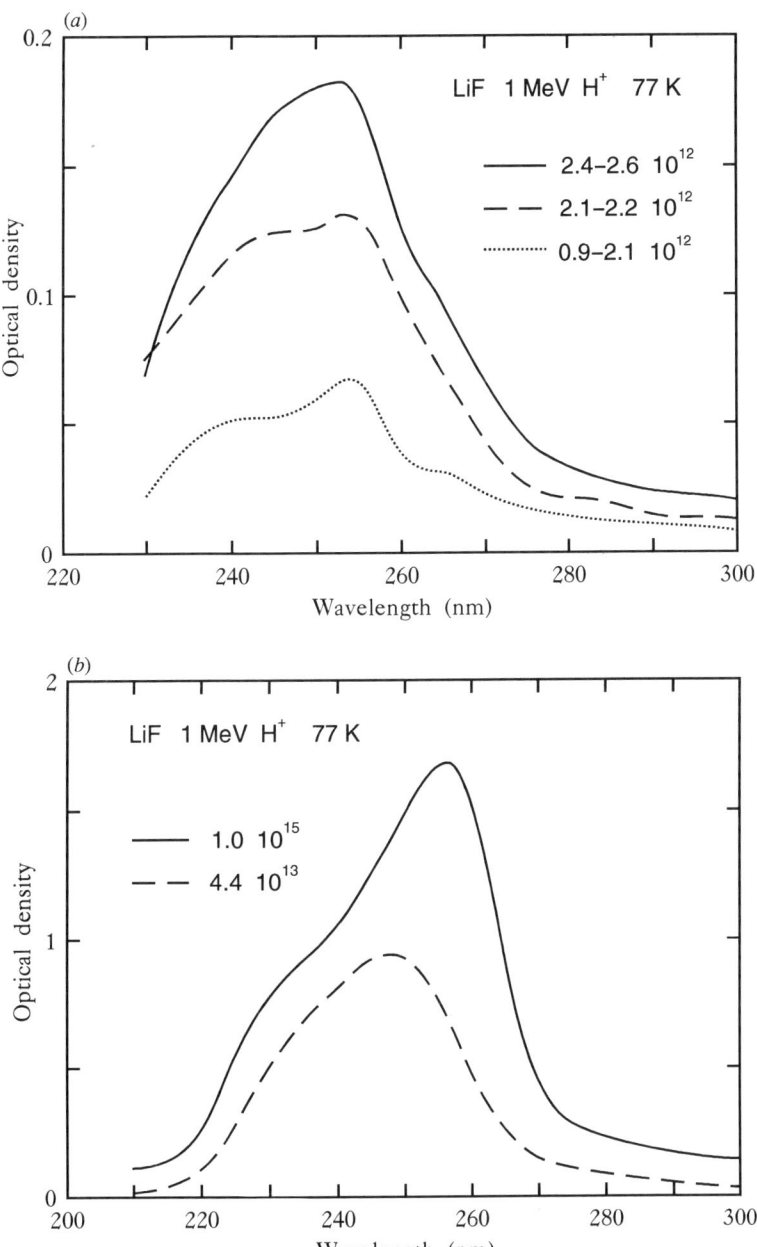

Fig. 3.12. (a) Absorption bands induced by 1 MeV proton implants at 77 K. Measurement was made during the implantation at 77 K so at these low doses each curve has some dose variation. Doses are in ions/cm^2. (b) Higher dose absorption curves following from (a).

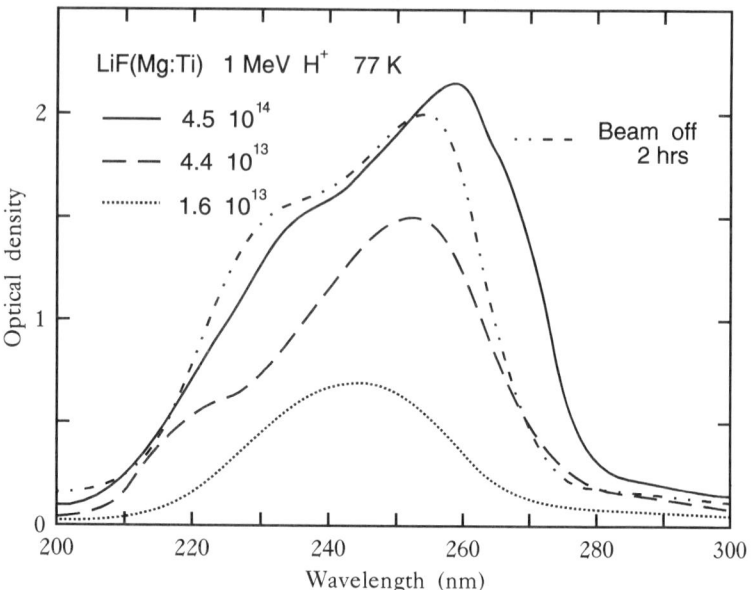

Fig. 3.13. Growth of absorption in the F band region with H^+ implantation at 77 K in LiF doped with 20 ppm Mg and ~ 0.5 ppm Ti. Note that further changes occur after implantation.

centres are influenced by the presence of the hydrogen, the F and F_2 colour centres, or intrinsic/impurity complexes. Figure 3.15 shows data for a 20 ppm Mg doped sample in the region of the F_2 band (Figure 3.15(a)) and the F band (Figure 3.15(b)). Heating from 77 K introduces a variety of changes. These depend on both the anneal temperature and the dopant level. In the pure sample the F_2 band develops a side band at 414 nm. In the region of the F band the doped samples similarly reveal a range of new features. In particular, it should be noted that the total optical absorption may increase with warming. Possible effects are that a break up of the F_2/impurity complexes may be expected to increase the fraction of F/impurity sites and hence alter the envelope of the F band absorption. Such changes are clearly apparent in samples which have been implanted for a second time, that is, after they have already undergone an initial annealing treatment. Figure 3.16 indicates an even wider range of related defect absorption bands which can contribute to the overall F or F_2 spectral regions. Detailed discussion of the point defect models is irrelevant here, but the key point to note is that, even for

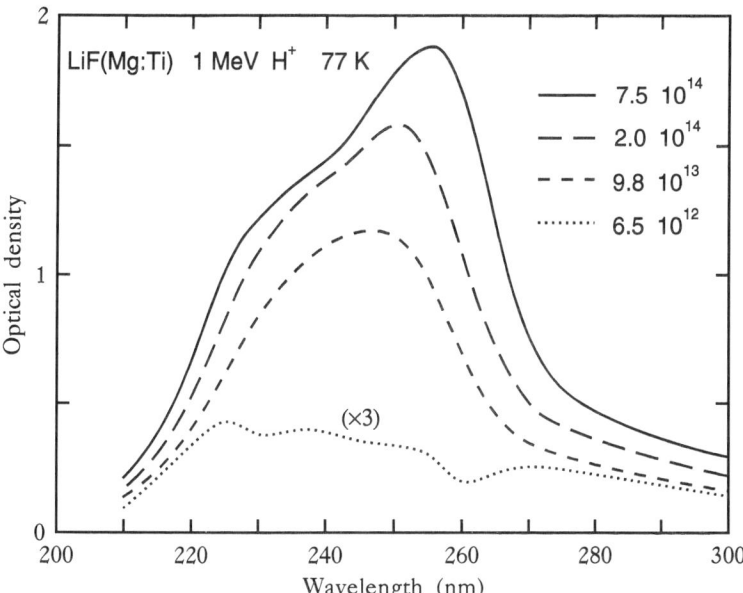

Fig. 3.14. Growth of absorption in the F band region with H^+ implantation at 77 K in LiF doped with 150 ppm Mg and $\sim$ 0.5 ppm Ti.

'simple' defects, the implants can reveal a wealth of complex associations of intrinsic and impurity features.

3.7 Growth curves

The growth curves, shown earlier in Figure 3.9, variously give apparent (and corrected) dose dependencies which are a function of the background impurities and the ion energy. They also indicate saturation and relaxation stages which may correspond to plastic flow of the damaged lattice. For example, the presence of ion implanted defects introduces strain which can be monitored by subtle optical absorption features such as changes in zero phonon lines. One such study was reported for inhomogeneous broadening of the 2.366 eV line from the F_4 defect in LiF at 4 K. Proton implants produce effects linked to the growth and collapse of colour centres (Blieden *et al.*, 1993). Products of such a collapse of the damaged lattice are platelets or clusters of F centres (e.g. Hobbs *et al.*, 1973). Similarly, excess metal, either from the result of metallic ion implantation or ions liberated from the target structure, may precipitate

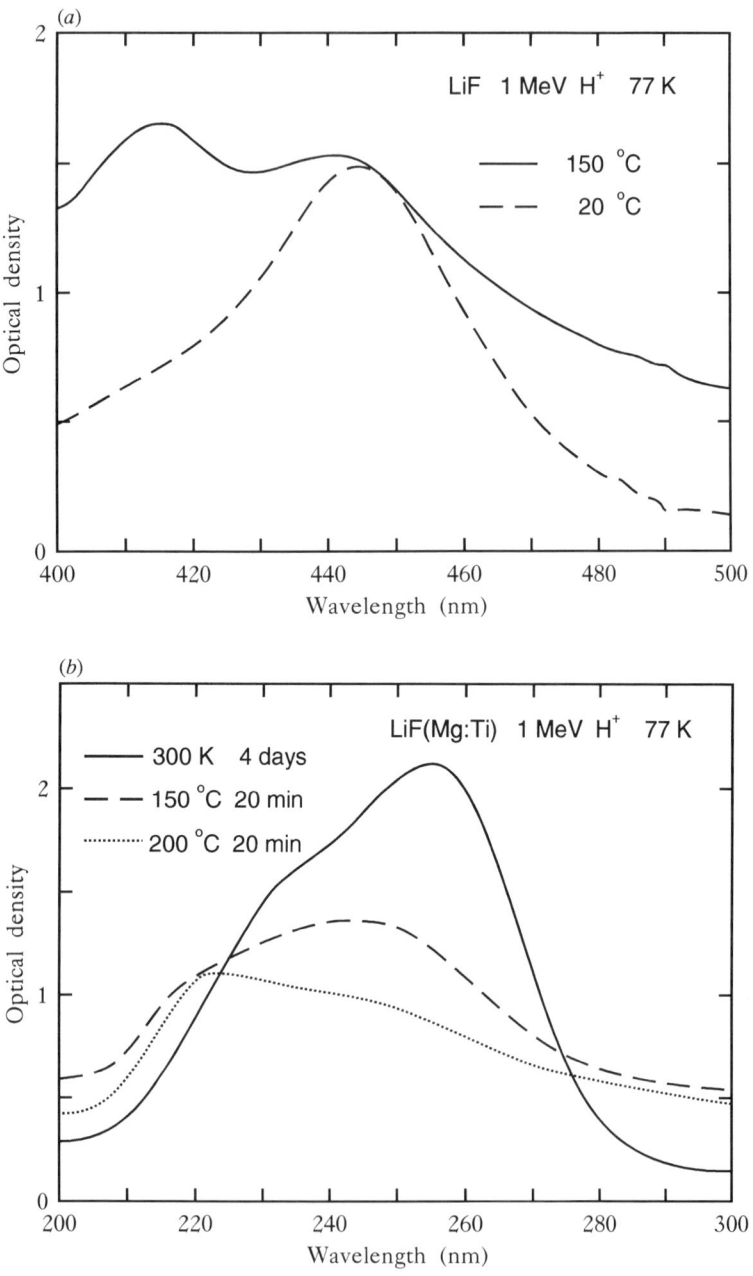

Fig. 3.15. Absorption curves in the region of (a) the F_2 band, and (b) the F band recorded at room temperature for a LiF sample implanted with protons at 77 K and then annealed to higher temperatures.

3.8 Molecular beam effects on absorption 95

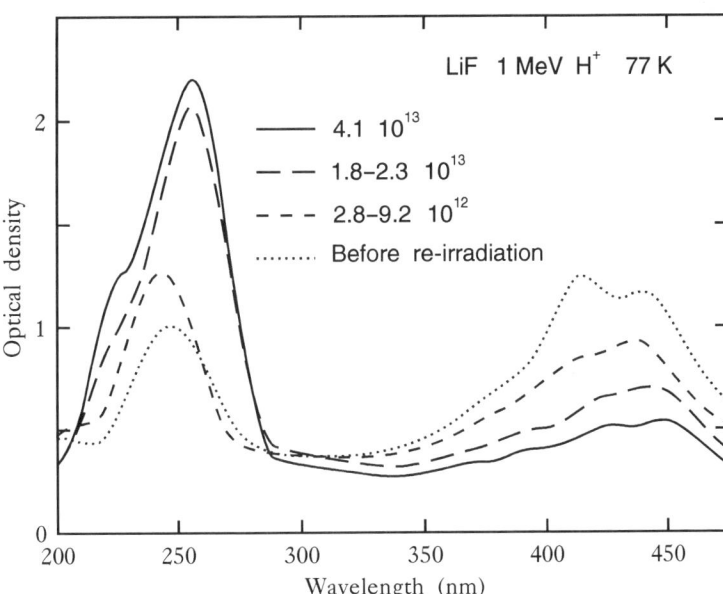

Fig. 3.16. Absorption curves recorded at room temperature for a LiF sample which had been implanted with protons at 77 K, annealed to higher temperatures and then implanted with further 1 MeV H^+ doses.

into colloids. For the alkali halides many colloid absorption bands have been recorded. Their growth is influenced by impurities, and LiF examples have been reviewed by Davidson *et al.* (1986) and Comins *et al.* (1988) and typified by their data shown in Figure 3.17. The diagram is particularly interesting as it is one of the few examples of optical absorption produced by ion beams which includes data taken near the UV band edge, although vacuum UV data are reported for the wide band gap material MgF_2 (Davidson *et al.*, 1993), and for silica (Antonini *et al.*, 1982) and sapphire, as mentioned in Section 3.4.

3.8 Molecular beam effects on absorption

As emphasised above, there are differences in the relative production rates of related colour centres which depend on the ionisation and defect production rates within an ion track. When such features are suspected, as between F and F_2 centres, then by changing ion mass, and hence the percentage of electronic and nuclear damage, it should be possible to emphasise interactions between related defects. An alternative route

96   *Optical absorption*

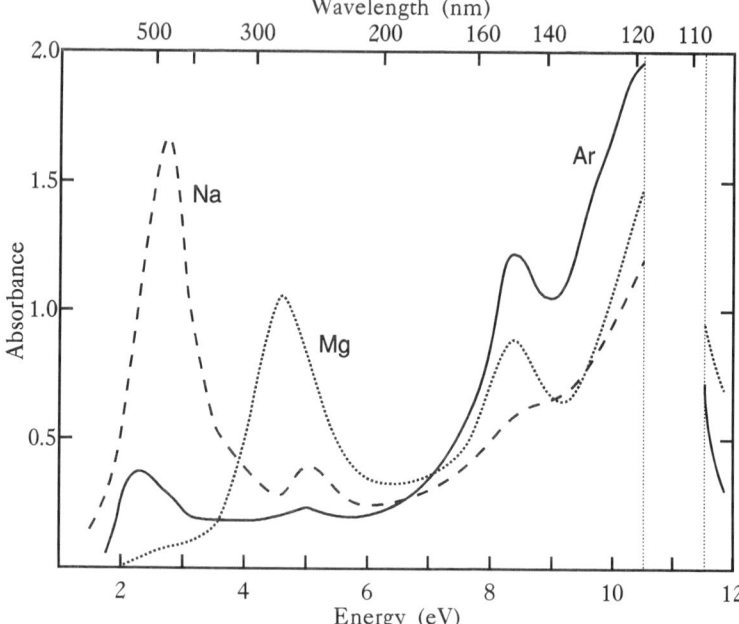

Fig. 3.17. Examples of the absorption spectra of LiF samples implanted with Na, Mg or Ar ions at 100 keV to doses of 10^{17} ions/cm^2.

to varying the rate of energy deposition in the damage track is to use molecular ion beams. This has the advantage that a comparison can be made between a singly charged ion M^+ at energy E and a molecular ion M_2^+ at energy $2E$. In both beams the particles will arrive at the surface with the same velocity and, after separation, the fragments of the molecule will have the same momentum and cover essentially the same projected range. The major difference is that the two components of the molecule will be both spatially and temporally coincident and so will generate a much higher energy density within the ion track. If the energy density effects are important, then, for normal ion beams, overlap of tracks during excitation are negligible because of the brief duration and small scale of any individual event. To avoid differences in input power, and hence temperature differences during the comparisons of the two beams, the higher energy beam should be at half the beam current (but equal particle flux). Variations on the theme have so far used simple gas sources such as hydrogen and nitrogen, but it should be noted that these may include metastable ions to give the sequence H^+, H_2^+ and H_3^+.

3.8 Molecular beam effects on absorption

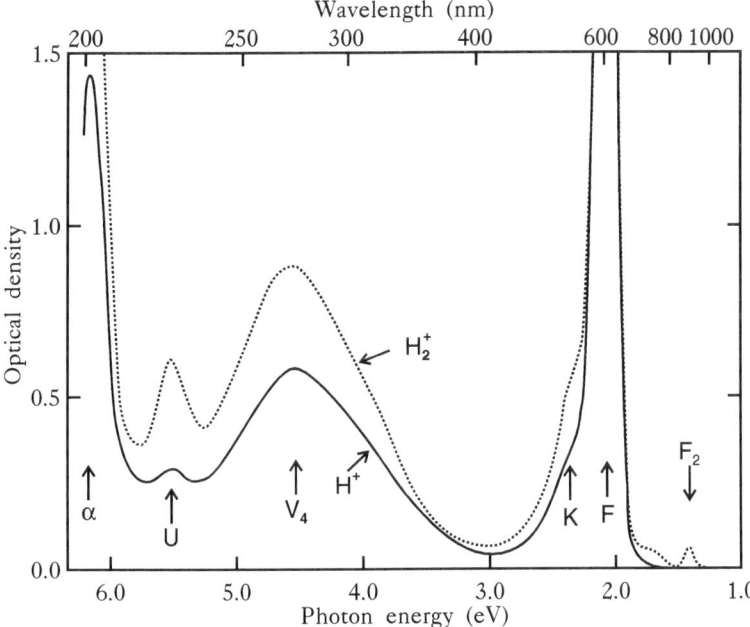

Fig. 3.18. Differences in optical absorption produced and measured at 40 K in KBr by 400 keV H^+ and 800 keV H_2^+ ion beams to the same particle dose. Note changes in the F_2, U and V_4 defect concentrations.

Examples of changes in the balance between alkali halide F and F_2 centres (i.e. vacancy and divacancy), and between H and V_4 centres (i.e. interstitial and di-interstitial) were demonstrated by Saidoh and Townsend (1977). Figure 3.18 shows the absorption spectra produced and measured at 40 K in KBr by beams of H^+ and H_2^+ ions. The total particle dose, beam power and ion ranges were maintained constant by using 400 keV H^+ and 800 keV H_2^+ beams which differ in current by a factor of two. Note that the interstitial (H) centre does not show a resolved absorption band in this example but the relative of the F_2 band is significantly altered. The effects are sensitive to the total damage level and Figure 3.19 presents the growth curves for the bands identified with F, F_2, U and V_4 centres. Both the divacancy and di-interstitial bands initially grow more rapidly during the molecular beam implant, whereas the F centre rate is suppressed. Although it was not the objective of the Saidoh and Townsend experiment, the use of hydrogen to generate the U band confirms that this is from a hydrogen related

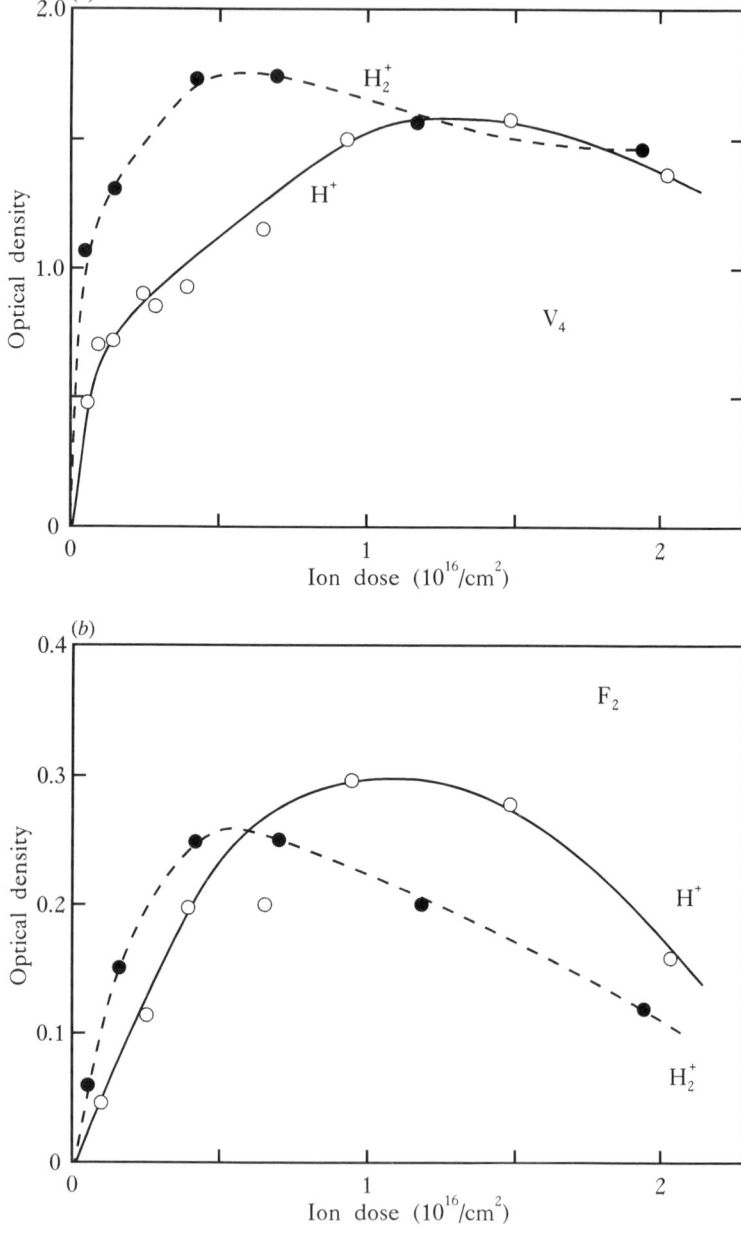

Fig. 3.19. For legend see facing page

3.8 Molecular beam effects on absorption

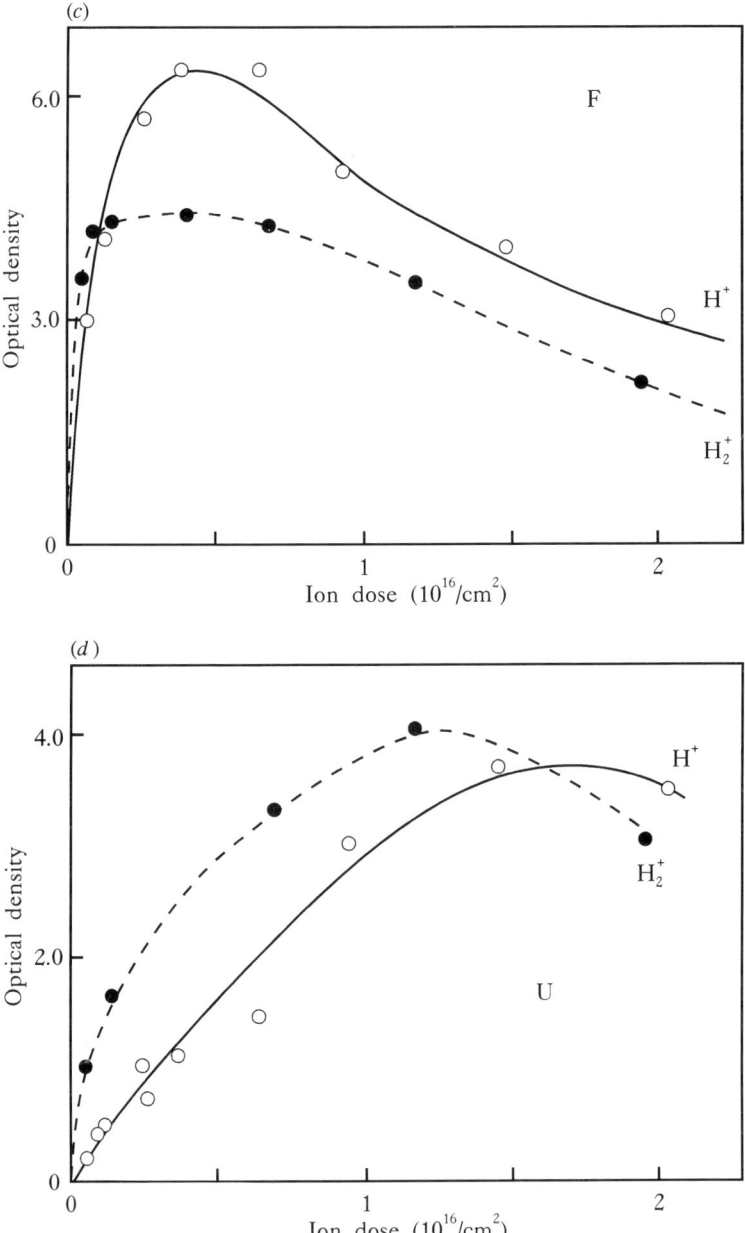

Fig. 3.19. Growth curves for the 40 K implants in KBr of H^+ and H_2^+ ion beams. The curves are for V_4, F_2, F and U bands.

defect. Ion/molecular beam comparisons have also been used in a search for bi-exciton luminescence (see Section 4.2).

3.9 Isotopic and ion species effects

The U centre production by hydrogen implantation may be viewed as a chemical method of confirming an impurity model of a defect. More subtle control of the implant ion will allow selection of specific isotopes so that both mass and hyperfine interactions can be varied. In studies of vibrational resonances examples exist for infra-red frequency shifts caused by changes between H^+ and D^+ implants in O—H vibrations in silica glasses (Mattern et al. 1976). The H and D variants have similarly been studied in rutile by Guermazi et al. (1987), but in this example such high impurity doses were implanted that chemical changes ensued and a new phase of Ti—O—OH developed and was detectable by high resolution microscopy and X-ray diffraction, as well as the infra-red optical absorption bands.

Differences between C—H and C—D bond modes were reported by Patrick and Choyke (1973) for SiC and examples of CO and CO_2 vibrational spectra are discussed by Bibring et al. (1984) for carbon implanted silica. In the latter example the relative intensity of the CO_2 signal falls with impurity doping whereas the CO signal rises rapidly.

Many implanted ions introduce infra-red absorption changes but, except for the lightest ions, isotopic separation of the vibrations is rarely possible. Magruder et al. (1990) compare the infra-red absorption features caused by implants of Ti, Cr, Mn, Fe, Cu, and Bi in silicate glasses. In particular they monitor two bands centred near 1232 and 1015 cm^{-1} (i.e. 8.117 and 9.852 microns) which they relate to the presence of a change in the degree of intermediate range order (i.e. over the scale of 2–5 nm), or non-bridging oxygen defects. The relative magnitudes of the two bands depend on the chemical properties of the implant and the consequent influence on the longitudinal vibrational modes. The observations are not simple to interpret as the sequence of relative effectiveness on the intermediate range order, by changing ion species, is different for low and high dose implants. One reason proposed is that the bismuth interacts chemically to form a different Si—O network, but the other ions only perturb the normal system.

Implantation in glass produces several changes including formation of new structures, as well as rearrangement of the chemical composition. Arnold et al. (1990) have combined visible, infra-red spectrometry with

3.10 Measurement of oscillator strength

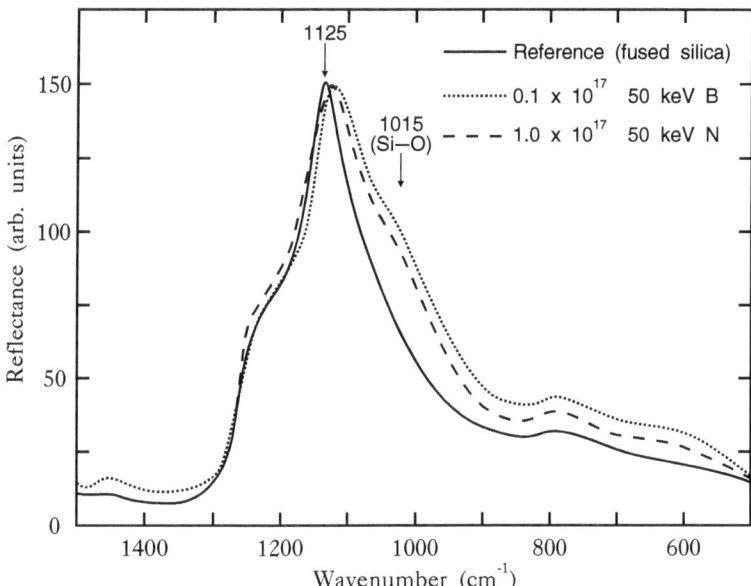

Fig. 3.20. Examples of changes in the infra-red vibrational spectra in silica implanted with B or N ions.

other analysis techniques for silica modified by H, Li, B, N, O and Si. Figure 3.20 shows a change in the infra-red as a result of B and N implants. For nitrogen the overall interpretation is that new phases of silicon oxy-nitride may occur, but these are not always fully compensated by existing ions. Therefore, secondary implants with silicon are needed to react with unbonded nitrogen to establish full stoichiometry of the new layer. Boron similarly bonds directly into the lattice.

Somewhat different behaviour occurs on annealing Li^+ implanted silica as a new crystalline layer emerges which, as measured from the infra-red reflectivity, resembles quartz, Figure 3.21. For this example the altered surface layer was achieved by using a 250 keV Li ion dose to 10^{17} ions/cm^2 and then annealing at 800 °C. Similar results for Li implanted silica and silicate glass are given by Arnold (1980) and Arnold and Peercy (1980).

3.10 Measurement of oscillator strength

The total dose delivered by an ion beam of implanted impurities is readily monitored by the beam current and deposited charge. There is

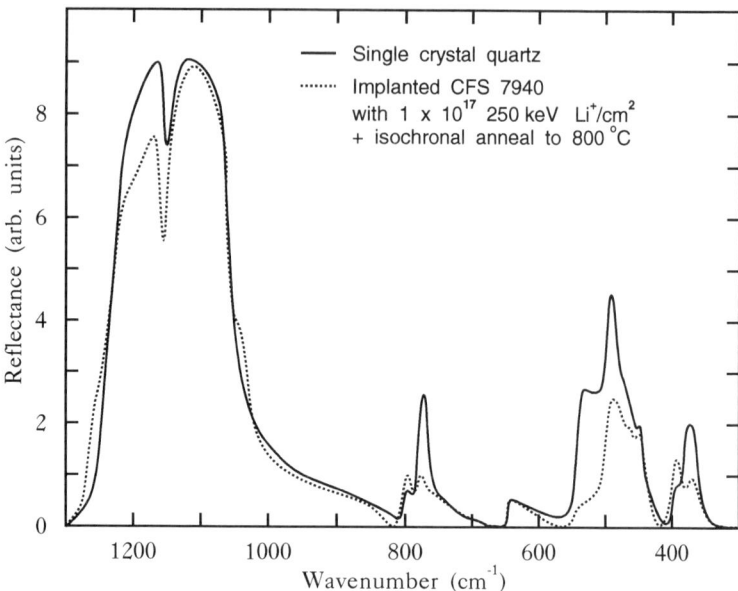

Fig. 3.21. A comparison of the infra-red absorption spectrum of quartz with that of material which has crystallised as a result of Li$^+$ implants and annealing in silica.

therefore an accurate control of the process, and an accurate measure of the dopant which is added to the target. When the optical absorption bands are clearly linked to these impurities, as in F_A-type defects, the known dopant value can be used to set limits on the value of the oscillator strength. Equation (3.3) showed that normally one can only assess the product term Nf in the Smakula expression, and separation of the two parameters is difficult. Henderson (1982) and Henderson and Garrison (1983) compared the production of defects in MgO, CaO and SrO as the result of Li$^+$ and Na$^+$ implantation. From the data they deduced an f value of 0.8 for the F$^+$ band observed at 250 nm. Having defined one f value it may then be possible to infer values for related defects such as the F_2 and F_2^{2+} centres which were generated by the ion beams. Henderson and Garrison made simultaneous measurements with electron spin resonance (ESR) but failed to detect Li ions, hence they assumed the Li did not substitute on to the cation sites. Instead, lithium and sodium are assumed to cluster into colloidal precipitates.

3.11 ESR and ENDOR

Isotopic control of the implant ions could be exploited in ESR and electron nuclear double resonance (ENDOR) to map out the range of defect interactions and defect symmetry. This possibility has been exploited by very few groups. One such is that of Devine and his collaborators (Devine et al., 1983, Devine and Fiori, 1986). From their work with ion implanted silica they noted a correlation between their ESR signal and an infra-red absorption band. This led them to ascribe the defects to complementary centres in which one was a hole trap and the other an electron trap. The damage profile of argon implanted silica has been determined by the ESR of the E_1' centre (an electron trapped at a bridging oxygen site).

Ion implantation in glasses generates the standard defect ESR signals (e.g. Weeks, 1991; Derryberry et al., 1991) such as E'-type centres in silica. The production rates are sensitive to the mass and the chemical nature of the implants. In an experiment to investigate defect formation in nitrogen implanted silica, Hosono et al. (1991) used a combination of vacuum UV absorption and ESR to identify the defects in OH free material. An absorption band at 7.5 eV (the E" centre from a neutral oxygen vacancy) was produced by N, Si or N plus Si implants. The relative intensities emphasise that it results from the Si—Si bonding. Approximately half of the N ions enter the glass network as N—Si bonds and the rest of the nitrogen appears as N_2 molecules. Detailed bond models were attempted using hyperfine interactions by implanting with either ^{14}N or ^{15}N. The ESR signals also revealed Si—N—Si hole traps; an e^- trapped N on a SiO_4^{4-} vacancy and a 17-electron NO_2 molecule.

Magnetic states which give ESR signals are identifiable with metallic impurity implants and the data prior to 1991 are reviewed by Weeks (1991).

3.12 High dose effects

3.12.1 Amorphisation

Amorphisation problems have already been introduced in Chapter 2. The simplest of these is amorphisation of the crystal lattice, or, in glasses, a relaxation to a new packing density. As indicated above, the energy deposited to amorphise a lattice is dependent on the type of bonding and the retention of relaxations and displacements of the target ions (Naguib and Kelly, 1975). Burnett and Page (1986) sketched a curve which relates these parameters, Figure 2.14 and this offers a useful first guide to the ion dose required, but it does not include temperature, directional or impurity effects. It should also be obvious that since amorphisation is associated with large lattice parameter changes, and therefore stress, the ion dose for amorphisation will vary with the depth of the amorphisation beneath the surface. The extensive review of amorphisation of surfaces of oxides by White *et al.* (1989) offers some scale for the ion doses required to amorphise the crystals, rather than the 'universal' units of energy/cm^3. For a particularly stable material, sapphire, the room temperature amorphisation dose is some 2×10^{17} ions/cm^2 for 150 keV Cr$^+$ ions, but this number falls to as low as 2×10^{15} ions/cm^2 when the implant is made at 77 K. This is probably an extreme example, in part because chromium has the same valence as the aluminium, and chromium ions only introduce a small degree of lattice mismatch if they move to lattice sites (i.e. as in ruby). Chemically stabilised disorder eases the process of amorphisation, and for sapphire implanted with group IV elements, C or Si, and also Zr, a factor of 10 less dose is required than when using comparable mass ions of different valency. One suggestion for this mechanism is that the carbon forms carbide bonds by substitution on to the oxide sublattice. These so distort the rhombohedral sapphire lattice that further damage is readily retained. Note also (Section 3.4) that data from Agnew (1992) suggests that even greater efficiency may be possible using boron, rather than group IV ions. In summary, the energy needed for amorphisation is primarily governed by retention of defects rather than difficulty in displacing lattice atoms.

Light ions such as helium should in principle require far greater ion doses as they deliver less nuclear collision energy than heavy ions. Hence to amorphise sapphire the calculated doses might be estimated to be some 10^{18} He$^+$ions/cm^2. However, at these dose levels the composition has so changed that the target is no longer sapphire. Further, with an inert gas, there will be bubble formation which will greatly distort the lattice. In attempts to amorphise sapphire with helium at high doses a problem occurs where the gas bubbles form a layer and cause the surface to peel, or eject entire surface blocks (this is termed blistering or exfoliation). In other crystals, such as LiNbO$_3$, light ion implantation never produces amorphisation, as the refractive indices within the damaged region do not reach a common value as they would for an amorphous material. This suggests that whilst the damage generates a highly disordered LiNbO$_3$/He layer, a combination of crystal field and/or stress terms drives a local recombination of the disorder which produces epitaxially regrowth aligned with the underlying lattice. Nevertheless, one should emphasise that with many materials helium ion doses of, say, 4×10^{16} ions/cm^2 readily induce complete amorphisation. Such examples certainly include natural minerals of quartz and zircon. Initial data indicate that many other materials behave similarly, for example paratellurite, calcite and $KNbO_3$. Not only are these materials less ionic but, at least for the first two, they naturally exist in both crystalline and amorphous phases. In mineralogy, samples which are amorphised by the radiation damage resulting from naturally occuring radioactive impurities are termed metamict. A survey of metamict materials may thus point the way to other candidates for ion beam amorphisation.

Amorphisation may not always be a desirable outcome of implantation. Diamond and silicon are covalently bonded and so are easily amorphised. In the case of silicon this is a useful step in the formation of semiconductor devices, since amorphisation with a silicon beam allows subsequent doping to be made into an amorphous layer. This avoids problems of channelling and lack of control on the dopant profile. For silicon the amorphous layer can be thermally recrystallised. By contrast, implants in diamond readily form amorphous carbon, possibly as graphite, and recrystallisation is not possible. There have been numerous attempts to avoid this difficulty, both to grow large diamonds by carbon implants into seeds, and to exploit semiconducting properties of diamond. Partial success has been reported for both objectives by implanting at high temperatures with careful control of the ion energy and dose rate. Rusbridge and Nelson (1980) used carbon implants between 600 and 900 °C, whereas Braunstein and Kalish (1981, 1983) went to higher implant temperatures followed by annealing to 1400 °C. For their semiconducting diamond they used dopants of Sb, P and Li. Optically the samples became darker if a graphite layer developed.

3.12.2 Colloids

Addition, or liberation, of metal ions frequently results in the formation of colloidal precipitates. These are characterised by strong optical absorption and reflectivity bands whose magnitude and peak positions change with colloid size. Earlier, but extensive, reviews of colloid properties have been made by Hughes and Jain (1979) and Hughes (1983). The band shape and position is a function of the metal, the dielectric in which it is embedded and the geometry of the atomic cluster. (An elliptical colloid may show two absorption bands characteristic of the two major diameters.) In a free electron gas model the dielectric parameters ϵ_1 and ϵ_2 are given by

$$\epsilon_1 = \epsilon_0 - \omega_p^2(\omega^2 + \omega_0^2)^{-2} \tag{3.8}$$

$$\epsilon_2 = \omega_p^2 \omega_0 (\omega(\omega^2 + \omega_0^2))^{-2} \tag{3.9}$$

where ω_p is the plasma frequency of the metal and ω_0 the collision frequency of free electrons. For small colloids of radius $r < 10$ nm embedded in a material of refractive index n_0 the peak wavelength, λ_m, of the absorption band, and halfwidth, $\Delta\lambda$, have been related from the Mie (1908) theory by Doyle (1958) by

$$\lambda_m = 2\pi c \omega_p^{-1}(\epsilon_0 + 2n_0^2)^{1/2} \tag{3.10}$$

$$\Delta\lambda = c(2\sigma)^{-1}(\epsilon_0 + n_0^2) \tag{3.11}$$

However, as the colloid size increases, the peaks move to longer wavelengths and the bands broaden. Colloid bands are therefore readily characterised by their changing position during colloid growth, whether from additional metal or during aggregation during annealing. Typically, there is colloid growth at temperatures up to, say, 300 °C, but evaporation into smaller clusters at more elevated temperatures. Colloids are frequently observed when an excess of monovalent metals or Ag, Cu and Au are present. In practice there is a relatively limited range of refractive indices so, independently of the host material, colloids from a particular metal give similarly positioned colloid bands. In extensive studies of colloids formed by ion implantation by the Lyon group, Thevenard et al. (1973a,b) used a set of alkali halides with additions of excess alkali metal to produce peaks at 680, 520 and 445 nm for K, Na and Li in LiF. Davidson et al. (1986) extended measurements into the UV (Figure 3.17) and saw colloid-type bands after Mg^+, Na^+ and Ar^+ implants. However, the Mg and earlier Li implants both gave a colloid peak near 275 nm. A similar band is initially shown in Figure 3.14 for proton implantation in one of the Mg doped LiF crystals, but it did not develop strongly in any of the proton damaged LiF. Consequently, there may be doubt as to whether the colloid is from the Li or Mg, particularly since Mg will readily substitute on to the Li lattice site.

Colloid formation is equally evident in ion implanted oxides. Since colloid formation is almost unavoidable for many ion beam conditions the literature includes a wide range of examples, for instance bands from Li^+ and Na^+ in MgO, CaO and SrO (Henderson and Garrison, 1983); various metals in MgO (Treilleux and Chassagne, 1980; Treilleux and

3.12 High dose effects

Thevenard, 1985; Perez et al., 1981; Canut et al., 1991). Silver and copper colloid examples include glass (Ke-Ming et al., 1985; Weeks, 1991; Hosono et al., 1992; Nistor et al., 1993; Wood et al., 1993) and $LiNbO_3$ (Rahmani et al., 1988, 1989).

Determination of colloid size is complicated by the facts that at small diameter the peak position is insensitive to the colloid size, and implantation does not form a uniform concentration of free metal and hence the colloid sizes within the peak of the damage distribution will differ from those at other depths. Standard depth analysis of implanted metal using RBS does not reveal that there can be a wide range in the colloid size distributions with depth. There is a further complication that radiation enhanced diffusion can allow colloids to cluster over a greater depth than expected from the ion range theory. Figure 3.22 shows data on the size distribution determined by electron microscopy of a glass implanted with silver ions (Nistor et al., 1993). A single figure is insufficient to convey all the information, as with depth beneath the surface one may specify the total silver content, which agrees with the RBS, the mean colloid size, and concentration of colloids. The example chosen is for 60 keV Ag ions at a dose level of 4×10^{16} ions/cm^2. RBS depth analysis is in broad agreement with the data of Figure 3.22(d).

Based on the optical measurements one cannot determine this level of complexity as the optical absorption bands from the various colloid sizes overlap in wavelength. Further, at some depths the colloids are in such close proximity that the simple analysis model, of isolated colloids, is inappropriate. The problem has been appreciated for a long time (Arnold and Borders, 1977). A simplistic analysis in terms of an average colloid size is equally difficult as both in transmission and reflection the apparent optical signals will be a summation of absorption and reflection terms in each of the colloid layers. Consequently, since the distribution pattern is not symmetric, the reflectivity determined from the ion implanted and rear faces of the sample may differ. The reflected intensity from a particular layer will depend on the optical absorption, and, thus, the intensity reaching and returning from the layer. A particularly clear example of such differences between front and rear reflections is given by Figure 3.23 for Cu^+ implanted in silica (Magruder et al., 1991). RBS depth analysis of the sample shows that the copper has become distributed around two depths, which, from the optical data, are interpreted as being of different size colloid distributions. Such bimodal distributions in the RBS were noted for silver (Arnold and Borders, 1977) and for copper (Magruder et al., 1990).

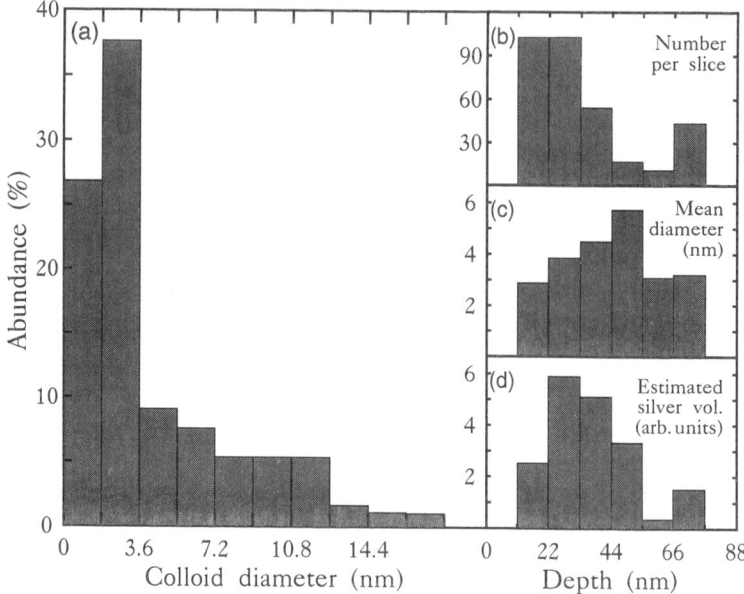

Fig. 3.22. Data on colloid size distributions resulting from Ag ion implantation into a silicate glass: (a) colloid size distribution as seen by TEM, (b) colloids per 11 nm depth interval, (c) mean diameter, (d) overall amount of Ag with depth.

There are now a number of examples of TEM studies of colloids formed by ion implantation, and in some cases, as for copper, the bimodal RBS data are mirrored by the TEM photographs (e.g. Hosono *et al.*, 1992). Nevertheless, there are other data which suggest that such features can be generated during the RBS by varying the analysing beam current. A more generally agreed result is that the RBS and TEM representations do not offer the same image of the depth distribution of the implanted metal, unless one integrates the total colloid distribution as in Figure 3.22(*d*). There is often evidence of colloids well beyond the expected implant distribution, suggesting that radiation enhanced diffusion transports material to the limits of the damage region. Finally, some authors note that the deepest colloids may form from metal ions of the bombarded glass rather than the implanted impurity.

3.12.3 Precipitate phases

Not only is there a range of colloid sizes, but also the metal may exist in several charge or structural states. This was mentioned earlier for Fe in

3.13 Summary of problems in interpretation

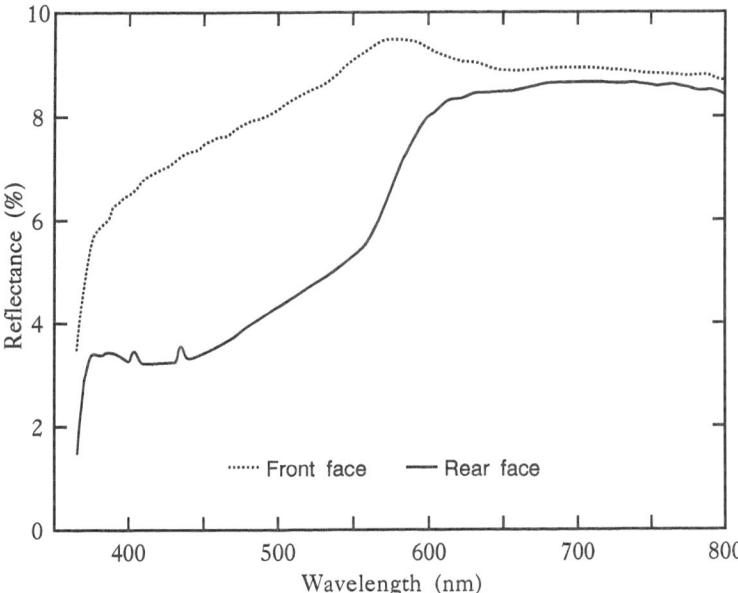

Fig. 3.23. Reflectivity spectra of silica implanted with copper, monitored from the implanted and rear faces.

sapphire (Figures 3.7 and 3.8). Similar examples using Mossbauer data were given by Kowalski *et al.* (1983) for Fe in LiF and Fe in MgO (Perez *et al.*, 1981). In the LiF the Fe^{3+} and Fe^{2+} were initially in proportions of 80 to 20%. The form of the iron content changed with total iron concentration to the charge states Fe^0 and Fe^{2+}, with some 30 and 40% in these forms. After annealing, further changes result in some 35% of the iron being precipitated out of the lattice as inclusions. Similarly, in the MgO example, annealing showed a range of iron compounds including Fe_2O_3, $MgFe_2O_4$ and Fe_3O_4.

3.13 Summary of problems in interpretation

It is clear that many factors influence the growth of optical absorption bands as a result of ion implantation. Because the ion dose is generally small compared with the number of intrinsic defects that are formed, one might at first sight ignore the possibility of intrinsic/impurity complexes resulting from the implant. However, most materials contain surprisingly large concentrations of trace impurities and, because of defect mobility, they associate with the intrinsic defects. Hence, a prediction of the rate of production of even the simple F-type centres in alkali halides is not possible; instead one must rely on observation. The LiF examples cited earlier had the advantage of relating to a material in which the defects had been well characterised so these provided a useful starting point for interpretation of ion implanted modifications. The results therefore indicate the inherent interpretive problems that one needs to face in other implanted materials. Weeks' (1991) review and current conference proceedings offer a wider perspective of materials in which ion implanted absorption studies are in progress. In many instances, as with silica (e.g. Antonini *et al.*, 1982; Magruder *et al.*, 1991; Polman *et al.*, 1991a; Polman and Poate, 1993), the optical absorption bands are either poorly resolved or dominated by absorption features which exist in non-implanted material. The presence of absorption does not of itself indicate future potential of the implanted material although it may be a necessary measurement in progress towards new applications such as lasers or non-linear properties (e.g. Becker *et al.*, 1991; Polman *et al.*, 1991b; and later sections of this text). Thus, although the problems are not specific to alkali halides, the factors which they underline are worth considering for interpretation of future studies include the following:

The F centre production rate varies in depth within the crystal.
The F_2 depth profile is not necessarily identical to that of the F.
Both F and F_2 production rates will be controlled by impurities.
The implant is also an impurity which may influence the growth.
Stress may suppress some types of defect.
The oscillator strength can alter with local disorder of the lattice.
Defects may interconvert.
Optical absorption bands overlap.
A single defect type may have several absorption bands.
Saturation levels will be a function of ion energy and impurities.
Collapse of a highly strained lattice can occur.
Colloidal defects and new phases may develop.
Saturation and plastic flow may occur.
Nuclear and electronic damage may not be independent.
Electronic stopping is a function of ion energy and ion mass.
Secondary effects occur within the ion track.
The diameter of the ion track is a strong function of energy.

Despite the resultant difficulties in interpretation it should be emphasised that ion defect absorption bands can provide new data on

3.13 Summary of problems in interpretation

implantation processes which may not be accessible by other routes. The full value of using optical measurements, particularly those taken during the implantation, for ion implanted insulating materials will only be apparent when there is a much larger data base involving many more materials.

References

Abu Hassan, L.H. and Townsend, P.D. (1986a). *J. Phys.*, **C19**, 99.
Abu Hassan, L.H. and Townsend, P.D. (1986b). *Rad. Effects*, **98**, 313.
Abu Hassan, L.H., Townsend, P.D. and Wood, R.A. 1988. *Nucl. Inst. Methods*, **B32**, 225.
Afonso, C.N., Ortiz, C. and Clark, G.J. (1985). *Phys. Stat. Sol.*, **(b)131**, 87.
Agnew, P. (1992). *Nucl. Inst. Methods*, **B65**, 305.
Agullo Lopez, F., Catlow, C.R.A. and Townsend, P.D. (1988). *Point Defects in Materials* (Academic Press, London).
Andrews, J.W., Hu, Y.Z. and Irene, E.A. (1989). Multichamber and in-situ processing of electronic materials *SPIE*, **1188**, 162.
Antonini, M., Camagni, P., Gibson, P.N. and Manara, A. (1982). *Rad. Effects*, **65**, 41.
Arnold, G.W. (1980). *Rad. Effects*, **47**, 15.
Arnold, G.W. and Borders, J.A. (1977). *J. Appl. Phys.*, **48**, 1488.
Arnold, G.W., Brow, R.K. and Myers, D.R. (1990). *J. Non-Cryst. Solids*, **120**, 234.
Arnold, G.W. and Mazzoldi, P. (1987). *Ion Beam Modification of Insulators* (Elsevier, Amsterdam).
Arnold, G.W. and Peercy, P.S. (1980). *J. Non-Cryst. Solids*, **41**, 359.
Becker, K., Yang, L., Haglund, R.F., Magruder, R.H., Weeks, R.A. and Zuhr, R.A. (1991). *Nucl. Inst. Methods*, **B59/60**, 1304.
Bibring, J.P., Chaumont, J. and Rocard, F. (1984). *Nucl. Inst. Methods*, **B1**, 628.
Blieden, G.S., Comins, J.D. and Derry, T.E. (1993). *Nucl. Inst. Methods*, **B80/81**, 1215.
Braunstein, G. and Kalish, R. (1981). *Nucl. Inst. Methods*, **182/183**, 691; ibid(1983), **209/210**, 387.
Burnett, P.J. and Page, T.F. (1984). *Proc. Brit. Ceram. Soc.*, **34**, 65.
Burnett, P.J. and Page, T.F. (1986). *Rad. Effects*, **97**, 283.
Burns, T.M., Chongsawangvirod, S., Andrews, J.W., Irene, E.A., McGuire, G. and Chevacharoeukul, S. (1991). *J. Vac. Sci. Tech.*, **9**, 41.
Canut, B., Dupin, J.P., Gea, L., Ramos, S.M.M. and Thevenard, P. (1991). *Nucl. Inst. Methods*, **B59/60**, 1211.
Chadderton, L. (1971). *Rad. Effects*, **8**, 77.
Chen, Y., Abraham, M.M. and Pedraza, D.F. (1991). *Nucl. Inst. Methods*, **B59/60**, 1163.
Comins, J.D., Davidson, A.T. and Terry, T.E. (1988). *Defect and Diffusion Forum*, **57/58**, 409.

References

Dalal, M.L., Rahmani, M. and Townsend, P.D. (1988). *Nucl. Inst. Methods*, **B32**, 61.
Davidson, A.T., Comins, J.D, Derry T.E. and Khumalo, F.S. (1986). *Rad. Effects*, **98**, 305.
Davidson, A.T., Raphuthi, A.M.J., Comins, J.D. and Derry, T.E. (1993). *Nucl. Inst. Methods*, **B80/81**, 1237.
Derryberry, S.L., Weeks, R.A., Weller, R.A. and Mendenhall, M. (1991). *Nucl. Inst. Methods*, **B59/60**, 1320.
Devine, R.A.B. and Debroux, M.H. (1983). *J. Appl. Phys.*, **54**, 4197.
Devine, R.A.B., Ferrieu, F. and Golanski, A. (1983). *Nucl. Inst. Methods*, **209/210**, 1201.
Devine, R.A.B. and Fiori, C. (1986). *Rad. Effects*, **99**, 191.
Doyle, W.T. (1958). *Phys. Rev.*, **111**, 1067.
Evans, B.D. and Hendricks, H.D. (1977). In *Ion Implantation in Semiconductors and Other Materials*, (Boulder, 1976) F. Chernow (ed.) (Plenum Press).
Guermazi, M., Thevenard, P., Brenier, R., Thomas, J.P. and Mackowski, J.M. (1987). *Nucl. Inst. Methods*, **B19/20**, 912.
Guillot, G. and Nouailhat, A. (1976). *Journ. de Phys.*, **37 C7**, 611.
Guillot, G. and Nouailhat, A. (1979) *Phys. Rev.*, **B19**, 2295.
Henderson, B. (1982). *Rad. Effects*, **64**, 35.
Henderson, B. and Garrison, A.K. (1983). *Rad. Effects*, **74**, 167.
Hodgson, E.R., Delgado, A. and Alvarez Rivas, J.L. (1979). *J. Phys.*, **C12**, 4393.
Hobbs, W.L., Hughes, A.E. and Pooley, D. (1973). *Proc. R. Soc.*, **A332**, 167.
Hosono, H., Abe, Y. and Matsunami, N. (1992). *Appl. Phys. Lett.*, **60**, 2613.
Hosono, H., Abe, Y., Oyashi, K. and Tanaka, S. (1991). *Phys. Rev.*, **B43**, 11966.
Hughes, A.E. (1983). *Rad. Effects*, **74**, 57.
Hughes, A.E. and Jain, S.C. (1979). *Adv. in Phys.*, **28**, 717.
Ke-Ming, W., Burman, C., Lanford, W. and Groleau, R. (1985). *Rad. Effec. Lett.*, **85**, 177.
Kowalski, J., Marest, G., Perez, A., Sawicka, B.D., Sawicka, J.A., Stanek, J. and Tyliszczak, T. (1983). *Nucl. Inst. Methods*, **209/210**, 1145.
Krefft, G.B. (1977). *J. Vac. Sci. Tech.*, **14**, 533.
Kristianpoller, N.N., Blieden, G.S. and Comins, J.D. (1992). *Nucl. Inst. Methods*, **B65**, 484.
Magruder. R.H., Morgan, S.H., Weeks, R.A. and Zuhr, R.A. (1990). *J. Non-Cryst. Solids*, **120**, 241.
Magruder, R.H., Zuhr, R.A. and Weeks, R.A. (1991). *Nucl. Inst. Methods*, **B59/60**, 1308.
Mattern, P.L., Thomas, G.J. and Bauer, W. (1976). *J. Vac. Sci. Tech.*, **13**, 430.
McHargue, C.J., Farlow, G.C., Sklad, P.S., White, C.W., Perez, A., Kornilios, N. and Marest, G. (1987). *Nucl. Inst. Methods*, **B19/20**, 813.
Mie, G. (1908). *Ann. Phys.*, **25**, 377.
Mouritz, A.P., Sood, D.K., St John, D.H., Swain, M.V. and Williams, J.S. (1987). *Nucl. Inst. Methods*, **B19/20**, 805.
Naguib, H.M. and Kelly, R. (1975). *Rad. Effects*, **25**, 1.
Nistor, L.C., van Landuyt, J., Barton, J.D., Hole, D.E., Skelland, N.D. and Townsend, P.D. (1993). *J. Non-Cryst. Solids*, **162**, 217.
Patrick, L. and Choyke, W.J. (1973). *Phys. Rev.*, **B8**, 1660.
Perez, A. (1974). PhD Thesis, Universite de Claude Bernard, Lyon, France.
Perez, A., Davenas, J. and Dupuy, C.H.S. (1976). *Nucl. Inst. Methods*, **132**, 219.

Perez, A., Treilleux, M., Fritsch, L. and Marest, G. (1981). *Nucl. Inst. Methods*, **182/183**, 747.
Polman, A., Jacobson, D.C., Lidgard, A. and Poate, J.M. (1991a). *Nucl. Inst. Methods*, **B59/60**, 1313.
Polman, A., Jacobson, D.C., Eaglesham, D.J., Kistler, R.C. and Poate, J.M. (1991b). *J. Appl. Phys.*, **70**, 3778.
Polman, A. and Poate, J.M. (1993). *J. Appl. Phys.*, **73**, 1669.
Pooley, D. (1966). *Brit. J. Appl. Phys.*, **17**, 855.
Rahmani, M., Abu Hassan, L.H., Townsend, P.D., Wilson, I.H. and Destefanis, G.L. (1988). *Nucl. Inst. Methods*, **B32**, 56.
Rahmani, M. and Townsend, P.D. (1989) *Vacuum*, **39**, 1157.
Rusbridge, K.L. and Nelson, R.S. (1980). Paper B1 presented at the 1980 IBMM Albany Conference (unpublished).
Saidoh, M. and Itoh, N. (1973). *J. Phys. Chem. Solids*, **34**, 1167.
Saidoh, M. and Townsend, P.D. (1977). *J. Phys.*, **C10**, 1541.
Saito, Y., Horie, H. and Suganomata, S. (1991). *Nucl. Inst. Methods*, **B59/60**, 1173.
Saito, Y., Kumagai, H. and Suganomata, S. (1985). *Japanese J. Appl. Phys.*, **24**, L880–2.
Sonder, E. and Templeton, L.C. (1972). *Rad. Effects*, **16**, 115.
Thevenard, P., Davenas, J., Perez, A. and Dupuy, C.H. (1973a). *Journ. de Phys.*, **34 C5**, 79.
Thevenard, P., Perez, A., Davenas, J. and Dupuy, C. H. (1973b). *Journ. de Phys.*, **34 C9**, 289.
Townsend, P.D. (1987). *Rept. Prog. Phys.*, **50**, 501.
Treilleux, M. and Chassagne, G. (1980). *Journ. de Phys.*, **41 C6**, 391.
Treilleux, M. and Thevenard, P. (1985). *Nucl. Inst. Methods*, **B7/8**, 601.
Weeks, R.A. (1991). *Mat. Sci. and Tech.*, **9**, 331.
White, C.W., McHargue, C.J., Sklad, P.S., Boatner, L.A. and Farlow, G.C. (1989). *Mat. Sci. Rept.*, **4**, 41.
Wood, R.A. and Townsend, P.D. (1991). *Nucl. Inst. Methods*, **B59/60**, 1331.
Wood, R.A. and Townsend, P.D. (1992). *Nucl. Inst. Methods*, **B65**, 502.
Wood, R.A., Townsend, P.D., Skelland, N.D., Hole, D.E., Barton, J. and Afonso, C.N. (1993). *J. Appl. Phys.*, **74**, 5754.

4
Luminescence

4.1 Luminescence processes

Luminescence transitions may occur within localised defect sites, for example to give the characteristic line emissions of rare earth ions, narrow emission bands of chromium in ruby, or they may produce broad bands from charge transfer between defects. Overall, emission bands may vary greatly in width, but nevertheless the luminescence spectra provide a measure of specific defect types, and even offer some quantitative measure of the changes in defect concentrations. Since the excitation energy for luminescence may be provided by many routes, ion implantation is no exception and it frequently produces strong luminescence from insulating targets. This feature is often used as a means of aligning the ion beam and it is common practice to have defining apertures with silica plates to check the ion beam position visually. Such intense luminescence can reveal a number of features relating to the changing defect state of the target. For example, in many of the materials used to form optical waveguides by ion implantation there is a decrease in luminescence intensity which approximately follows the amorphisation in the crystal. Hence, one has a visual estimate of the progress of the amorphisation. Quantitative recording of the wavelength dependence of the signal, in terms of luminescence efficiency and spectral changes, should provide details on not only the defects pre-existing in the material but also the ion beam induced changes. Consequently, a number of research groups have used the luminescence produced during implantation to follow such modifications.

Lattices modified by the ion beam will show changes in their subsequent luminescence performance and the effects of implantation have been recorded in photoluminescence, laser emission, cathodoluminescence and thermoluminescence. Even for light ions with MeV energy the ions only penetrate a few microns into the surface. Therefore, these changes are difficult to resolve from bulk features if the luminescence is subsequently studied using an excitation energy which penetrates the entire sample, as do X-rays. The most obvious changes due to implantation occur during the ion beam excited measurements, or with cathodoluminescence, in which the electrons only penetrate a comparable depth into the surface as the ion beam. Subsequent applications, such as waveguide lasers, will also reveal changes in energy levels, lifetime and line widths resulting from perturbation of the luminescence sites.

Although rarely studied intentionally, contaminants on the surface of the material may contribute significantly to both the ion beam luminescence and cathodoluminescence. Such impurity effects can easily be misinterpreted as arising from the implanted layer of the target since the presence of the surface contaminants may only be apparent, or indeed may only adhere to the surface, in the presence of the ion beam. For example, the implantation may selectively sputter one species from a compound target, and thus allow a new range of surface reactions. An example in which the original interpretation of the luminescence signals was ascribed to defects within the $LiNbO_3$ target, rather than to surface contaminants, is mentioned in Section 4.4.8. The detection of small quantities of surface contaminants is possible because luminescence techniques, e.g. using photon counting, are extremely sensitive. Therefore, even submonolayer concentrations of active sites can directly generate a detectable signal. However, confusion may arise because the observed signals may be relatively strong despite the fact that the contaminant level is low. In part this may be because the energy deposited from the ion beam is transported via excitons (i.e. bound e^-h^+ pairs) or free charges to the luminescence sites. Impurities and surfaces both act as exciton traps and luminescence sites so energy release can be concentrated at special sites. This implies that the emission spectra are biased in favour of defect sites. At the other extreme of high concentrations of dopants, luminescent materials formed by implantation can lead to waveguide laser production, new phosphors for TV screens and more sensitive radiation dosimeters.

4.2 Luminescence during ion implantation

Not only can the ion beam induced luminescence provide a convenient aid to beam positioning and an indication of the changes in the defect types and concentrations which are taking place in the target crystal, but additionally it may provide information on transient defect states formed during the implantation process. The transient features may include transitions at defects or ionised states which only exist within the ion track during the initial energy deposition. Similarly, transient signals may be generated by ionisation of defects in the cascade regions of the nuclear collision damage. In this respect luminescence is far more sensitive than optical absorption since one can use photon counting and, inherently, the signal to noise for luminescence signals is superior to that for absorption since luminescence is seen against a zero background whereas absorption is measured as a decrease in a strong optical signal. Post-implantation excitation may therefore show different emission spectra from those recorded during implantation.

It should be recalled that luminescence may arise from either simple electron–hole recombination, recombination after lattice relaxation at a specific defect or exciton site, or charge transfer processes between spatially linked defect sites. Finally, light may be generated after long range charge transport in the conduction band. The luminescence signal may therefore contain a wide variety of features, some of which are characteristic of the intrinsic defects, whilst others are monitors of impurity related states.

4.3 Effects of implantation temperature

The implantation generates a non-thermodynamic set of conditions within the ion track and hence it is appropriate to question whether the effective lattice temperature is the same as the bulk temperature as recorded by thermocouples. This is particularly relevant for ion beam bombardment of insulators as the deposition of several watts/cm^2 into a micron-thick layer of a poor thermal conductor is likely to develop very strong temperature gradients. The luminescence itself may in principle provide a probe of the implant volume. One such example of the luminescence approach was made during MeV electron bombardment damage of ruby by Townsend (1961) and Mitchell and Townsend (1963) and re-evaluated for ion beam damage by Chandler and Townsend (1979a). The characteristic red emission lines from the chromium impurity ions in the Al$_2$O$_3$ lattice result from an inner shell transition within the chromium energy levels, but the crystal field coupling produces a splitting of the levels and a temperature dependence. These features can be determined under equilibrium conditions by optical excitation of the luminescence,

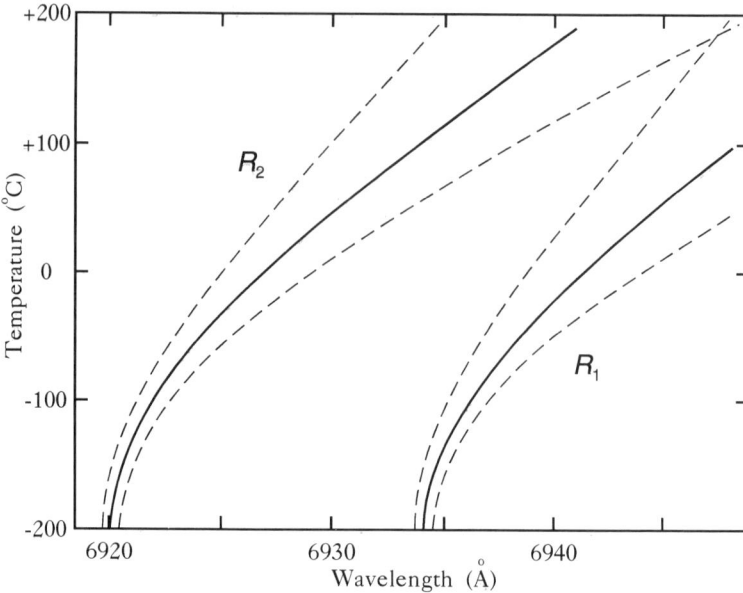

Fig. 4.1. The temperature dependence of the R_1 and R_2 emission lines of ruby. The dashed lines indicate the halfwidth values.

as shown in Figure 4.1. Further, because the excited lifetimes of the states are milliseconds, the emission in the ion beam excited case is likely to be representative of the pseudo-equilibrium temperature conditions, rather than dynamic events within a collision cascade. Chandler and Townsend found that the emission lines broadened, as expected for a temperature gradient in the layer, and underwent wavelength shifts corresponding to differences in temperature between the implant layer and the bulk material. Ruby is particularly interesting in this example as the thermal conductivity of the crystal at low temperature (e.g. at 50 K) is greater than that for copper, but this conductivity falls by a factor of 100 by room temperature. Consequently, implants at equal power in cooled and ambient temperature substrates produce quite different effective heating. The results reported were that for a 0.5 mm-thick ruby crystal at 77 K an input power of $0.2 \, \text{W/cm}^2$ caused a local temperature rise of <10 K. In a less favourable condition for temperature accuracy, implantation of a 2 mm-thick sample at room temperature with a power of $3 \, \text{W/cm}^2$ produced a temperature rise of 120 K within the implant layer. In both cases the important feature was that such temperature excursions were not recorded by thermocouples clamped to the crystals. Both examples

4.4 In situ luminescence

indicate that the temperature of the implant layer differs from the bulk measurement but the precise temperature rise is difficult to estimate.

4.4 *In situ* luminescence

Luminescence may be generated in a variety of ways, with familiar routes being by optical or electron beam excitation. However, ion implantation into an insulating target material is invariably accompanied by visible luminescence. The intensity, and even the spectrum, will often vary visibly with ion beam dose or target temperature. Since these variations reflect the changing defect state of the target, one may monitor them in an effort to understand the damage processes which occur. As mentioned earlier, the high excitation density within the ion beam track may reveal transient defects which could not be sensed by subsequent excitation methods. Hence, for both reasons, it is useful to record the ion beam induced luminescence. Nevertheless, it is important to note that, as will become apparent, the simple monitor of light intensity associated with a specific type of defect may not reflect changes in concentration of that defect type correctly. Despite this caveat, there is still considerable information to be gained from the study of ion beam excited luminescence. Examples of several such studies will now be discussed.

4.4.1 Alkali halides – excitons

There have been extensive studies of point defects in alkali halides and many of the defect types are very well characterised. One well documented feature is the exciton. The exciton absorption occurs close to the absorption band edge but there is a large Stokes shift for excitons in alkali halides as they relax, and their emission occurs nearer mid-gap energies. Because of the scale of the relaxation, and the fact that exciton decay is strongly perturbed by coupling of the bound electron–hole pair to impurity sites, the precise emission wavelengths of the exciton decay vary according to the impurity or other defect sites which are perturbing the luminescence transition. In NaCl the sigma and pi emission bands are expected, theoretically, near 3.4 and 6.22 eV, respectively, with so-called, M-like levels, at lower energies. Experimentally, the highest energy band is variously seen from 5.1 to 5.8 eV and the pi band from 2.9 to 4.1 eV depending on the dopants and method of excitation. A possible interpretation of these variations is that the luminescence emission peak is strongly influenced by defect states which induce relaxed exciton decay.

120　　　　　　　　　　　Luminescence

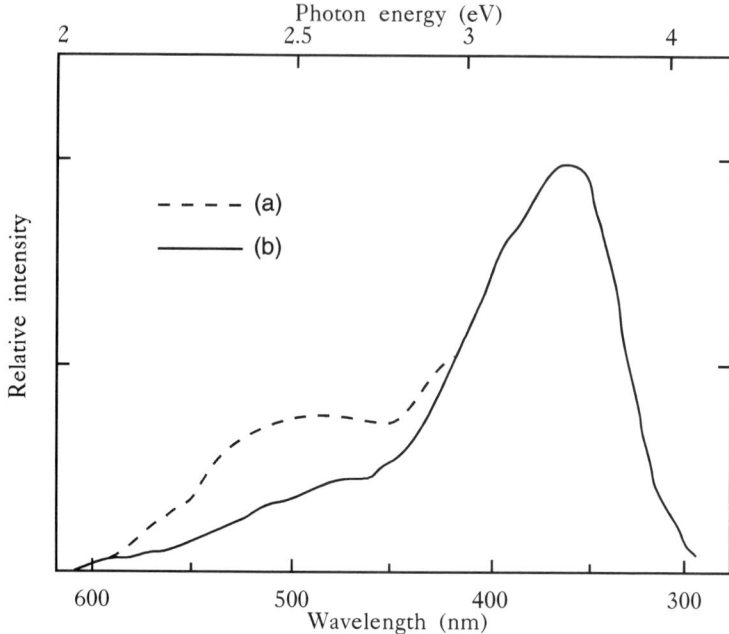

Fig. 4.2. Luminescence at 77 K from NaCl during 1 MeV N⁺ implantation. (a) is for a continuous beam and (b) rejects long lifetime components by using a PSD and 200 Hz beam modulation.

In NaCl such a multiplicity of perturbed sites have been separated during cathodoluminescence and ion beam excitation (Aguilar et al., 1979). With simple continuous beam excitation one observes an envelope of emission bands generated by decay at the various defect sites. Lifetime analysis can provide some separation of the features as the use of beam modulation and phase sensitive detection selectively rejects signals from processes with lifetimes which are long compared with the modulation period.

Figure 4.2 shows data for NaCl in which 200 Hz modulation has suppressed the M-type emission near 2.4 eV (517 nm) but retained the pi signal near 3.4 eV (364 nm). Ion beam excitation produces both an excitation density, and a dynamic defect concentration within the ion track which is not normally feasible with electron or X-ray excitation. This increase in local defect densities influences the exciton decay paths, and so in turn changes the ratio of the component bands. The interactions include both intrinsic defects and extrinsic trace impurities; consequently, emission spectra from 'pure' NaCl from different manufacturers are

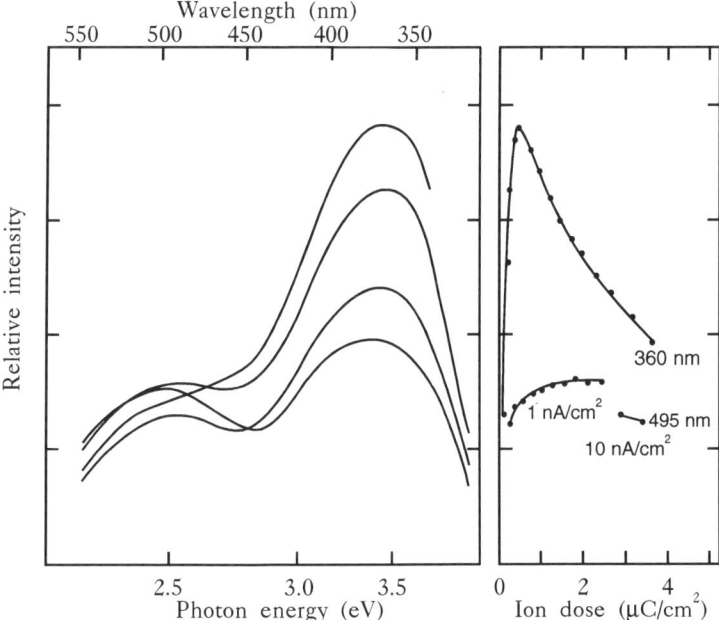

Fig. 4.3. Luminescence at 77 K from NaCl produced by Rank Industries, showing dose dependence and current density effects.

noticeably different. The data presented in Figures 4.3 and 4.4 are for exciton luminescence generated by nitrogen implantation in NaCl at 77 K for two pure crystals from different suppliers. It is immediately obvious that there are minor shifts in peak position, and the ratio of the peak intensities is inverted. In principle, the samples only differ in impurity content at the parts per million level and thus, at first sight, one would not expect significant differences for emission associated with an intrinsic excitonic decay route. However, the excitons are mobile and during diffusion will sample many lattice sites; consequently, the presence of imperfections will cause changes in the mode of decay and also introduce perturbations in the energy of the transitions. Further, the ratios of the peak intensities are not constant but change with total ion beam dose, because of the change in the number of defects in the layer. The equilibrium distribution between alternative forms of the defects is equally important, as when the ion beam current density was altered from 1 to 10 nA/cm^2, the intensities of the component emission bands moved immediately to a new equilibrium condition. Such dose rate effects

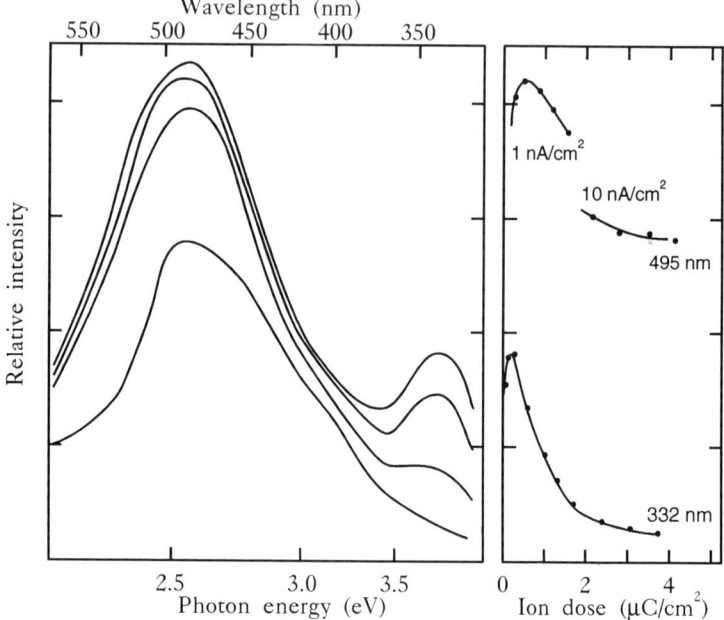

Fig. 4.4. Luminescence at 77 K from NaCl produced by the Harshaw company, showing dose dependence and current density effects.

are known to cause conversion of alkali halide defects, e.g. between the F and F_2 halogen vacancy centres.

4.4.2 Alkali halides – a search for bi-excitons

During luminescence measurements of Ne implanted NaCl, Ukai et al. (1976) proposed that the 2.4 eV emission might originate from a bi-exciton luminescence. Their data base was somewhat limited and they made their suggestion on changes in relative intensity of the 2.4 and 3.1 eV emission during the course of their implant sequence. However, as indicated above in Figures 4.3 and 4.4, the emission intensities are not constant with accumulated ion beam damage. Thus a comparison of intensity ratios is not reliable. In later work by Chandler and Townsend (1979b), H^+, H_2^+, N^+ and N_2^+ beams were used and the results showed that the decay, or growth, of the two emissions was a function of the total ion dose and the energy density in the ion track. The data suggested that the bi-exciton model for the 2.4 eV emission was not correct. The requirement for both excitons to be formed at the same time and to be spatially interacting is particularly easily tested in ion beam induced luminescence, since by converting from single ions to molecular ion bombardment one can change the local energy density deposited along the ion tracks. Molecular ions dissociate at the surface but their separate damage tracks are temporally and spatially linked. Figure 4.5(a) and (b) shows that for both N^+ and N_2^+ beams at different current densities, the ratio of the two luminescence bands cannot be used to detect bi-exciton effects. Further, on emphasising electronic excitation effects by converting from an H^+ to an H_2^+ beam (Figure 4.5(c)), the dose dependence curves are quite smooth, without any discontinuity at the point where bi-exciton yields should have altered as a result of changing excitation track density.

4.4.3 CaF_2

Aono et al. (1988) have recorded the luminescence from europium implanted into CaF_2. Emission was seen both during implantation and during subsequent excitation with other ion beams. The colour of the signal changed as a function of ion dose, changing from blue to pink to orange over the dose range 0.2 to 2×10^{15} ions/cm². The colour changes result from a mixture of emission bands centred near 360, 450, 610 and 710 nm. The violet signal at 360 nm arises from the CaF_2 host. From the spectroscopy of europium the 450 and 610 nm bands are related to the presence of Eu^{2+} and Eu^{3+} ions respectively. For charge stability one assumes that the divalent Eu corresponds to substitution of the Eu on to the Ca lattice sites. Movement of implanted material on to the lattice sites is thus accompanied by changes in the spectra. Aono et al. suggested that the europium rapidly moves substitutionally on to the CaF_2 lattice sites, but the ratio of the two charge states varies with the damage and at the higher concentrations the Eu^{2+} signal decreases.

4.4.4 Silica

Quartz and silica both produce an intense blue emission centred near 460 nm (2.7 eV) during ion beam bombardment. Weaker emissions occur near 640 nm (1.94 eV) and 280 nm (4.45 eV). These are quite specimen dependent. For example, in some silica material the red emission is initially intense but is rapidly reduced by the bombardment (although the details of the defect site have not been confirmed). As for NaCl, the blue emission is thought to arise from decay of a relaxed exciton. Additional details are reported which include some structure in the main blue band

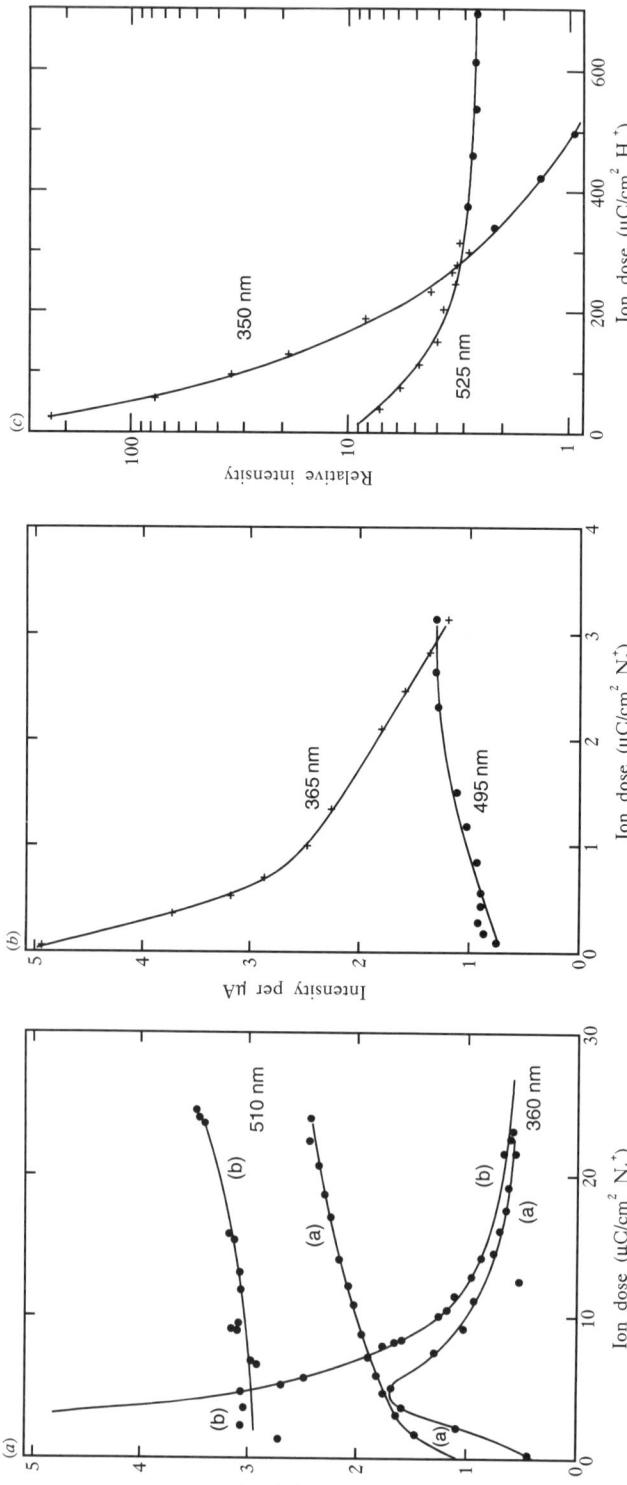

Fig. 4.5. (a) Intensity variations versus ion dose, from the region of the emission peaks using 1 MeV N^+ bombarded NaCl for two current densities of a: 230 nA/cm² and b: 16 nA/cm². (b) Signal changes in NaCl with dose using 2 MeV N_2^+ beams at 0.5 nA/cm² only. (c) 77 K luminescence from NaCl at different ionisation densities using 1 MeV H^+ ions at 60 nA to a dose of 300 μC, followed by a 2 MeV H_2^+ beam.

and evidence for line spectra from silicon that is being sputtered. Such an example is included in the data of Jaque and Townsend (1981) when using 180 keV argon ion excitation. If one concentrates on the signals originating within the implant region, then the most obvious effects are the wide range of intensity dependence with ion dose, ion energy and ion mass. This complexity has been studied by several authors (including Jaque and Townsend, 1981; Chandler et al., 1979; Abu-Hassan et al., 1987; Abu-Hassan and Townsend, 1988; Wang et al., 1991). The overall conclusion is that the luminescence is generated by exciton decay but the efficiency of such exciton formation and decay depends on the method of exciton generation, the diffusion of the excitons and, finally, the damage state of the region where the decay occurs. In order to assess the contributions to the observed total light intensity it is necessary to sum the effects from the regions in which there is (a) electronic excitation of the pure material, (b) nuclear collision damage, (c) electronic excitation of the cascade damaged regions and (d) exciton transport beyond the implant. Further, silica is particularly interesting in that electronic energy deposition can slowly generate new point defects. There are probably saturation limits to both the concentration of isolated point defects and complex clusters. Electronic excitation of nuclear collision damage (i.e. which generates the more complex defect clusters) can induce relaxation of clusters into simpler point defects, or even total recovery of the damage. One must therefore integrate effects resulting from the overlap of electronic and nuclear collisions. To a first approximation this gives terms from electronic excitation of undamaged material, light produced by nuclear collisions, and electronic excitation of collision damaged regions. Account must be taken of annealing of the nuclear damage by electronic excitation, plus the light formed by exciton diffusion beyond the implant region. The light is the result of a summation of terms such as those of Equation (4.1):

$$I = \int K_p \left[\frac{dE}{dx}\right]_{el} dx + \int K_p \left[\frac{dE}{dx}\right]_{nuc} dx$$
$$+ \int K_d \left(\left[\frac{dE}{dx}\right]_{nuc} - K_c \left[\frac{dE}{dx}\right]_{nuc} \left[\frac{dE}{dx}\right]_{el}\right) \left[\frac{dE}{dx}\right]_{el} dx \quad (4.1)$$

in which K_p, K_d relate to light production in pure and damaged silica and K_c is a measure of cascade annealing by electronic processes. Note also that light may be formed as a result of exciton diffusion and decay. Therefore, the volume influenced will not be limited to the immediate region in which nuclear and electronic energy are deposited. Although all

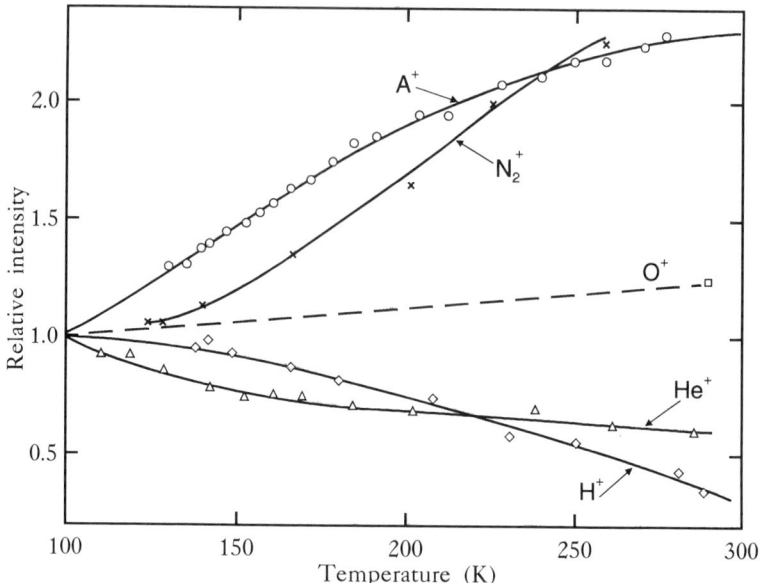

Fig. 4.6. Normalised blue luminescence intensity produced in silica at 77 K using beams of Ar^+ at 180 keV, N_2^+ at 90 keV, and O^+, He^+ and H^+ at 1 MeV.

these factors are quite simple in isolation, the overall effect is to produce a complex signal dependence on ion dose and ion species, which is of course temperature dependent. For example, in the Jaque and Townsend paper the luminescence signal with protons increased smoothly with dose, whereas for oxygen there was a steady decline at 40 K. Separation of predominantly electronic or predominantly nuclear collision effects is immediately obvious for the temperature dependence of the silica luminescence intensity as a function of ion mass (Figure 4.6).

An unfortunate result of these changing exciton luminescence efficiencies with the defect state of the silica is that if one attempts to study the energy dependence of the light production it is essential to use a new sample for each energy. The simpler approach of using one sample and varying the energy can produce a number of artefacts in the form of the curve. For example, for the low energy (<60 keV) nitrogen beam data of Figure 4.7 there is apparently a strong peak in luminescence efficiency with 40 keV ions, together with a shoulder near 10 keV. The data are not obviously sensitive to the sequence of implants. However, this low energy efficiency 'peak' vanishes on extending the range of ion energies, as given

4.4 In situ luminescence

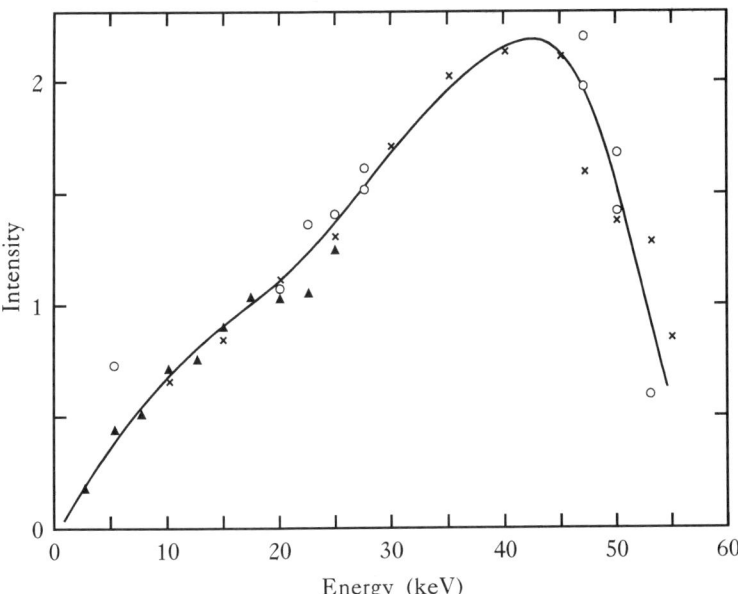

Fig. 4.7. Apparent luminescence intensity during N^+ and N_2^+ bombardment of silica. Crosses are for an initial decreasing N^+ energy sequence and open circles are for subsequent selected energies. Triangles are for N_2^+ data.

in Figure 4.8. Instead, new higher energy efficiency maxima appear but their position is dependent on the direction of the energy sequence, the previous dose and the maximum energy used. Hence Figures 4.7 and 4.8 emphasise the fact that the position of the maximum is an artefact of the energy sequence used in repeated measurements on a single silica sample. In these examples with nitrogen implants there is an additional factor arising from chemical effects leading to the formation of a silicon oxy-nitride compound. The changes in composition will also alter the diffusion rate of excitons. The analysis also suggested that excitons can diffuse over path lengths of at least a micron, which for shallow implants indicates that light can be produced at greater depths than either the projected range or the extent of the ion beam damage.

4.4.5 Sapphire

A similar study of luminescence produced during ion implantation of sapphire shows many of the same problems. Whilst clearly resolved emission bands are apparent at 336, 414 and 390 nm, which are charac-

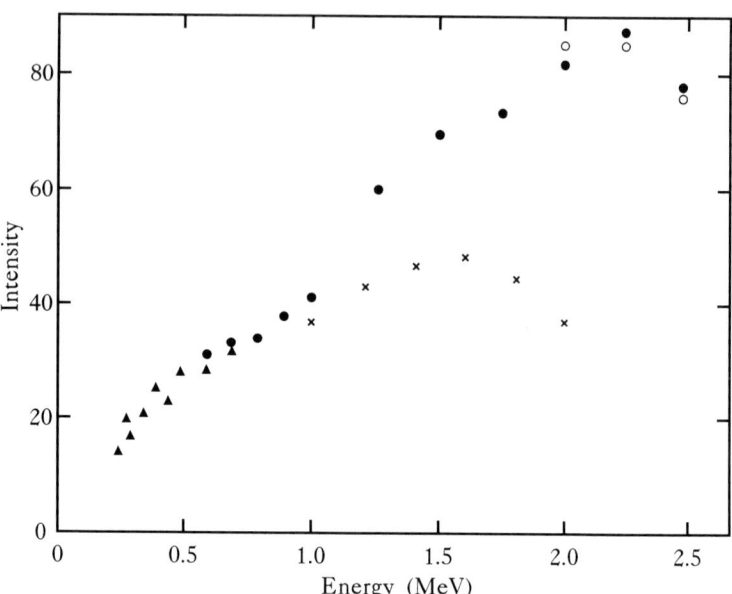

Fig. 4.8. High energy data for luminescence during N^+ implants in silica. Filled circles are a sequence from 2.5 down to 0.6 MeV, open circles from 2.0 up to 2.5 MeV. Crosses and triangles are for a different sample from 2.0 down to 0.2 MeV.

teristic of electronic relaxation at F^+, F and F_2 centres, the intensity of these features is not an accurate measure of their concentrations within the sample. With light ion excitation (e.g. H^+ and He^+), the apparent signal intensities are greater because the energy loss is primarily by electronic excitation. This does not form new defects, but does offer the energy in a form which can relax back via excited states of pre-existing growth defects, i.e. luminescence. Heavy ion (e.g. Ar^+) beam damage produces many new defects and an even higher concentration of transient atomic displacements in the lattice. The presence of such defect structures both provides non-radiative decay paths and, by distortion of the lattice, influences defect transitions which have extended wave functions. The diffuse excited states of the F centre are perturbed and so the F centre emission is totally suppressed by ion beam bombardment, whereas that of the more tightly bound states of the F^+ centre are only reduced in intensity.

As for silica, one cannot determine the energy dependence of the luminescence intensity by repeated measurements on the same sample.

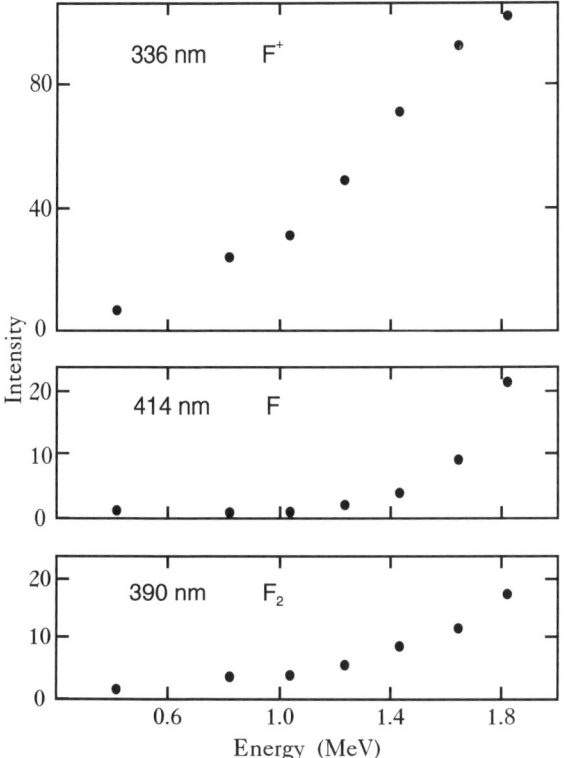

Fig. 4.9. Sapphire luminescence intensities of three defect species produced during He$^+$ implantation using successively lower ion beam energies.

Attempts to do so merely introduce a 'peak' in the energy dependence curve, which is adjustable according to the range and sequence of energies used. The presence of such an artefact was neatly demonstrated by Al Ghamdi *et al.* (1990), who contrasted a decreasing energy sequence (Figure 4.9) with one in which they alternately used high and low energies (Figure 4.10). The steps were sufficiently large that there was no overlap of the damage in the second case. Hence light could be produced at high efficiency as excitons diffused into the central undamaged layer. The main conclusions were, firstly, that for sapphire the light emission originates by exciton decay within the undamaged crystal, and, with minor exceptions, ion implantation reduces the luminescence efficiency. Secondly, the initial light intensity produced when the implant commences, reflects an intrinsic defect concentration which is unexpected by reference to the optical

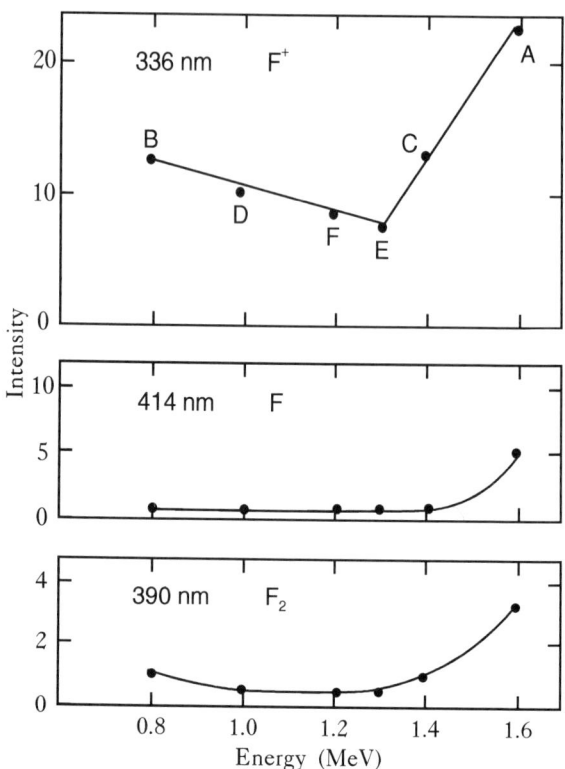

Fig. 4.10. As for Figure 4.9 but with the alternating energy sequence given by A–F.

absorption data. The same conclusion was also reached by Chen *et al.* (1991).

4.4.6 *LiNbO$_3$ – impurity and stoichiometric effects*

LiNbO$_3$ is a key material for electro-optic devices and has been studied extensively for ion implanted optical waveguides. During ion implantation it emits a weak blue luminesence which fades during the destruction of the lattice. Since the characterisation of niobate defects is not very advanced, at least as compared with the halides and silica, the spectra have been recorded in an effort to gain more understanding of the basic defect features. A variety of emission features have been observed including broad luminescence bands attributed to intrinsic defects, line features

4.4 In situ luminescence

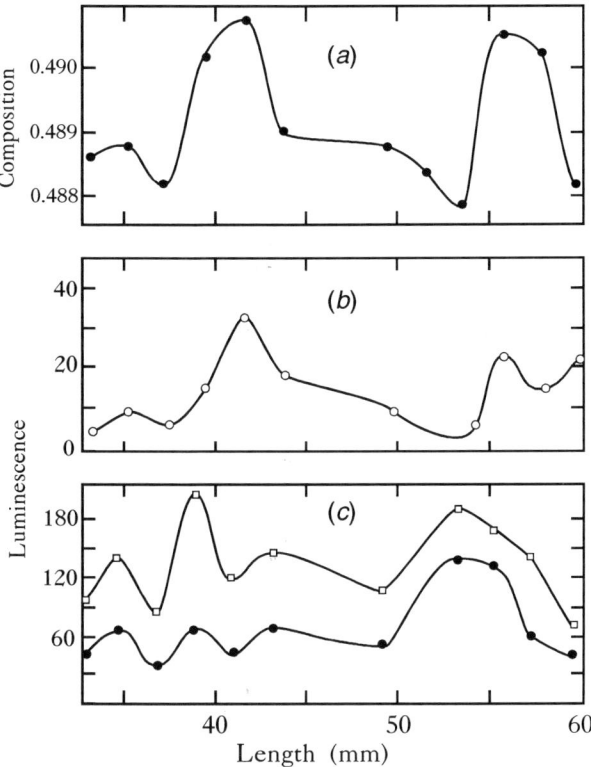

Fig. 4.11. Signals from sections of a boule of LiNbO$_3$ doped with Mg and Ti. (a) % of Li$_2$O as derived from the absorption edge, with luminescence signals recorded at (b) 370 nm, and in (c) 450 nm (squares) and 350 nm (circles).

considered as evidence for exciton decay, emission from impurity atoms, and a set of lines which are linked to surface impurities. Related comparisons of ion beam induced luminescence and cathodoluminescence (Al Ghamdi et al., 1988) show that the emission band intensities are indicative of the stoichiometry of the LiNbO$_3$, with up to 100 times more emission intensity from stoichiometric material than from congruent growth crystals, as also noted in X-ray excitation by Garcia-Cabanes et al. (1988). The ion and electron beam induced variations in luminescence spectra and intensity measured along the length of a doped boule have been used to record fluctuations in stoichiometry and dopant concentrations, (Figure 4.11). In the Al Ghamdi et al. paper, the stoichiometry, Mg and Ti features were separately associated with emission bands at 370,

350 and 450 nm respectively. An interesting feature was that the three variables showed a similar intensity pattern along the boule but were laterally displaced, suggesting that variations in growth conditions sequentially influence the stoichiometry and the takeup of impurities from the melt. The similarity of the ion and electron beam excitation is to be expected since proton excitation delivers some 95% of the energy deposition via electronic processes and there are only minor effects of damage production from nuclear collisions.

Impurity related signals have been noted as a sharp emission line at 671 nm in Mg doped material (Haycock and Townsend, 1986). However, this prominent line feature is from free lithium metal. This suggests that such free metal can be liberated by substitution of Mg on to Li sites during bombardment.

4.4.7 LiNbO$_3$ – excitons

Some of the line emission seen during implantation of LiNbO$_3$ with protons has been assigned to exciton transitions. This was the first direct evidence of excitons in this material. The spectra shown in the composite Figure 4.12 give examples of broad emission bands plus a variety of narrow bands or lines. The intensity of the broad bands and some lines, e.g. the one labelled I$_7$, are sensitive to the crystal face which is excited and show polarisation effects.

This is not unexpected as the niobate lattice is approximately rhombohedral and highly anisotropic, leading to polarised emission features. A bound electron–hole pair closely resembles a hydrogen atom, hence the energies associated with unrelaxed excitonic transitions normally follow a hydrogenic type set of spacings given by

$$E_n = E_g - (E_g - E_1)/n^2 \qquad (4.2)$$

in which E_g is the energy level at the bottom of the conduction band (i.e. equivalent to the vacuum level for a hydrogen atom). A value of 4.19 eV is used based on the ultra-violet reflectivity data of Mamedov et al. (1984). The series limit, E_1, is chosen from the observed value of 3.37 eV, and n is the excitonic state label. The large dielectric constant of 5.7 suggests that there will be changes in the effective index depending on the scale of the wavefunctions. Hence the simple prediction of a series of lines at 302, 311 and 368 nm may be distorted, both by a changing dielectric constant and by crystal field splitting from the anisotropic LiNbO$_3$. The lines at positions I$_7$ and I$_8$ are believed to be the result of such a splitting of the

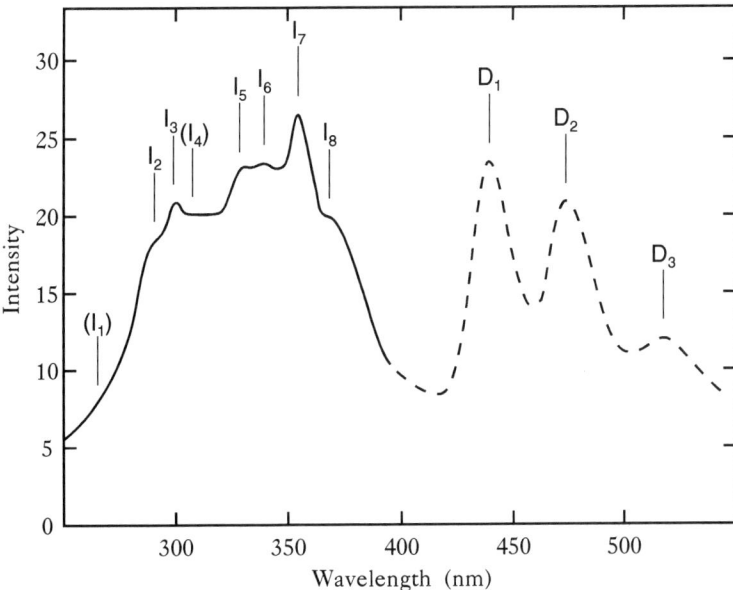

Fig. 4.12. Luminescence bands seen in LiNbO$_3$ during proton implants at 77 K. Initial signals are labelled I$_1$–I$_8$ and later damage generated bands as D$_1$–D$_3$.

368 nm line. The I$_7$, I$_8$, I$_4$ and I$_3$ lines assigned to exciton transitions are detectable throughout the temperature range 60–300 K. The behaviour of the other lines is different in character. For example line I$_5$ shows a clear temperature dependence with an activation energy of 0.35 eV. Other lines are sample, dose and beam dependent. The separate set of lines labelled D$_{1-3}$ will now be discussed.

4.4.8 Surface impurity emission

Haycock and Townsend (1987a,b) frequently observed a number of line features during implantation of LiNbO$_3$ at the positions labelled D$_{1-3}$ and originally thought that they were characteristic of trace impurities in the material. Their arguments for this were that the lines were quite commonly observable, but the signal intensities changed with ion dose in a fashion which was dependent on the supplier and dopants added to the crystal. Additionally, there was a strong temperature dependence on emission intensities, and repeatable differences between the growth of luminescence intensity curves for X, Y and Z cut crystals. Haycock and Townsend did not consider the possibility of surface contaminants condensed from the vacuum system as comparisons with other crystal luminescence made during their data collection did not show such line features. However, Luff (1989) found precisely the same emission lines on a wide range of surfaces during cathodoluminescence studies. Luff traced their origin to a vacuum problem and noted that these, and a variety of other sharp line features, were identifiable to surface adsorbed CO and CO_2. There is therefore a cautionary note that such lines may indeed show dose dependence, definable activation energies assessed by intensity changes with temperature, and dependence on crystal face, dopants and manufacturer, but this is not proof that the emission is produced within the crystal. Luminescence is highly sensitive to energy transported to specialised recombination or active sites. With ion beam excitation the energy is often deposited less than a micron in depth from the surface. This is within the diffusion range of both excitons and free carriers. Hence, surface active sites can dominate the light production. The Haycock and Townsend example of the D_{1-3} type emission merely emphasises that surface bonding of gas impurities is highly sensitive to surface states which are controlled by stoichiometry, trace impurities, defects and temperature. In this sense Haycock and Townsend were indirectly measuring properties of their different $LiNbO_3$ samples. Later work under much higher vacuum conditions suppressed these emission bands. This is not only an error which can occur with this material but, rather, it is a general problem. A more recent cathodoluminescence example (Eremenko *et al.*, 1988) has been discussed by Luff and Townsend (1990).

For materials such as the oxides mentioned above, and of course halides, there is the possibility of electronically driven decomposition of surface layers, particularly the loss of oxygen or halogen. Such changes are clearly demonstrated for ion beam sputtering of silica in which the changes of the oxygen in the background vacuum conditions from 10^{-5} to 10^{-8} mbar change the sputtering rate by a factor of nearly two (Holmen and Jacobsson, 1990). Oxygen sputtering is greater than that for the silicon, hence there is preferential loss of oxygen but this oxygen sputtering rate determines the net rate of the SiO_2. Sputtering in the presence of a higher partial pressure of oxygen causes regrowth of SiO_2 by reaction with the exposed silicon. Other background ions are not effective in altering the sputtering rate. Sputtering is not the only property which is likely to be modified by the oxygen partial pressure and luminescence data are also likely to be influenced by such problems.

4.4.9 Solid argon

Luminescence production is not limited to conventional inorganic solids and a more exotic example is the production of light from solid inert gases. Although these materials have a very wide energy gap, of near 10 eV, the atoms in the ground state are only weakly bonded with bond strengths typically of 0.08 eV. Hence, ion bombardment produces sputtering, both from nuclear collisions and electronic excitation. Light is produced within the solid by normal exciton relaxation processes, but because of the very large sputtering yields a significant amount of light originates from the ejected particles. The data for argon from the Bell work have been summarised by Brown (1988). The strong feature at 9.8 eV arises from the decay of the Ar_2^* excimer as the lattice relaxes in the process of decomposition. The relaxation alters the transition energies sufficiently that the light is not re-absorbed. Weaker emission bands are seen at higher energies of 11.2 and 11.7 eV and are ascribed to excited dimers and isolated argon atoms which decay after they have been ejected into the vacuum system

4.4.10 Summary of in situ luminescence effects

The preceding examples show that because luminescence measurements are highly sensitive they record ion beam excited features associated with impurities, stoichiometry, ion beam damage and surface contaminants. Although such information should in principle be valuable, the observed data must be treated with caution as there is no guarantee that either the intensity, or the spectra, yield unequivocal details of the defect types and their concentrations. Indeed, not all types of potentially luminescent defect site participate in the radiative decay. Nevertheless, as exemplified by the lithium niobate data, ion beam luminescence gave the first evidence for the exciton levels, showed that Li metal was liberated in Mg doped material, and showed that there are a variety of subtly controlled reactions with surface contaminants. With greater experience of such data one assumes that the *in situ* analyses can continue to provide information which is not otherwise readily accessible.

4.5 Photoluminescence

After ion beam implantation the changes resulting from new defects, added impurities, lattice relaxations and/or amorphisation can all be detected by luminescence phenomena. In the simplest examples such modifications are detected by photoluminescence but, with increasing interest in forming waveguide lasers by implantation, one must also consider the differences in luminescence efficiency and allowed transitions of the laser guide regions. Effects are apparent both on the threshold power for lasing and the preferred resonant wavelength for multi-line laser dopants. Numerous examples are given in Chapters 6 and 7.

4.5.1 Luminescence of NaF

Among the simpler features of photoluminescence are those such as have been reported by Afonso et al. (1987) where they implanted NaF with a variety of dopants including H, B and Ne. The implants both form and, because of their impurity effects, stabilise F_2^+ centres. These are interesting as such intrinsic defects have been used to form the basis of tunable infra-red lasers (Mollenaur, 1980, 1985; Schneider, 1981). In the NaF results the defects were stable and, when pumped at 488 nm, gave broad emission bands centred at 580, 650 and 750 nm, from F_2, F_3^+ and F_2^+ centres. In particular, the 750 nm feature was strongest for the H implants. Such an emission band was only produced by ion implantation and did not appear in neutron or electron beam irradiated crystals. Presumably, this implies that isolated intrinsic F_2-type defects are not responsible, but instead the light is generated at a defect (possibly an F_2 centre) associated with the chemical impurity of hydrogen. Note that later data by Kristianpoller et al. (1992) showed rather different spectra but with traces of the same components' emission bands. They concluded that more complex defects formed with trace impurities may be relevant.

4.5.2 Synthesis of new semiconductor alloys

Photoluminescence measurements are frequently made to study impurity effects in semiconducting compounds as they show shifts in the band gap and spectral changes related to annealing of damage or relocation of impurities. One such example, by Braunstein et al. (1987), was seen after the implantation of Mn into CdTe. After Mn implantation they observed loss of the original photoluminescence signal near 780 nm. Instead, even in samples annealed to 730 °C, a new spectrum appeared which was characteristic of a wider energy gap. Braunstein et al. were thus able to deduce that the Mn impurities caused the formation of a wider bandgap alloy $Cd_{1-x}Mn_xTe$. The new alloy was sensitive to applied magnetic fields and the optical signals moved to lower energies under the application of a field.

4.5.3 Divacancies in sapphire

Detection of divacancy-type defects has proved difficult by optical absorption or ESR in Al_2O_3, but less so using luminescence. In studies of sapphire implanted with MeV energy Fe^{2+} the generation of the divacancies, and their subsequent growth at the expense of single lattice defects, was demonstrated from the photoluminescence data (Chen et al., 1991). The luminescence bands from the component defects were resolved by selective excitation in the F, F^+ and F_2 regions of the absorption spectrum. However, the absorption bands overlap rather more than the luminescence features so greater discrimination was possible during luminescence.

4.6 Waveguide lasers

A major potential application of ion implantation is to form optical waveguide lasers in insulating crystals. At present two approaches are being considered. In the first of these, standard lasing materials have been structured into waveguides by using ion beam effects to either increase the refractive index within the guide, or by using index decreases associated with amorphisation at the end of the ion range to form a barrier region (see Chapter 5). The first such laser waveguide formed by ion implantation to define the refractive index was in Nd:YAG (Chandler et al., 1989; Field et al., 1991a; Zhang et al., 1991). In this example the index enhancement was generated by nuclear collision damage in the waveguide region, whereas in later lasers formed in Nd:YAP or Nd:LiNbO$_3$ (Field et al., 1990, 1991b), the approach was to define a low index optical barrier by a high energy ion beam. A further method was used in the fourth example of Nd:$Bi_4Ge_3O_{12}$ in which the damage only enhances the index, most probably as a consequence of relaxation into a variant of the original $Bi_4Ge_3O_{12}$ lattice (Mahdavi and Townsend, 1990a,b). Further discussion of the refractive index profiles, and the factors which influence them, will be given in Chapters 6 and 7.

In general, the arguments in favour of such guide structures are that they may require very low pump powers and the pump source can be a semiconductor laser array. Pumping may be via the end of the

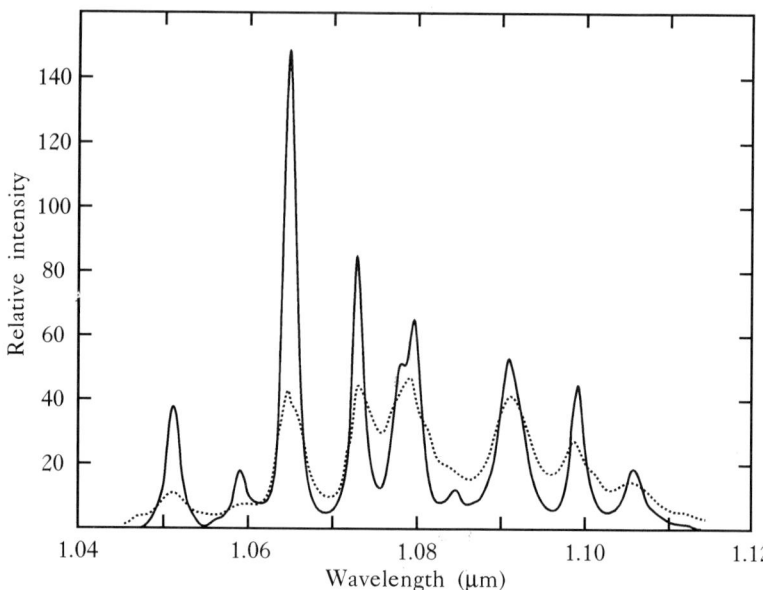

Fig. 4.13. A comparison of the photoluminescence spectra of bulk Nd:YAG (solid line) with light generated within a waveguide formed by He implantation (dotted line).

guide, or if the pump beam is strongly absorbed, from the side. Less obvious is that the ion implantation may influence the range of suitable transitions and the line widths. Figure 4.13 shows signals from He^+ implanted Nd:YAP, and other examples are given in Chapter 7. After implantation the relative intensities of the lines have altered and there is some luminescence line broadening prior to lasing. These features are a function of the anisotropy of the crystal lattice and consequently also vary with the choice of ion beam direction relative to the crystal axes. Note that the separation of vacancies and interstitials is sensitive to the variations in diffusion rate along different axes.

Rather than define a laser waveguide from a bulk laser material, a second approach to the problem is to commence with an undoped crystal and ion implant the impurity ions which determine the laser action. In principle this may have advantages, as one could combine, on the same surface, regions of both passive and lasing material. Also, it should be possible to introduce dopants which are not readily accepted by the host lattice.

One such example is to implant Er ions into SiO_2 (Polman *et al.*,

4.6 Waveguide lasers

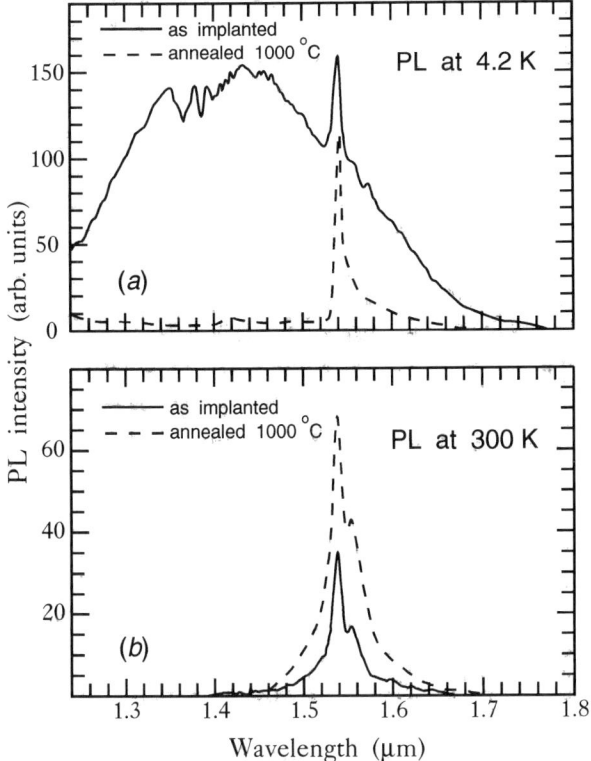

Fig. 4.14. Photoluminescence spectra of Er implanted silica before (solid) and after (dashed) annealing at 1000 °C. Data were taken at (a) 4.2 or (b) 300 K.

1991a,b). Erbium is particularly attractive as the 1.54 μm transition is ideally matched for amplifiers and lasers in optical fibre transmission. Diffusion doping is extremely difficult for this example since in silica and many materials Er is effectively immobile for reasonable temperatures and timescales. As shown by Figure 4.14 the photoluminescence spectra are significantly enhanced in intensity, and there is an increase in excited state lifetime as a result of post-implantation annealing. Whilst this is encouraging it has not yet lead to the demonstration of waveguide Er lasers. In part this may be because the impurity ions are unable to diffuse or substitute on the silica lattice sites; instead, the heat treatments could result in ions being precipitated as colloids or other phase separations.

Buchal and Mohr (1991) and Buchal (1991) approached the problem rather differently in their attempts to make an erbium doped $LiNbO_3$

waveguide laser. Introducing such heavy ions over the micron depth scale of the waveguide would imply accelerator energies of many MeV. Therefore, Buchal implanted both Ti and Er and then annealed the material at a sufficiently high temperature that both ions diffused together to make a doped waveguide. Unfortunately, such implants cause major disorder and loss of oxygen in the niobate but this problem is mostly surmounted by reoxidation, although there is some evidence that the implanted and diffused material is more snsitive to optical damage than material doped by other routes.

4.7 Thermoluminescence

Among the most sensitive luminescence techniques for the detection of defects is thermoluminescence (TL). One may separate trapping levels by their appearance at different temperatures, and, by the magnitude of the glow peaks, compare changes in trap population. Additionally, in those examples where the emission spectra have been recorded, information is available on the recombination centres. Consequently, defect effects resulting from ion implantation may produce distinguishable features despite the fact that the implant will generally only modify a very small percentage of the sample thickness.

4.7.1 Silica and quartz

Evidence that implantation into a layer only microns thick can influence the TL of a millimetre-thick sample is provided by the data of T. Hedde (unpublished 1985 and cited by Townsend, 1987) shown in Figure 4.15. This presents an isometric plot of $I(\lambda, T)$, together with a contour map, of the glow curves obtained by ion beam amorphisation of the face of a quartz crystal. Prior to implantion the pure quartz sample produced thermoluminescence after X-ray excitation which displayed glow peak emission only near 500 nm, and at temperatures of 55, 95, 175 and 240 °C. Amorphisation of quartz was produced by implantation over the upper few microns of the crystal. However, as shown by the diagram, the new glow curve contains not only the original green glow peaks, but also additional strong blue signals at 340 nm. The changes in peak temperature and emission band show that both new trapping centres, and recombination centres, have been generated. Point defects in silica and quartz are very similar (e.g. Griscom, 1980, 1985; Halliburton, 1985; McKeever, 1984; Yang and McKeever, 1988, 1990), and small shifts in

4.7 Thermoluminescence

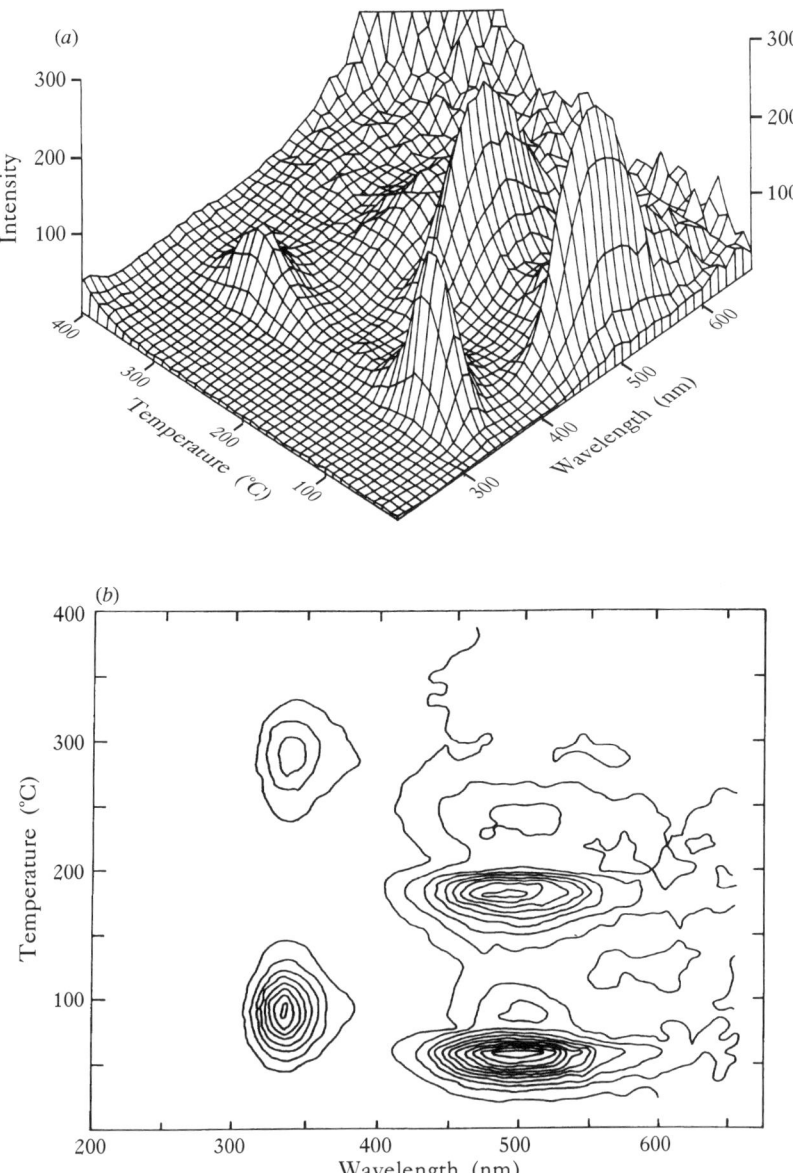

Fig. 4.15. Thermoluminescence spectra of nitrogen implanted quartz. Prior to implantation the crystal only gave the 500 nm emission peaks.

defect stability might simply mean that the new colour centres are modified versions of the traps which occurred in the original crystal. However, for the present discussion it is not important to know if such defects are solely the result of the amorphisation to silica, or whether they include implant impurity effects. The more general comments are, firstly, that the presence of defects in a micron-thick layer can be seen against the background of pre-existing defects. Secondly, the glow peak pattern seen at 340 nm shows that new trapping levels are produced which are expressed as peaks at new glow peak temperatures, and recombination occurs at different sites from those in the original material. The observation that the emission occurs either in the blue or in the green, and that there are no glow peaks which emit at the same temperature at both wavelengths, may imply that charge released in the bulk does not significantly diffuse from the interior to the surface layer, or vice versa.

The glow peaks in quartz are of considerable interest as they form the basis of the age determinations of pottery and soils (Aitken, 1985), despite the fact that the detailed structures of the sites which release and capture the charge are still the subject of discussion (Yang and McKeever 1988, 1990). One should note that there are many problems in comparing TL signals from different laboratories (Betts et al., 1993). For example, in dating measurements it is customary to heat the samples at much higher rates than was possible in the acquisition of the data of Figure 4.15. For the spectral measurement a heating rate of only some 20 °C per minute was possible, whereas more conventional polychromatic data are recorded at 150 °C per minute. The peak temperature, T_{max}, heating rate, β, detrapping energy, E, and frequency of attempt to escape, Y, are related by

$$Y T_{max}^2 = \beta E \exp(E/kT_{max}) \qquad (4.3)$$

Increasing the heating rate causes the traps to empty at higher temperatures, typically near 100, 150, 275 and 350 °C. The 100 °C peak is a complex association of an intrinsic defect and an impurity but the 150 °C peak is less clearly understood.

In an attempt to test models suggested for the 150 °C trap, Arnold (1977, 1982) used a range of ion species from H^+ to Xe^+ to form defects under different irradiation fields in which the ratio of electronic to nuclear collision damage was altered. He also varied the type of silica and used a range of commercial fused silicas, r–f sputtered silica films and crystalline quartz. As previously mentioned, the amorphised silica is sensitive in its luminescence to electronic excitation of damage and/or relaxation of complex defects. Similar trends have been recorded on many occasions for silica for compaction, refractive index and stress (e.g. EerNisse, 1974, 1977; Dellin et al., 1977). Arnold interpreted his data to show that for the silica there was greater enhancement of the 150 °C glow peak in regions of nuclear collision damage. In fact, the intensity of the signal grew linearly with energy deposited in nuclear collisions. Ion species effects only appeared at high dose levels when the impurity effects overrode those from the intrinsic damage. Arnold thus suggested that the trapping level responsible is a defect complex which certainly involves several intrinsic point defects. The glow peak seen at 330 °C was prominent in silica glasses containing traces of Al. The 100 °C peak was further linked to an intrinsic oxygen vacancy centre; however, more recent views include association with impurities (Yang and McKeever, 1988, 1990). Specific examples include luminescence bands near 380–390 nm (E' and H_3O_4 ions); 400 nm (Ge); 460 nm (excitons); 470 nm (AlO_4) and 580–590 nm (E' variants) from 'pure' silica. Other emission bands are even less clearly identified or are attributed to other dopants.

4.7.2 CaF_2

Thermoluminescence trapping and recombination sites are frequently a consequence of complex composite defects made up from intrinsic and impurity components. These exist in chemically and ion beam doped CaF_2, and the signals from the complexes dominate the signals recorded by ion beam impurity doping of CaF_2 by Bangert et al. (1982). Doped CaF_2 is potentially a useful radiation dosimeter and conventional chemical doping of the impurities was compared with their addition by ion implantion. Impurity ions of Mn, Dy, and O were implanted as well as the self-ions of F and/or Ca. Chemical doping can in principle suffer from cross-contamination as the light production may be sensitive to trace impurities at less than the ppm level. Implants provide a cleaner doping route. The implanted crystals gave emission spectra which altered with the dopant, with peaks at 360, 460 or 480 nm from Ce, Mn and Dy. Figure 4.16 shows examples for the Ce and Mn implants. The data confirmed the indications from the chemical doping work, which was particularly valuable since Ce is often a trace impurity and separation of Ce from other dopant effects was difficult.

The association of the Ce with intrinsic defects became obvious during self-ion implants and as a result of adding impurity ions by implantation. Also, in CaF_2 there is the further possibility of generating localised im-

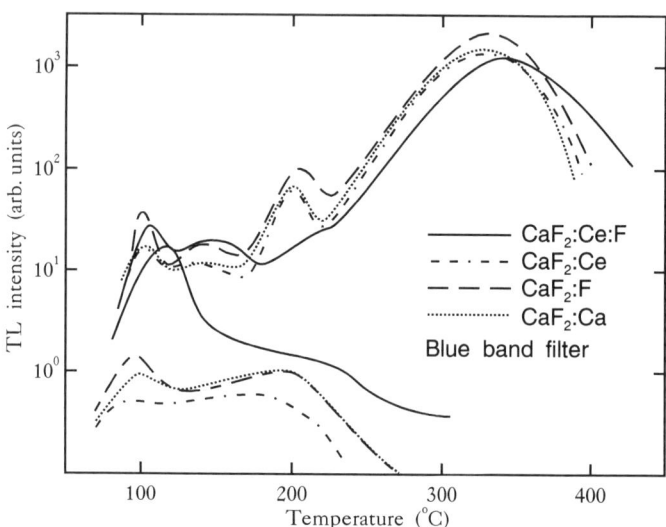

Fig. 4.16. TL glow curves produced by beta irradiation of ion implanted CaF_2 after post-implantation annealing to 500 °C. The upper curves were for slow cooled samples, and the lower set for rapidly cooled samples. Data are for implants of Ca (·····), F (————), Ce (·———·), Ce plus F (———).

perfections by thermal treatments. Therefore, annealing of ion damage can in fact increase the concentration of intrinsic defects. Furthermore, the survival of the defects and their association into impurity/intrinsic defect complexes is a function of the cooling rate. Rapid cooling can freeze in complexes which were thermodynamically feasible at high temperature. Figure 4.16 clearly demonstrates the differences between slow and rapid cooling on the glow curve patterns. In these data the TL signals were produced by post-implantation excitation of the material with electrons, and then the TL was recorded with a blue band filter to emphasise the signals generated at Ce sites. The set of curves indicate that there are relatively minor differences between the Ce, F, and Ca implants, hence the blue emission is strongly influenced by traces of Ce in the original samples. Defect models for the incorporation of Ce into the CaF_2 lattice include the suggestion of a substitutional Ce^{3+} ion adjacent to an interstitial F^- ion. In particular, this is thought to control the intensity of the peak near 100 °C. Clearly, one route to an enhancement of such a defect is the double implantation of Ce and F. The data of Figure 4.16 (bottom spectra) for the quenched material show that the 100 °C peak is indeed enhanced after such double implants. A combination of implants and annealing cycles suggested that the high temperature peaks near 180, 220 and 350 °C all involve intrinsic defect clusters. This assertion was made from the dependence of the changes with the size of the collision cascades for ions of different mass.

4.7.3 LiF dosimeters

An even more commonly used thermoluminescence dosimeter material is LiF doped with Mg and Ti as the primary impurities. Optimisation of the radiation sensitivity includes several processing steps. In the first the Mg impurities are atomically dispersed by heating to 400 °C, then aggregation into trimers built up from three (Mg^{2+} + Li^{+}_{vac}) units. These complex defects act as the charge traps. During the heating stage, luminescence is produced at the Ti sites. The Ti exists in the Ti^{4+} state and, in order to have charge compensation, models have been proposed in which the Ti^{4+} ions cluster with three fluorine vacancies or, more probably, associate with O^{2-} ions substituted on halogen sites. The oxygen is not introduced intentionally but is assumed to enter the melt during growth of the LiF boule. Under such a model it is difficult to demonstrate the role of oxygen in the sensitisation, or to ensure that adequate gas has entered the crystal to compensate all the titanium ions. Wintersgill *et al.* (1977, 1978) tested the oxygen model by using LiF with a range of Ti dopings, into which they implanted oxygen. For samples with a high Ti content the growth conditions had limited the number of available compensating oxygen ions, hence the implants of oxygen strongly enhanced the emission in these samples. To ensure that the changes were not an artefact of ion beam damage comparisons were made with Ne implants, since they are chemically inert but of nominally the same mass and, hence, track dimension. Their data therefore strongly support the model of the (Ti^{4+}+$3O^{2-}$) luminescent centres.

4.8 Impurity doping of CaO

Ion beams were used in a similar luminescence problem which occurred in CaO doped with Bi (Hughes and Pells, 1974). Although doped with Bi the phosphor showed sharp luminescence lines characteristic of Gd. The most probable explanation was that the Gd had entered the CaO during the high temperature diffusion of the Bi, but to refine the starting materials and crucible further was difficult. A simple approach was to implant Bi or Gd separately into pure CaO which was initially free of the fluorescence signal. The results showed that the diffused trace impurities entered the CaO during the Bi diffusion stage.

4.9 Cathodoluminescence

In many respects the emission spectra obtained during electron beam irradiation are similar to those produced during light ion bombardment as in both cases the energy transfer is primarily electronic. A useful guide for the depth of sample which is generating the light is that a 10 keV electron penetrates one micron in unit density material, whereas the same depth requires about 200 keV for a He ion. For luminescence applications the guide is a little simplistic as light may be produced at some distance from the point of excitation if energy is transported via exciton or charge migration. Since such diffusion lengths are often of a micron scale, the spectrum of the light will be biased in favour of special recombination sites, such as surfaces, impurities, dislocations or other defects.

Luminescence is always a probe of such defect sites and as such can be used to follow the history of defects as they change their structural arrangement during annealing. An early example of this approach was made by Yu and Bryant (1979) in a study of the location of erbium ions implanted into ZnS. The rare earth ions have sufficiently sharp line spectra that the modifications from the local crystal field reveal the interactions, with neighbouring atoms and the local site symmetry. In the ZnS example the implants were made at low temperature. Hence the range of defects could include free interstitial ions of erbium. Interstitials are very mobile, so that by 50 K free interstitials will have disappeared. The cathodoluminescence spectra showed that this was also the case for the Er ions and they moved on to substitutional sites. A similar study by Bryant and Nahum (1977) followed the passage of Yb implanted in CdTe from sites which were primarily interstital, to substitution on the CdTe lattice. The detailed symmetry studies revealed, however, that there are many more complex variations than the idealised interstititial and substitutional sites for the impurities. Bryant (1982) reviewed several examples from their group of such cathodoluminescence studies of rare earth implants. It should be noted that not all impurities act as luminescence centres, even if the first observation is that the light intensity increases as a result of implantation. For example, in the case of ZnTe there is light production from trace impurities of oxygen linked to Te vacancies. Hence any ion beam damage which increases the concentration of Te vacancies can enhance the luminescence signal, and this should not be confused with the introduction of active impurity ions.

A later study of Er implanted ZnSe (Abolhassani and Bryant, 1987) clearly identified at least three distinct defects of Er^{3+}: at a Zn substitutional site; at a Se substitutional site; or with a Zn interstitial. The analysis benefits from the use of a single Er isotope which results in improved line resolution compared with diffusion doped material. The various sites involve up to 16 close neighbours with upwards of 26% bond length relaxations.

Cathodoluminescence has been used to study the depth distribution of the damage and impurity sites which were formed by ion implantation, by probing with electrons of various energies. Much of the early work was by Norris and his collaborators (1973,1978,1980,1982). An earlier survey was given by Yacobi and Holt (1990). The materials studied were semiconductors such as CdTe, ZnTe, CdSe, ZnS and GaAs implanted with either inert gases or active impurities. In many cases the component luminescence bands were resolved and so the separation of impurity and defect sites was simple. The energy separated profiles have been confirmed by surface etching.

4.10 Depth effects

More important than the detailed spectra is the observation that in many ion implanted systems the damage effects extend a considerable distance beyond the projected ion range. For semiconductors the cathodoluminescence analyses are more complex because of charge compensation, as well as carrier transport to regions beyond the implant. Nevertheless, the presence of modified material to depths of 3–10 times the ion range has been claimed from refractive index changes in ZnTe and silica; chemical etching of silica is influenced well beyond the implant depth, and transmission electron microscope studies of colloids formed by metal implants in glass show not only the colloids at the projected ion range, but also void structures buried much deeper. The overall conclusion is that not only is there charge migration, as detected by the cathodoluminescence studies, but, probably because of a combination of stress, temperature gradients and radiation enhanced diffusion, many point defects migrate significantly beyond the region of the implantation. From the viewpoint of luminescence studies this means that there is a wealth of information concealed within the optical signal.

References

Abolhassani, N. and Bryant, F.J. (1987). *J. Phys.*, **C20**, 207.
Abu Hassan, L.H. and Townsend, P.D. (1988). *Nucl. Inst. Methods*, **B32**, 293.
Abu Hassan, L.H., Townsend, P.D. and Webb, R.P. (1987). *Nucl. Inst. Methods*, **B19/20**, 927.
Afonso, C.N., Gomez, I.M. and Ortiz, C. (1987). *Phys. Stat. Sol.*, **(b)143**, 777.
Aguilar, M., Chandler, P.J. and Townsend, P.D. (1979). *Rad. Effects*, **40**, 1.
Aitken, M.J. (1985). *Thermoluminescence Dating* (Academic Press, New York).
Al Ghamdi, A., Jones, G., Luff, B.J., Townsend, P.D. and Polgar, K. (1988). *Nucl. Inst. Methods*, **B32**, 51.
Al Ghamdi, A. and Townsend, P.D. (1990). *Rad. Effects and Defects in Solids*, **115**, 73.
Aono, K., Iwaki, M. and Namba, S. (1988). *Nucl. Inst. Methods*, **B32**, 231.
Arnold, G.W. (1975). In *Ion Implantation in Semiconductors*, p.275, F. Chernow, J.A. Borders and D.K. Brice (eds.) (Plenum, New York).
Arnold, G.W. (1982). *Rad. Effects*, **65**, 17.
Bangert, U., Thiel, K., Ahmed, K. and Townsend, P.D. (1982). *Rad. Effects*, **64**, 143.
Betts, D.S., Couturier, L., Khayrat, A.H., Luff, B.J. and Townsend, P.D. (1993). *J. Phys. D*, **26**, 843; *ibid* **26**, 849.
Braunstein, G.H., Dresselhaus, G. and Withrow, S.P. (1987). *Nucl. Inst. Methods*, **B19/20**, 851.
Brown, W.L. (1988). *Nucl. Inst. Methods*, **B32**, 1.
Bryant, F.J. (1982). *Rad. Effects*, **65**, 81.
Bryant, F.J. and Nahum, J. (1977). *Rad. Effects*, **31**, 106.
Buchal, C. (1991). *Nucl. Inst. Methods*, **B59/60**, 1142.
Buchal, C. and Mohr, S. (1991). *J. Mater. Res.*, **6**, 134.
Chandler, P.J. and Townsend, P.D. (1979a). *Rad. Effects Lett.*, **43**, 61.
Chandler, P.J. and Townsend, P.D. (1979b). *Rad. Effects*, **42**, 65.
Chandler, P.J., Field, S.J., Hanna, D.C., Shepherd, D.P., Townsend, P.D., Tropper, A.C. and Zhang, L. (1989). *Elec. Lett.*, **25**, 985.
Chandler, P.J., Jaque, F. and Townsend, P.D. (1979). *Rad. Effects*, **42**, 45.
Chen, Y., Abraham, M.M. and Pedraza, D.F. (1991). *Nucl. Inst. Methods*, **B59/60**, 1163.
Dellin, T.A., Tichenor, D.A. and Barsis, E.H. (1977). *J. Appl. Phys.*, **48**, 1131.
EerNisse, E.P. (1974). *J. Appl. Phys.*, **45**, 167; *ibid* (1977), **48**, 3337.

Eremenko, V.V., Fugol, I.Ya., Samovarov, V.N., Demirskii, V.M., Zhuraviev, V.M. and Uyutnov, S.A. (1988). *Sov. J. Low Temp. Phys.*, **14**, 882.
Field, S.J., Hanna, D.C., Shepherd, D.P., Tropper, A.C., Chandler, P.J., Townsend, P.D. and Zhang, L. (1990). *Elect. Lett.*, **26**, 1826.
Field, S.J., Hanna, D.C., Shepherd, D.P., Tropper, A.C., Chandler, P.J., Townsend, P.D. and Zhang, L. (1991a). *IEEE J. Quant. Elect.*, **27**, 428.
Field, S.J., Hanna, D.C., Shepherd, D.P., Tropper, A.C., Chandler, P.J., Townsend, P.D. and Zhang, L. (1991b). *Opt. Lett.*, **16**, 481.
Garcia Cabanes, A., Zaldo, C., Sanz-Garcia, J.A., Cabrera, J.M., Agullo-Lopez, F., Pareja, R., Polgar, K., Raksanyi, K. and Foldvari, I. (1988). *Phys. Rev.*, **B37**, 6085.
Griscom, D.L. (1980). *J. Non-Cryst. Solids*, **40**, 211; *ibid* (1985), **73**, 51.
Halliburton, L.E. (1985). *Cryst. Latt. Defects and Amorph. Mats.*, **12**, 163.
Haycock, P.W. and Townsend, P.D. (1986). *Rad. Effects*, **98**, 243.
Haycock, P.W. and Townsend, P.D. (1987a). *Cryst. Latt. Defects and Amorph. Mats.*, **15**, 143.
Haycock, P.W. and Townsend, P.D. (1987b). *J. Phys.*, **C20**, 319.
Haycock, P.W. (1986). D. Phil. Thesis, Sussex, (unpublished).
Hedde, T. (1985). BSc. Dissertation, Sussex, (unpublished).
Holmen, G. and Jacobsson, H. (1990). *J. Appl. Phys.*, **68**, 2962.
Hughes, A.E. and Pells, G.P. (1974). *J. Phys.*, **C7**, 3997.
Jaque, F. and Townsend, P.D. (1981). *Nucl. Inst. Methods*, **182/183**, 781.
Kristianpoller, N.N., Blieden, G.S. and Comins, J.D. (1992). *Nucl. Inst. Methods*, **B65**, 484.
Luff, B.J. (1989). D. Phil. Thesis, Sussex, (unpublished).
Luff, B.J. and Townsend, P.D. (1990). *J. Sov. Low Temp. Phys.*, **16**, 124.
Mamedov, A.M., Osman, M.A. and Hajieva, L.C. (1984). *Appl. Phys.*, **A34**, 189.
McKeever, S.W.S. (1984). *Rad. Protect. Dosimetry*, **8**, 81.
Mitchell, E.W.J. and Townsend, P.D. (1963). *Proc. Phys. Soc.*, **81**, 12.
Mollenauer, L.F. (1980). *Opt. Lett.*, **5**, 188.
Mollenauer, L.F. (1985). *Laser Handbook*, p.143 M.L. Stitch and M. Bass (eds.) (North Holland, Amsterdam).
Norris, C.B., Barnes, C.E. and Beezehold, W. (1973). *J. Appl. Phys.*, **44**, 3209.
Norris, C.B., Barnes, C.E. and Zanio, K.R. (1978). *J. Appl. Phys.*, **48**, 1659.
Norris, C.B. (1980). *J. Appl. Phys.*, **51**, 1998; *ibid* (1982), **53**, 5172, 5177.
Polman, A., Lidgard, A., Jacobson, D.C., Becker, P.C., Kistler, R.C., Blonder, G.E. and Poate, J.M. (1991a). *Elect. Lett.*, **27**, 993.
Polman, A., Jacobson, D.C., Eaglesham, D.J., Kistler, R.C. and Poate, J.M. (1991b). *J. Appl. Phys.*, **70**, 37780.
Schneider, I. (1981). *Opt. Lett.*, **6**, 157.
Townsend, P.D. (1961). PhD Thesis, Reading, (unpublished).
Townsend, P.D. (1987). *Rept. Prog. Phys.*, **50**, 501.
Ukai, T., Matsunami, N., Morita, K. and Itoh, N. (1976). *Phys. Lett.*, **56A**, 127.
Wang, P.W., Aldbridge, R.G., Kinser, D.L., Weeks, R.A. and Tolk, N.H. (1991). *Nucl. Inst. Methods*, **B59/60**, 1317.
Wintersgill, M.C. Townsend, P.D. and Cusso-Perez, F. (1977). *Journ. de Phys.*, **38-C7**, 123.
Wintersgill, M.C. and Townsend, P.D. (1978). *Phys. Stat. Sol.*, **(a)47**, K67.
Yacobi, B.G. and Holt, D.B. (1990). *Cathodoluminescence Microscopy of Inorganic Solids* (Plenum, New York).
Yang, X.H. and McKeever, S.W.S. (1988). *Nucl. Tracks*, **14**, 75.

Yang, X.H. and McKeever, S.W.S. (1990). *J. Phys.*, **D23**, 237. *ibid Rad. Prot. Dosimetry*, **33**, 27
Yu, C.C. and Bryant, J.F. (1979). *J. Lumin.*, **18/19**, 841.
Zhang, L., Chandler, P.J. and Townsend, P.D. (1991). *J. Appl. Phys.*, **69**, 3440.

5
Ion implanted waveguide analysis

5.1 Characteristics of ion implanted waveguides

A waveguide is characterised by a region of high refractive index bounded by regions of lower index. The confinement of the light, as well as the spatial distribution of optical energy inside the guiding layer depends on the refractive index profile. There are several conventional techniques for fabricating optical waveguides. These techniques, including epitaxial growth, metal diffusion and ion exchange, increase the refractive index of the surface layer for a few microns (Figure 5.1), and this high index layer is surrounded by the low index of air and substrate to form an optical waveguide. Ion implantation, as a surface modification technique, can modify the optical properties of an insulator surface. However, when light ions are used, particularly when dealing with crystals, instead of changing the refractive index of the surface layer, a low index optical barrier is built up at the end of the ions' track due to elastic energy deposition from ions to the lattice. Therefore, the surface layer, ideally the same as the substrate, is surrounded by the low index of air and this optical barrier (Figure 5.1(c)). During the ion implantation, some point defects may be produced in the surface layer due to ionisation and excitation when the ions are travelling fast. These simple defects will change the properties of the material, and induce absorption and scattering loss. In practice, it has been found that post-implant annealing at a moderate temperature can either reduce or aggregate these defects depending on the material in question. In many materials, a low loss optical waveguide ($\sim 0.5\,\mathrm{dB/cm}$) can be produced by ion implantation and subsequent annealing.

Compared with other waveguide fabrication methods, ion implantation has some unique advantages. It appears to be a universal technique for

152   Ion implanted waveguide analysis

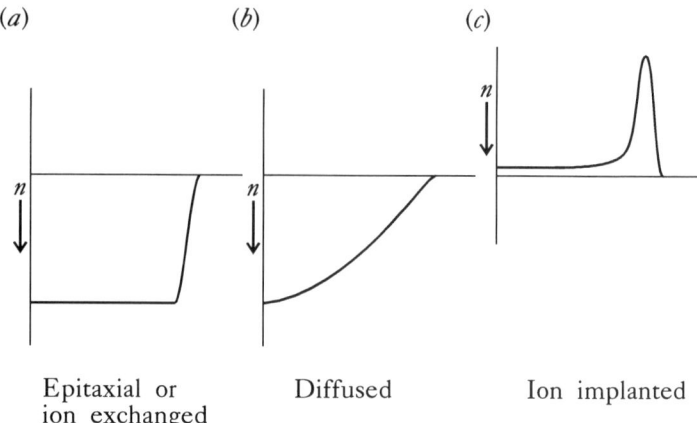

Fig. 5.1. Refractive index profiles of waveguides fabricated by various techniques.

producing waveguide structures in most optical materials. It has a superior controllability and reproducibility to the other techniques. The ambient or low temperature at which it is usually performed can be chosen such that all ferroelectric materials have stable phases. As there is no established standard technique for fabricating optical waveguides, ion implantation, with its many advantages, will compete against all other techniques to find a role in the development of integrated optics.

As implanted waveguides in general are optical barrier waveguides, during the past few years a special analysis method, the reflectivity calculation method (RCM), has been developed by Chandler and Lama (1986a) for characterising the refractive index profiles of implanted waveguides. This chapter will introduce and discuss this measurement technique and analysis method.

5.2 Waveguide mode theory

The dielectric slab is the simplest form of optical waveguide. It is formed by the combination of three different refractive index dielectric media (usually airspace n_0, waveguide n_1 and substrate n_2) in a sandwich structure as shown in Figure 5.2. For confining electromagnetic waves it is necessary that the central medium has a higher refractive index than either of the two sides ($n_1 > n_0, n_2$). Within the waveguide structure, because of optical interference effects, the trapped light can only travel at certain angles, called 'mode angles' ($\theta_m, m = 0, 1, 2...$), the lower ones

5.2 Waveguide mode theory

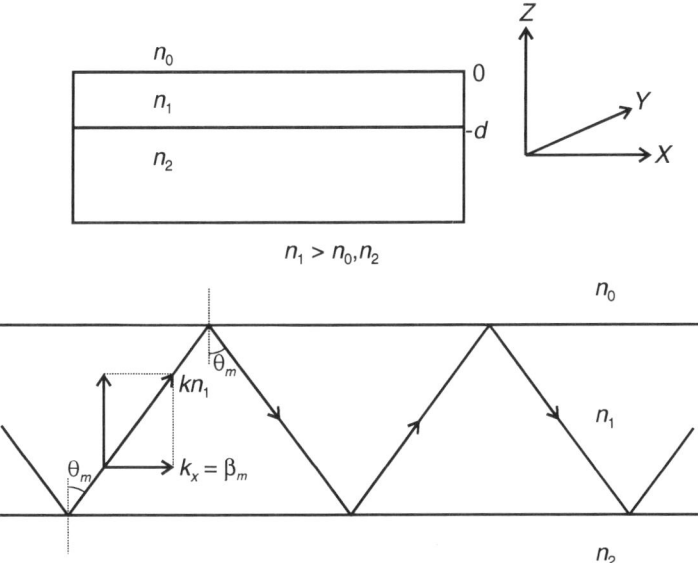

Fig. 5.2. Configuration of a slab waveguide. It is formed by three different refractive index dielectric media in a sandwich structure. Geometrical optics approach, in which the field distribution can be analysed as a superposition of two plane waves: $K_x = Kn_1 \sin\theta_m$, $K_z = Kn_1\cos\theta_m$.

being almost parallel to the guide axis, and the higher ones zig-zag until they exceed the critical angle criteria ('cut-off'). The velocity of a particular mode along the guide will be given by its effective mode index n_m which depends on the mode angle $n_m = n_1 \sin\theta_m$. The number and spacing of these modes in any particular guide structure will depend on its width and refractive index profile.

A proper mathematical treatment of the slab waveguide is to set up an eigenvalue problem from Maxwell's equations, and proceed to solve it, in order to obtain the normal allowed modes (eigenstates) corresponding to the light propagating directions in the guide, and their associated field distributions (eigenfunctions). However, it is possible to analyse the light propagation in a waveguide from several other points of view. One of these is the use of geometrical optics, where a ray of light is considered to be reflected back and forth between the two interfaces, following a zig-zag path in the waveguide. From simple interference theory, for a mode angle to be allowed, the total optical path length per zig-zag must be a multiple of λ. This includes the phase changes on reflection at the two interfaces ($2\varphi_{10}$ and $2\varphi_{12}$) given by the Fresnel equations, leading to a synchronous equation (Tien, 1977):

$$2k_z d + 2\varphi_{10} + 2\varphi_{12} = 2m\pi \qquad (5.1)$$

where $k_z = k(n_1^2 - n_m^2)^{\frac{1}{2}}$, k is the wave vector, d the guide width and m the mode order ($m = 0, 1, 2...$). This can be solved transcendentally to give the allowed values of n_m for a given slab profile (n_0, n_1, n_2, d) and wavevector (k).

An alternative method is a quantum mechanics analogy treatment, which utilises the fact that light waves in a waveguide have properties similar to those of an electron trapped in a potential well. This model is useful because it may be developed by the use of standard quantum mechanical approximations, such as perturbation theory and the WKB method (Hocker and Burns, 1975; White and Heidrich, 1976), into a solution for a graded index waveguide, which is usually the case in actual practice. These models are not necessarily rigorous, but they do provide an insight into a complex problem. These approaches will be discussed in the following sections.

5.2.1 Maxwell equation approach

In order to obtain a complete description of the waveguide modes and field distributions associated with these modes, Maxwell's equations must be solved so that they satisfy the waveguide boundary conditions. In rectangular coordinates, the wave equation can be derived from Maxwell's equations as follows:

$$\left[\frac{\partial^2}{\partial x^2} + \frac{\partial^2}{\partial y^2} + \frac{\partial^2}{\partial z^2}\right] E_j = \left[\frac{n_j^2}{c^2}\right]\left[\frac{\partial^2 E_j}{\partial t^2}\right] \qquad (5.2)$$

where $j = 0, 1, 2$ denotes airspace, waveguide and substrate respectively. The solution to the above equation in a planar waveguide will propagate as:

$$E_j = A \exp i(\omega t \pm k_{xj}x_j \pm k_{yj}y_j \pm k_{zj}z_j) \qquad (5.3)$$

where k_{xj}, k_{yj} and k_{zj} are the components of the wave vector. Since a slab waveguide structure consists of three media, the wave in each medium should be considered separately. Because the ratio of thickness to width of a practical slab waveguide is of the order of $< 1\%$, the wave travelling in the guide can be considered to be distributed uniformly along the wavefront of the light beam, which lies along the y direction in Figure 5.2. This leads to:

$$\frac{\partial}{\partial y} = \pm ik_{y0} = \pm ik_{y1} = \pm ik_{y2} = 0 \tag{5.4}$$

In addition, since the waves in the guide, substrate and airspace are matched along their interfaces, the components of the wave vectors parallel to the propagating plane (yz), i.e. k_{xj}, must be the same in all three media It is customary to denote these three equal vectors by the symbol β. Thus: $\beta = k_{x0} = k_{x1} = k_{x2} = kn_1 \sin\theta$, where θ is the propagation angle in the guide. The allowed values of β (determined from the boundary conditions) are going to be the eigenvalue solutions (normal modes) from Maxwell's equations ($\beta_m = kn_m$, $m = 0, 1, 2....$), where n_m are the 'effective' mode indices of the eigensolutions.

By assuming $\partial/\partial y = 0$ (Equation (5.4)), Maxwell's wave equation can be separated into two independent sets. One, which involves the field components E_y, H_x and H_z only, is called the TE (transverse electric) wave, and the other, which involves H_y, E_x and E_z only, is called the TM (transverse magnetic) wave. The wave equations applying to the TE wave can be written as:

$$\begin{aligned} \frac{\partial^2 E_{yj}}{\partial z^2} &= -(k^2 n_j^2 - \beta^2) E_{yj} \\ H_{xj} &= -\frac{i}{k}\frac{\partial}{\partial z} E_{yj} \\ H_{zj} &= \frac{i}{k}\frac{\partial}{\partial x} E_{yj} \end{aligned} \tag{5.5}$$

The solutions of the electric field component (E_y) for Equations (5.5) in the three different regions are

$$\begin{aligned} E_{y0} &= A e^{-\delta z} & \text{(air)} \\ E_{y1} &= A \cos\epsilon z + B \sin\epsilon z & \text{(waveguide)} \\ E_{y2} &= (A \cos\epsilon d - B \sin\epsilon d) e^{\gamma(z+d)} & \text{(substrate)} \end{aligned} \tag{5.6}$$

where

$$\begin{aligned} \delta &= (\beta^2 - n_0^2 k^2)^{\frac{1}{2}} \\ \epsilon &= (n_1^2 k^2 - \beta^2)^{\frac{1}{2}} \\ \gamma &= (\beta^2 - n_2^2 k^2)^{\frac{1}{2}} \end{aligned} \tag{5.7}$$

and it should be noted that the first TE boundary condition $E_{y0} = E_{y1}$ (at $z = 0$) and $E_{y1} = E_{y2}$ (at $z = -d$) has been implemented.

From Equations (5.6) it can be clearly seen that in the z direction the electric field in the guide is a standing wave, which means that the energy carried is confined to the waveguide, but it decays exponentially as an evanescent wave in both side regions. The unknown constants A and B in Equations (5.6) can be determined by matching the second boundary condition at the two interfaces (i.e. for the tangential magnetic field H_x, or in the TE case the derivative of E_y i.e. $\partial E_y/\partial z$). This results in:

$$\begin{aligned}\delta A + \epsilon B &= 0 \\ (\epsilon \sin \epsilon d - \gamma \cos \epsilon d)A + (\epsilon \cos \epsilon d + \gamma \sin \epsilon d)B &= 0\end{aligned} \quad (5.8)$$

By eliminating the unknowns A and B, we have

$$\tan \epsilon d = \frac{\epsilon(\gamma + \delta)}{(\epsilon^2 - \gamma \delta)} \quad (5.9)$$

This is the eigenvalue (mode) equation, which gives the allowed eigenvalues (β_m) or mode indices (angles θ_m) for TE light to propagate in the guide. In terms of the definitions of ϵ, γ, δ (Equations (5.7)) the final form of the mode eigenvalue equation can be written as follows:

$$k_z d = m\pi + \tan^{-1}\left[\frac{n_m^2 - n_0^2}{n_1^2 - n_m^2}\right]^{\frac{1}{2}} + \tan^{-1}\left[\frac{n_m^2 - n_2^2}{n_1^2 - n_m^2}\right]^{\frac{1}{2}} \quad (5.10)$$

where $n_m = \beta/k = n_1 \sin \theta_m$ represents the mode index. It specifies the ratio of the wave velocity in a vacuum to that in the propagating direction (x) of the waveguide, and thus plays the role of the effective refractive index. It is obvious that Equation (5.10) is a transcendental equation, which can only be solved by a numerical method. The solution thus obtained is multi-valued, and each value of n_m represents a waveguide mode. The number of possible modes depends on the thickness and the refractive index of the waveguide and the wavelength of the light. Different waveguide modes are identified by their mode order $m = 0, 1, 2, ...$, respectively. The field distributions of the $m = 0, 1, 2, ...$ modes have $1, 2, 3, ...$ maxima, respectively, across the waveguide width, as shown in Figure 5.3.

A parallel discussion can be applied for TM waves, by suitably adjusting the second boundary condition. In that case the mode eigenvalue equation can be derived as

$$k_z d = m\pi + \tan^{-1}\left[\frac{n_1^2}{n_0^2}\right]\left[\frac{n_m^2 - n_0^2}{n_1^2 - n_m^2}\right]^{\frac{1}{2}} + \tan^{-1}\left[\frac{n_1^2}{n_2^2}\right]\left[\frac{n_m^2 - n_2^2}{n_1^2 - n_m^2}\right]^{\frac{1}{2}}$$

It is important to realise that the set of discrete eigenvalues of the waveguide modes is not a complete solution of the waveguide equation. It is the solution only for when the condition ($kn_2 < \beta = kn_1 \sin \theta_m < kn_1$) is satisfied. The complete set should also include the continuous solutions which lie in the ranges (1) $kn_0 < \beta < kn_2$, and (2) $0 < \beta < kn_0$. These continuous solutions indicate that other types of 'mode' exist in the

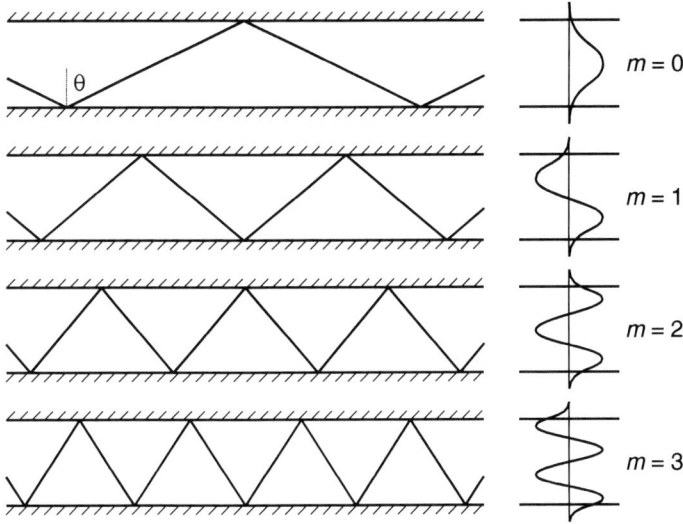

Fig. 5.3. Different waveguide modes are identified by their mode order $m = 0, 1, 2, \ldots$. The $m = 0$ mode has the largest mode angle. The field distributions of the $m = 0, 1, 2, \ldots$ modes have $1, 2, 3, \ldots$ maxima, respectively, across the waveguide width.

waveguide system, which are called 'substrate modes' and 'air modes' respectively. Substrate modes are only travelling in the waveguide and substrate regions. This is because in this case it can be seen from Equations (5.7) that when $\gamma^2 = (\beta^2 - k^2 n_2^2) < 0$, then γ becomes imaginary, so that the evanescent field in the substrate region turns into a radiation field. In other words, as the mode angle θ_m decreases further into the range of $\sin^{-1}(n_0/n_1) < \theta_m < \sin^{-1}(n_2/n_1)$, a plane wave will be totally reflected only at the upper interface (air-waveguide), but no longer so at the lower interface (guide-substrate). Air modes are travelling in all three air, guide and substrate regions. In this case the coefficients γ and δ in Equations (5.6) both become imaginary, so the evanescent fields in air and substrate regions turn into radiation fields, and therefore the plane wave is no longer totally reflected at either of the interfaces. All the types of mode are summarised in Figure 5.4.

In the above analysis, the wave in each layer is represented by two parameters (effectively, amplitude and phase), and the waves in two adjacent layers are related by two boundary condition equations (for continuity of electric and magnetic fields). Thus for a simple slab waveguide with a total of three layers (two boundaries) there are four unknowns related by

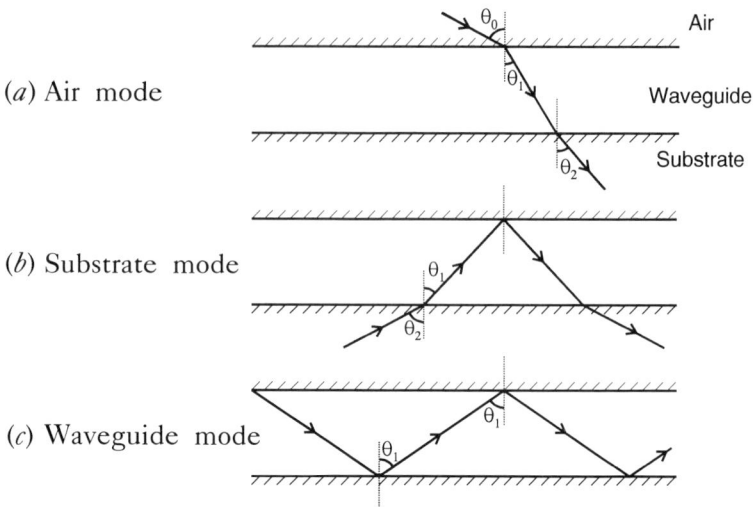

Fig. 5.4. Summary of the three types of mode: (*a*) air mode – travelling in all three air, guide and substrate regions; (*b*) substrate mode – travelling only in the waveguide and substrate regions; (*c*) waveguide mode – travelling only in the waveguide region.

two pairs of simultaneous equations, and the solution is elementary. As an extension of this method, the solution for a waveguide with a graded index profile may be achieved by dividing the guide into a series of thin uniform layers. For each interface the two boundary condition equations can be represented as a 2×2 matrix, and so the whole solution can be achieved by a multiplication of this series of matrices, one for each interface. The number of layers may be increased until the solutions converge to an acceptable precision.

5.2.2 Quantum mechanics analogy

A useful alternative model for the waveguide problem may be constructed by considering the light waves in it as though they were analogous to electrons trapped in a potential well. This analogy allows the direct application of quantum mechanics approximation techniques, such as WKB (Wentzel–Kramers–Brillouin) and perturbation methods, to the calculation of mode indices (Hocker, 1975; White, 1976; Lama and Chandler, 1988). This is especially useful for graded index waveguides where direct solutions are not possible. The Schrödinger equation for an

5.2 Waveguide mode theory

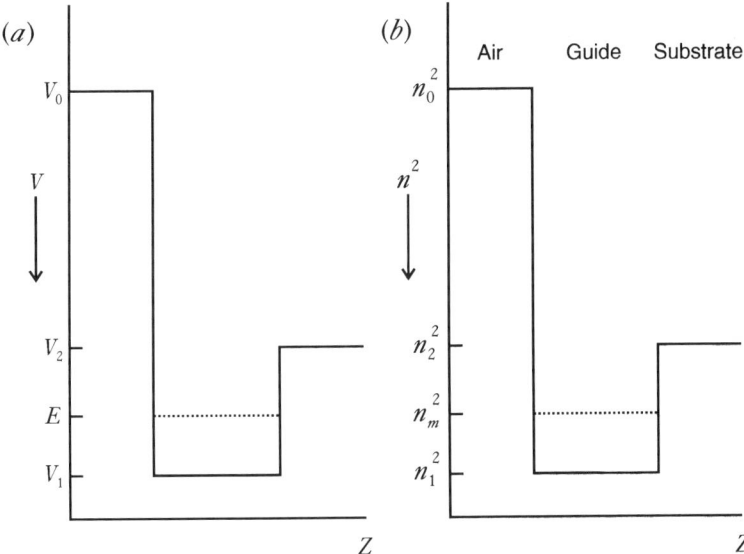

Fig. 5.5. Analogy between a planar waveguide and a quantum potential well: (a) electron potential well – characterized by energy levels; (b) planar waveguide – characterized by mode refractive indices.

electron with energy E trapped in a square well (Figure 5.5(a)) is

$$\frac{\partial^2 \varphi_j}{\partial z^2} + \frac{2m}{\hbar^2}(E - V_j)\varphi_j = 0 \quad j = 0, 1, 2 \ldots \quad (5.11)$$

where $V_1 > V_2 > V_0$. In the case of a waveguide, the wave equation for a uniform refractive index slab waveguide is (see Equation (5.5))

$$\frac{\partial^2 \varphi_j}{\partial z^2} + k^2\left(-n_m^2 + n_j^2\right)\varphi_j = 0 \quad j = 0, 1, 2 \ldots \quad (5.12)$$

If we then make the associations

$$\frac{2mE}{\hbar^2} \rightarrow -k^2 n_m^2$$
$$\frac{2mV_j}{\hbar^2} \rightarrow -k^2 n_j^2$$

Equations (5.11) and (5.12) will be identical. Here, in this model, we compare the bound state of an electron in a potential well with the waveguide modes in the guide. It follows, then, that we may represent a uniform waveguide by the square potential well as shown in Figure 5.5(b). It can be seen from the diagram that the allowed energy levels E_m of the electron correspond to allowed n_m^2 values of the waveguide modes.

5.3 Waveguide coupling

It is very important to be able to efficiently couple light into the various waveguide modes for applications and for measuring the refractive index profile. In the characterisation of optical waveguides it is convenient to study wide optical wells which confine many modes. But for many practical applications the monomode guide is essential. In the latter it is still necessary to measure mode index and loss. In general there are three coupling methods to access waveguide modes. They are fibre-coupling, end-coupling and prism-coupling. Fibre-coupling is used in device applications to couple the light into and out from waveguides, whilst end-coupling and prism-coupling are used for experimental measurements. To detect electric field profiles and insertion loss, the end-coupling method is commonly applicable. For analysing refractive index profiles, the prism-coupling method is widely used. Coupling in the light via a prism is necessary whenever separation of modes is needed for multimode guides.

5.3.1 End coupling

The end-coupling experiment has been designed for observing field profiles of waveguide modes as an indication of the mode confinement. The experiment requires the samples to be end-polished with an optically smooth finish. No chips should be observed with a ×500 magnification microscope on the edges of both ends of the sample. The field pattern of the waveguide modes can be seen on a screen or a TV camera by end-coupling-out employing a ×40 microscope objective. The laser beam coupling into the waveguide can be achieved either by end-coupling for a single-mode waveguide which uses another ×40 microscope objective, or by prism-coupling, which is useful for selecting a single mode in a multi-mode waveguide.

The end-coupling-in and end-coupling-out experimental set-up is shown in Figure 5.6(a). Initially, a screen takes the place of the TV camera to view the image. The microscope objective lenses have X and Z direction adjustment for focusing on the sample ends. The sample is mounted horizontally on a Y and Z adjustable table in order to view different parts of the guide. A set of horizontal interference fringes tend to appear on the screen, in the bottom half of the display when the surface of the sample falls below the objective lens axis, and in the top half when the reverse is true, as Figure 5.7 shows. The fringes result from interference

5.3 *Waveguide coupling* 161

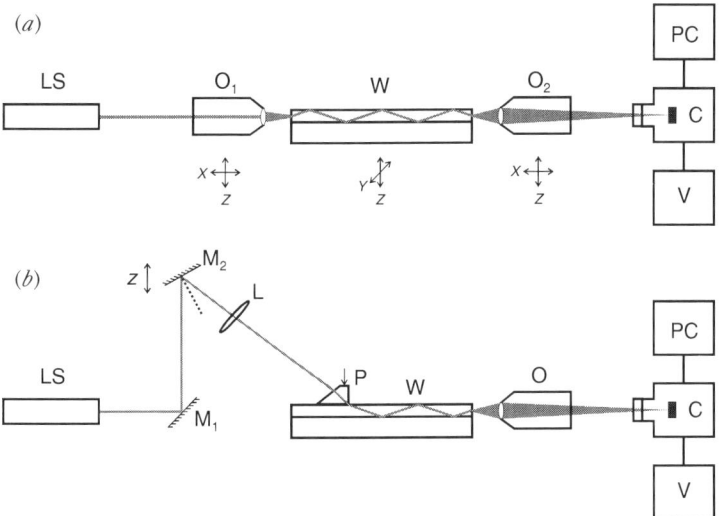

Fig. 5.6. End-coupling experimental arrangement: (*a*) end-coupling-in and end-coupling-out; (*b*) prism-coupling-in and end-coupling-out. LS – laser; O, O_1, O_2 – microscope objectives; L – lens; W – waveguide; M_1, M_2 – reflecting mirrors; P – prism; C – CCTV camera; PC – computer; V – video monitor.

between the two light sources – a direct laser beam from the first microscope objective, and a reflection of it from the surface of the sample. The fringe spacing increases as the waveguide surface of the sample gets close to the optical axis of the first objective lens. Eventually, the fringes will disappear and in their place on the screen there will be a very bright line running horizontally through the middle. The bright line is the field pattern of the waveguide modes. Finally, the distant screen is replaced by a TV camera without a lens, to let the light impinge directly on to the sensor element (512×512 pixels over an area $\sim 1 \text{ cm}^2$). The magnified field pattern of the mode can be displayed on the video screen. An image processing computer system then analyses the intensity distribution of the field pattern of the mode, and fits this to the refractive index profile of the waveguide.

The alternative prism-coupling-in experimental set-up used to observe field profiles of individual modes is illustrated in Figure 5.6(*b*). A pair of mirrors is placed between the laser and the sample instead of the first microscope objective. The laser beam hits the bottom mirror placed at 45° to the horizontal direction and is reflected on to the top mirror, adjustable for height and angle. A lens is inserted between the top mirror

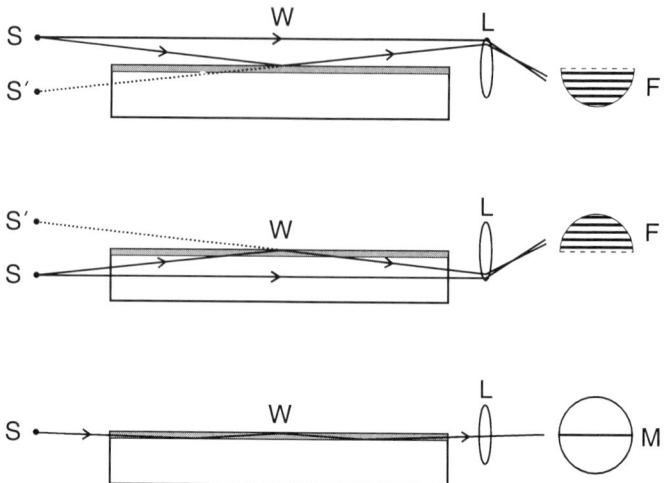

Fig. 5.7. Illustration of the end-coupling method for displaying the near field mode pattern. S – light source; S' – image of light source; W – waveguide; L – lens; F – interference fringes; M – waveguide mode.

and the prism at a focal length from the prism, which ensures that the laser beam through the lens always impinges on the same spot on the prism where the contact position should be. The rest of the set-up is the same as explained in the first part. Altering the height of the top mirror couples-in different modes, and the field pattern of each mode can be shown on the video screen and analysed by the computer.

Confirmation of the validity of the assignments of modes can be obtained by direct observation of the intensity patterns emitted from the end of a waveguide. If light is coupled into the waveguide at one end, and emitted at the other, then the field pattern is the sum of all mode contributions. However, selective excitation by a prism allows access to a specific mode. As shown in Figure 5.6(b) a combination of prism-input-coupling and end-emission gives the field distributions of each mode. Figure 5.8 gives examples of intensity profiles detected by using prism-coupling-in and end-coupling-out for a multiple energy He^+ implanted quartz waveguide. This waveguide has a square optical well bounded by air on the upper surface, and a low index amorphous silica barrier on the substrate side. From the index profile it is clear that there are six modes confined in this well, therefore only six real modes can be monitored by end-coupling. This classic square well is reflected in the symmetry and intensity patterns of the light from each mode. By

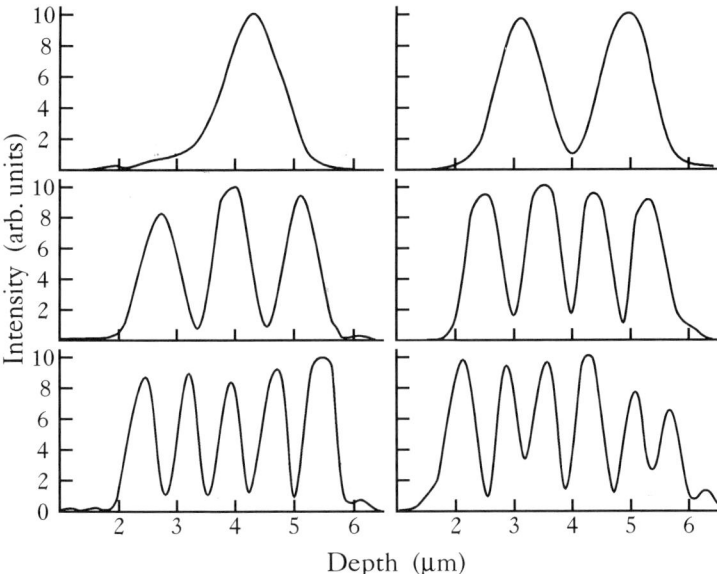

Fig. 5.8. Field intensity profiles for the six propagating modes observed in a multiple energy implanted sample of quartz. The depth scales have arbitrary zeros and are accurate to ±5%.

contrast, $Bi_4Ge_3O_{12}$ implantation, for example, increases the refractive index in the surface layer, forming a buried well, as will be mentioned later. The index increases from the surface with depth, so it is not possible to prism-couple to the $m = 0$ mode. Due to its asymmetric waveguide structure the intensity distribution will not be centred in the well.

5.3.2 Prism coupling

Tien et al. (1969, 1970) and Harris et al. (1970) first studied the use of the prism for excitation of waveguide modes in planar waveguides. Since then, prism-coupling has become an important technique in integrated optics studies. A prism which has a higher index than the waveguide is pressed against it, leaving a small air gap between the base of the prism and the top surface of the waveguide. The air gap is of the order of 0.1 μm, which is about one-quarter of the wavelength of visible light. A light beam from a laser enters the prism and is totally reflected at its base. The total reflection leaves only evanescent fields in the air gap, and it is

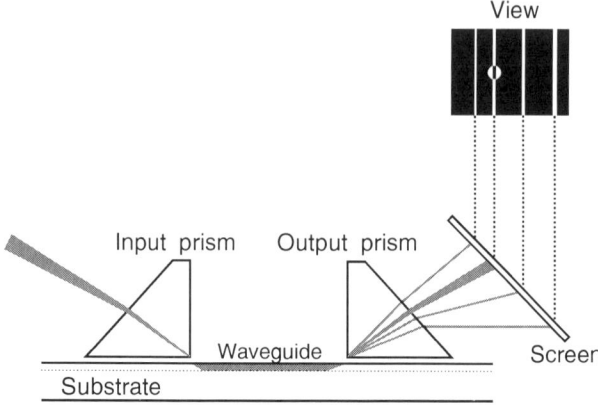

Fig. 5.9. The experimental arrangement commonly used to observe a spectrum of bright mode lines.

these evanescent fields which couple the light wave from the prism into the waveguide.

A direct calculation of the fields in the air gap suggests that the coupling strength between the prism and the waveguide should be very small. But a maximum coupling efficiency of more than 90% has been observed (Ulrich, 1971) A simple explanation is that the light energy in the prism is continuously fed into the waveguide at the successive reflection points under the laser beam, and these contributions are in phase.

The mode line spectrum which is a consequence of the prism-coupling technique is easy to measure experimentally. It consists of the angular positions of all the excited modes in the waveguide. These angular positions can be used to calculate the wave velocities of all the normal guiding modes and hence the refractive index and thickness of the waveguide. For a graded index waveguide, it is very useful to determine its refractive index profile by this method. This coupling technique is also widely used to measure the loss in waveguides, as will be described later.

Figure 5.9 shows the experimental arrangement commonly used to observe a spectrum of bright mode lines. In this experiment, two prism couplers are used spaced a distance d apart, to couple a laser beam into and out of the planar waveguide. A screen at the right is used to display light emerging from the output prism. In the actual experimental arrangement, the coupling system is mounted on a rotatable table in order that the incoming laser beam can be oriented into any synchronous mode

5.3 Waveguide coupling

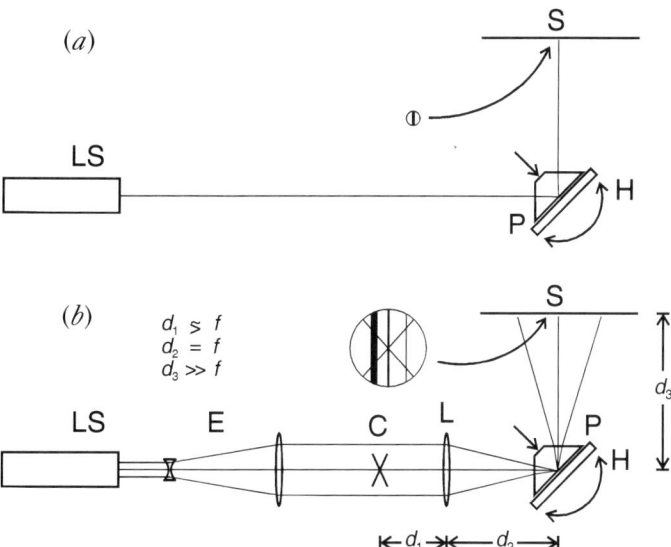

Fig. 5.10. Experimental arrangement for observation of coupling and intermode scattering. This setup is also used for the determination of the mode indices by the 'dark mode' method. LS – laser; E – beam expander; C – crosswires; L – focusing lens; P – prism; H – sample holder; S – screen.

angle, so that successive modes can be excited in the waveguide. As the light wave in the mode is propagated from the input prism to the output prism, part of the light is scattered. This scattering can be divided into two types. One is in-plane scattering, and the other is out-of-plane scattering (the plane being the plane of incidence). The former causes light to be scattered into other guiding modes. This scattered light is also propagated in the waveguide, and as it reaches the output prism, it is also coupled out of the prism into several different directions. This occurs because the scattered light involves many waveguide modes and each of them has its own output angle satisfying the synchronous condition. Thus a series of bright spots with gradually decreasing intensity should be observed on the screen. The brightest one represents the coupled-in mode. In addition, the out-of-plane scattering causes light to be scattered out of the plane of incidence. As a consequence of this scattering, instead of only seeing bright spots, a series of parallel bright lines are observed on the screen (with a spot at the centre of the coupled-in line). These lines constitute the bright mode spectrum.

Figure 5.10 illustrates an alternative arrangement for observing the mode line spectrum, especially for ion implanted waveguides, in which only one isosceles prism is used (Tien et al., 1969, 1972). By contrast with the two-prism technique, which gives bright mode lines, this is called the dark mode line spectrum method. This method relies on an expanded laser beam being focused by a converging lens, through the isosceles prism and on to the guide itself. The picture seen on the distant screen is a large bright background circle crossed by several dark lines. The size of the circle is inversely proportional to the focal length of the lens. In this arrangement, due to the use of an expander, the laser light does not only couple into one waveguide mode, but, being a cone of light, it covers several mode angles. The synchronous condition is therefore satisfied for several modes at the same time, and so the light travels into the waveguide in each of these several modes. This coupling-in of light results in a lack of reflected light at the prism base, and consequently the dark mode line spectrum is displayed on the screen. Note that the first few dark mode lines appear at positions coincident with the m bright lines in the two-prism method. It can be proved that these correspond to the real guiding waveguide modes (Lama and Chandler, 1988). But many further dark lines can also be seen in the dark mode line technique. These extra dark lines gradually broaden for higher mode number. A rather non-rigorous explanation for these extra non-guiding dark lines is that they are the result of optical interference between the multiple reflections occurring at the interfaces in the waveguide. They are strongly related to the structure of the waveguide. The measurement of these extra dark lines has been shown to provide very useful information for the analysis of the planar optical waveguide profiles (Chandler and Lama, 1986a). This will be discussed in more detail in Section 5.4.3 concerning the reflectivity calculation method for determining index profiles from the complete dark mode spectrum.

If a waveguide increases in refractive index with depth from the surface then the prism-coupling method may not detect the lowest order modes. It is a problem which arises very obviously in $LiNbO_3$ and $Bi_4Ge_3O_{12}$ (Chandler et al., 1989a; Mahdavi et al., 1990). To overcome this problem, an analysis made with several wavelengths is necessary. The details of this problem have been discussed by Chandler et al. (1989a) and Zhang et al. (1991) for $LiNbO_3$. They used several analysing wavelengths and careful thinning of the guide to enable greater coupling from the prism into the lower modes.

5.4 Index profile determination

The theoretical considerations of Section 5.2 dealt with various methods for calculating the mode spacings of a given waveguide profile. In the analysis of a practical waveguide, the reverse calculation is necessary in order to deduce the index profile from a given set of experimental modes. Only in the simplest of cases is this possible by a direct analytic method (e.g. an infinitely deep square well giving quadratically spaced modes, or a parabolic well giving uniformly spaced modes). In general, the mode spectrum has a more complex spacing. In this case it is usual to set up a trial index profile shape described by several parameters and calculate the theoretical mode curve of this trial profile. A parameter optimisation routine (least-squares fit) is then needed to adjust the profile shape until the theoretical modes match the experimental ones. This is necessary because in any rigorous analysis, all the modes are affected by all the parameters. A non-rigorous approximation which is an exception to this rule is the 'direct' WKB method. In this analysis, any particular mode is assumed only to be affected by the shape of the well beneath it, and not by the higher profile. This method therefore 'builds' a profile from the bottom upwards using one mode value at a time, and its implementation is consequently speedy.

5.4.1 WKB approximation for a graded index profile

In practical cases planar waveguides are frequently non-uniformly distributed in index. It is in general impossible to calculate the mode indices of these waveguides, but the WKB method from quantum mechanics can be adapted to deal with these non-uniform planar waveguides. Because the refractive index, in this case, varies along the z direction, according to the WKB approximation, the mode eigenvalue equation (Equation (5.1)) can be rewritten as a phase integral equation as follows:

$$\int_0^{z_t} k_z(z)dz + \varphi_{10} + \varphi_{12} = m\pi \quad m = 0, 1, 2\ldots \quad (5.13)$$

where, from the wave vector relation (Figure 5.2), $k_z^2(z)$ can be written as

$$k_z^2(z) = k^2 n_1^2(z) - k_x^2 \quad (5.14)$$

where $k_x = \beta = kn_1(z)\sin\theta_m(z)$. The limits of the integral $z = 0$ and $z = z_t$ are the classical turning points (at which the standing wave becomes an evanescent wave) of the wave function in the potential well. The turning point z_t satisfies the condition $n_m = n_1(z_t)$. To apply Equation

(5.13) the most difficult problem is to determine the phase changes ($2\varphi_{10}$ and $2\varphi_{12}$) in a given graded waveguide. It is not possible to employ Fresnel's law directly, but a reasonable assumption can be made for most common cases. Comparing the index of the most dense waveguide ($n_1(z) \sim 2.0$) with that of air ($n_0 \sim 1.0$), the ratio of the indices is about 2:1, and thus for all practical purposes, the approximation $2\varphi_{10} \sim \pi$ can be taken. In addition, since the quantity $2\varphi_{12}$ is the phase change which occurs in the wave function's reflection at the turning point z_t, another linear approximation can be taken for φ_{12} as

$$\tan\varphi_{12} = \lim \left[\frac{n_m^2 - n_2^2}{n_1^2 - n_m^2}\right]^{\frac{1}{2}} \longrightarrow 1 \quad \text{(as } n_1 \to n_m \text{ and } n_2 \to n_m\text{)}$$

and hence, $\varphi_{12} = \pi/4$.

Knowing φ_{10} and φ_{12}, and the shape of the index distribution in the z direction, Equation (5.13) can be used to solve for the mode indices. In actual applications, the WKB method has been applied in more cases than any other technique to analyse the index profiles of graded index waveguides. It turns out to be remarkably accurate in some cases, such as in- or out-diffused waveguides in which the index is gradually changing with depth. But it is less suitable for other cases, such as proton exchange, epitaxial growth, and ion implanted waveguides, in which the profile has a steep index function, or an optical barrier.

5.4.2 Ion implanted optical barrier waveguides

As discussed earlier, the WKB method which is adapted for the determination of refractive index profiles can only be applied to the study of perfectly bound eigenstates where a unique determination of the discrete solutions is possible. Therefore, it appears to have a major limitation in the case of waveguide profiles as generated by ion implantation. Ion implanted waveguides in crystalline materials normally rely on a reduced index barrier at the end of the implanted region. The guiding layer in these devices is therefore located between the surface and this barrier. The barrier will satisfactorily confine the main non-stationary (leaky) 'modes' which have a low tunnelling probability. For this case, the problem is no longer an eigenvalue (stationary) one, and so approximations deriving from such a formalism are invalid. A more obvious shortcoming of the WKB method arises when attempting to account for the modes which are observed, not as resonant bright lines from a coupling-out prism, but by the so-called 'dark mode' method. These observed modes

5.4 Index profile determination

include not only the real resonant modes which are confined in the actual optical well (guide), but also the 'substrate' modes which are beyond the limit expected from the well depth (barrier height). These 'substrate' modes are closely related to the refractive index distribution, so that they form part of the available useful information for determining the index profile of the waveguide. It is therefore hoped that their inclusion in any calculation can assist in a more accurate characterisation of the guide profile. But the WKB method does not model these 'substrate' modes, and so it does not use all of the available data. It is certainly not able to determine the full index profile of a barrier waveguide.

To sum up, it can be seen that for non-bound states, the time independent Schrödinger approach is not applicable, as the wave solutions are not stationary. It is more realistic and accurate to characterise the wave which exists in a barrier waveguide as a travelling wave with component vectors in two directions at right-angles (k_x, k_z), one in the propagating direction as $\exp i(k_x x - \omega t)$, and the other in the transverse z direction. The latter is the sum of waves with two differing amplitudes, $A \exp i(k_z z - \omega t)$ and $B \exp i(k_z z + \omega t)$, representing the positive and negative travelling transverse wave caused by multiple reflections at the boundaries. The solution for the z direction will have a partly stationary and partly 'leaky' nature. Chandler and Lama (1986a) developed a new method using this principle, a 'reflectivity' calculation method, in order to characterise the non-stationary case waveguides. It has been demonstrated that this method can analyse ion-implanted waveguides remarkably well, and also several other types of guide, such as epitaxial, proton exchange (Chandler *et al.*, 1990), and also more complex structures such as the double barrier (Chandler *et al.*, 1988, 1989b). This method has been used so far to characterise ion implanted guides in many substances. The model, detailed theory and applications of this method will be discussed in the following sections.

5.4.3 Reflectivity calculation method (RCM)

The reflectivity calculation method attempts to model the actual technique used for the experimental observation of mode lines by the dark mode method. The phenomenon of 'dark modes' is in fact an interference effect produced by the reflectivity of light at a single point on the coupling prism interface (Figure 5.11). It does not depend on any 'guiding' of light. If resonance occurs at a certain coupling angle, more light will be absorbed/scattered/tunnelled than at non-resonance, and so less light

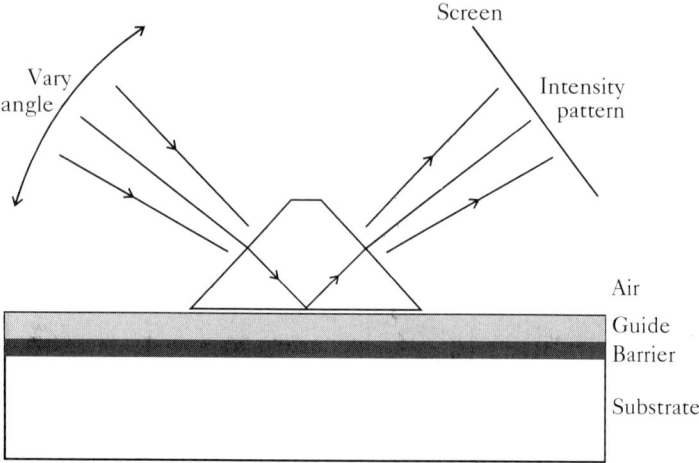

Fig. 5.11. Experimental method for the observation of a back-reflected intensity pattern containing 'dark mode' lines.

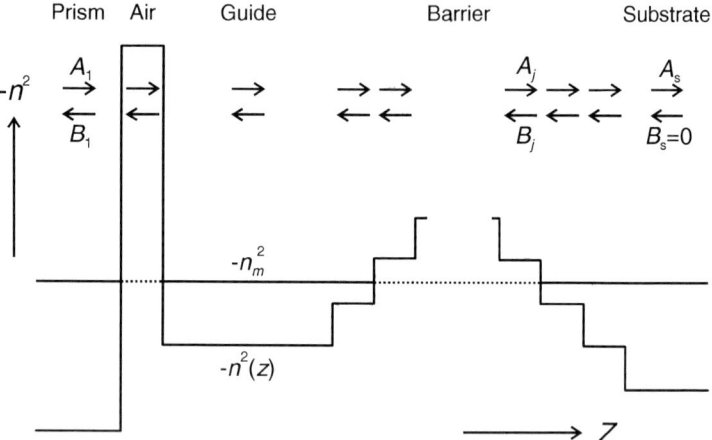

Fig. 5.12. Travelling wave components in the model for a non-bound waveguide system. In the calculation the profile is represented by up to a hundred steps.

will be reflected back on to the screen – resulting in a dark line. The resonance angles, of which the lower ones correspond to the bright mode positions, can therefore be determined from the minima in the calculated reflectivity as a function of incident angle.

Figure 5.12 gives a model of the refractive index profile of a typical barrier waveguide. The transverse light component (k_z) is considered

5.4 Index profile determination

incident on the system in the positive z direction from the coupling prism (via an air gap) into the guiding region, and then to partially leak through the barrier into the substrate. The barrier shape is simulated by a set of discrete steps (up to 50 or 100). Because of partial reflections at all the step boundaries, there is also a travelling wave in the negative z direction. These two opposite travelling waves can be represented mathematically as $\exp i(\pm k_z z - \omega t)$, where k_z is the z component of the wave vector in the medium. According to the discussion in a previous section (Equation (5.14)), we have

$$k_z = k[n^2(z) - n_m^2]^{\frac{1}{2}} \qquad (5.15)$$

k being the wave vector in free space, and n_m the effective mode index $n(z)\sin\theta(z)$. It is to be emphasised that in the present model, n_m is no longer an eigenvalue of a bound mode, but is continuously variable. It will have certain values at which the wavefunction exhibits strong resonance. The sign of k_z^2 indicates that the wave in different regions behaves in different manners, either in an oscillating manner when k_z is real ($n_m < n(z)$), e.g. in the coupling prism, guide, substrate, or the barrier region with index greater than the n_m level; or in an evanescent manner when k_z is imaginary ($n_m > n(z)$), e.g. in the air gap and barrier regions with index less than the n_m level. Let A and B represent the amplitudes of the two opposite travelling waves in any region. Hence the general wave function in the z direction can be written as

$$E(z) = A\exp(+ik_z z) + B\exp(-ik_z z) \qquad (5.16)$$

where $k_z = k\cos\theta$. A and B are complex to allow for phase matching at the step boundaries and k_z can be either real, representing an oscillating wave, or imaginary representing an evanescent wave. In the former case the wavefunction can be rewritten as

$$E(z) = A'\cos(k_z z) + B'\sin(k_z z) \quad \text{(real)} \qquad (5.17)$$

where $k_z = k[n^2(z) - n_m^2]^{\frac{1}{2}}$, and in the latter case the wavefunction can be rewritten as

$$E(z) = A\exp(-k'_z z) + B\exp(k'_z z) \quad \text{(imag.)} \qquad (5.18)$$

where $k'_z = k[n_m^2 - n^2(z)]^{\frac{1}{2}}$. The use of this terminology has the advantage that in all the regions, only the amplitudes are complex, and not the exponents, resulting in simplified and more rapid computation. As Figure 5.12 shows, the whole non-bound waveguide system is divided into many layers, and hence the two complex amplitudes A and B in one region are related to those in an adjacent region by the two boundary condition requirements. These are

$$
\begin{array}{lll}
(1)\ H_j(z) = H_{j+1}(z) & \text{(at the } j:j+1 \text{ boundary)} \\
(2)\ E_j(z) = E_{j+1}(z) & \text{(at the } j:j+1 \text{ boundary)}
\end{array} \quad (5.19)
$$

For TE polarisation (1) can be rewritten as

$$\frac{\partial E_j(z)}{\partial z} = \frac{\partial E_{j+1}(z)}{\partial z}$$

and for TM polarisation (2) can be rewritten as

$$\frac{1}{n_j^2}\frac{\partial H_j(z)}{\partial z} = \frac{1}{n_{j+1}^2}\frac{\partial H_{j+1}(z)}{\partial z}$$

For simplicity the z origin of each region can be taken at its first boundary. In general, these boundary conditions for a barrier waveguide system can be divided into four different cases, according to the index distribution regions. The wavefunctions have different representations in the two types of region (k real, k imaginary) as shown above, and so the four boundary cases are oscillating to oscillating, evanescent to evanescent, oscillating to evanescent and evanescent to oscillating, as described below.

(a) Oscillating to oscillating (e.g. within the guide, substrate or prism). In this case both wave functions have oscillating natures, and so according to Equations (5.17), (5.18) and (5.19) the boundary conditions lead to the two relations

$$
\begin{aligned}
A_j\cos(k_{zj}T_j) + B_j\sin(k_{zj}T_j) &= A_{j+1} \\
k_{zj}A_j\sin(k_{zj}T_j) - k_{zj}B_j\cos(k_{zj}T_j) &= -k_{z(j+1)}B_{j+1}
\end{aligned}
$$

These two relations can be represented by a 2×2 matrix, which after inversion becomes

$$
\begin{bmatrix} A_j \\ B_j \end{bmatrix} = \begin{bmatrix} \cos(k_{zj}T_j) & -\frac{k_{z(j+1)}}{k_{zj}}\sin(k_{zj}T_j) \\ \sin(k_{zj}T_j) & \frac{k_{z(j+1)}}{k_{zj}}\cos(k_{zj}T_j) \end{bmatrix} \begin{bmatrix} A_{j+1} \\ B_{j+1} \end{bmatrix}
$$

(b) Evanescent to evanescent (e.g. within the barrier). In this case both wavefunctions have evanescent natures. By a similar treatment to that above, we have

$$
\begin{aligned}
A_j\exp(-k'_{zj}T_j) + B_j\exp(k'_{zj}T_j) &= A_{j+1} + B_{j+1} \\
k'_{zj}A_j\exp(-k'_{zj}T_j) - k'_{zj}B_j\exp(k'_{zj}T_j) &= k'_{z(j+1)}A_{j+1} - k'_{z(j+1)}B_{j+1}
\end{aligned}
$$

5.4 Index profile determination

which gives the matrix

$$\begin{bmatrix} A_j \\ B_j \end{bmatrix} = \begin{bmatrix} \frac{E}{2}\left(1 + \frac{k'_{z(j+1)}}{k'_{zj}}\right) & \frac{E}{2}\left(1 - \frac{k'_{z(j+1)}}{k'_{zj}}\right) \\ \frac{1}{2E}\left(1 - \frac{k'_{z(j+1)}}{k'_{zj}}\right) & \frac{1}{2E}\left(1 + \frac{k'_{z(j+1)}}{k'_{zj}}\right) \end{bmatrix} \begin{bmatrix} A_{j+1} \\ B_{j+1} \end{bmatrix}$$

where $E = \exp(k'_{zj}T_j)$.

(c) Evanescent to oscillating (e.g. from air gap to guide, or from barrier to substrate). Here we have

$$A_j \exp(-k'_{zj}T_j) + B_j \exp(k'_{zj}T_j) = A_{j+1}$$
$$k'_{zj}A_j \exp(-k'_{zj}T_j) - k'_{zj}B_j \exp(k'_{zj}T_j) = -k_{z(j+1)}B_{j+1}$$

giving the matrix

$$\begin{bmatrix} A_j \\ B_j \end{bmatrix} = \begin{bmatrix} \frac{E}{2} & -\frac{k_{z(j+1)}}{k'_{zj}}\frac{E}{2} \\ \frac{1}{2E} & \frac{k_{z(j+1)}}{k'_{zj}}\frac{1}{2E} \end{bmatrix} \begin{bmatrix} A_{j+1} \\ B_{j+1} \end{bmatrix}$$

where, also, $E = \exp(k'_{zj}T_j)$.

(d) Oscillating to evanescent (e.g. from prism to air gap, or from guide to barrier). Here we have

$$A_j \cos(k_{zj}T_j) + B_j \sin(k_{zj}T_j) = A_{j+1} + B_{j+1}$$
$$k_{zj}A_j \sin(k_{zj}T_j) - k_{zj}B_j \cos(k_{zj}T_j) = k'_{z(j+1)}A_{j+1} - k'_{z(j+1)}B_{j+1}$$

giving the matrix

$$\begin{bmatrix} A_j \\ B_j \end{bmatrix} = \begin{bmatrix} C + \frac{k'_{z(j+1)}}{k_{zj}}S & C - \frac{k'_{z(j+1)}}{k_{zj}}S \\ S - \frac{k'_{z(j+1)}}{k_{zj}}C & S + \frac{k'_{z(j+1)}}{k_{zj}}C \end{bmatrix} \begin{bmatrix} A_{j+1} \\ B_{j+1} \end{bmatrix}$$

where $C = \cos(k_{zj}T_j)$, $S = \sin(k_{zj}T_j)$.

The incident and reflected amplitudes at the coupling prism (A_1 and B_1 in Figure 5.12) may therefore be related to the two amplitudes in the substrate region beyond the optical barrier (A_s and B_s) by the product of $s-1$ of the above 2×2 real matrices (M_j), each representing an interface within the s regions of the guide and barrier. Thus

$$\begin{bmatrix} A_1 \\ B_1 \end{bmatrix} = \prod_{j=1}^{s-1} M_j \begin{bmatrix} A_s \\ B_s \end{bmatrix} = \begin{bmatrix} P_{11} & P_{12} \\ P_{21} & P_{22} \end{bmatrix} \begin{bmatrix} A_s \\ B_s \end{bmatrix}$$

From Figure 5.12 it can be seen that there is no light travelling back from the substrate to the waveguide, i.e. $B_s = 0$. Hence the reflection from the prism interface is given by

$$R = \left|\frac{B_1}{A_1}\right|^2 = \left|\frac{P_{21}}{P_{11}}\right|^2$$

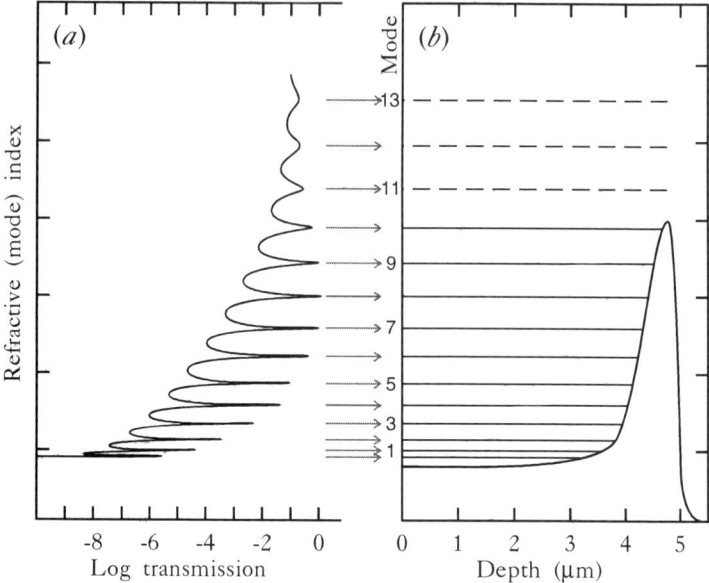

Fig. 5.13. (a) Example of the calculated transmission coefficient $(1 - R)$, using the reflectivity calculation method for the waveguide profile shown in (b).

The above discussion for TE propagation differs in the case of TM because of the different boundary conditions. This introduces the factors $(n_j/n_{j+1})^2$, multiplying the (k_{j+1}/k_j) terms of the M_j matrices.

Figure 5.13(a) shows an example in which the reflection coefficient R has been calculated in this manner for a chosen barrier waveguide profile as in Figure 5.13(b). By continuously varying the mode index n_m (actually the mode angle θ_m), one can find that there are a series of minima in the values of R occurring at certain angles. The distribution of $(1 - R)$ has been plotted logarithmically in Figure 5.13(a) against n_m. Sharp dips in the reflectivity – spikes in $(1 - R)$ – occur at certain resonant values of the mode index and these would be seen as dark mode lines on the screen. As can be seen from the diagram, near the bottom of the well (high index value) where the barrier is thick and hence tunnelling unlikely, the spikes are very sharp, and here they correspond closely to the stationary state eigensolutions for n_m. Nearer the top of the barrier where tunnelling can occur, the spikes become broader which corresponds to broad dark modes seen on the screen. Beyond the top of the barrier the reflectivity function continues to show fluctuations. These oscillations in reflectivity correspond to very broad dark mode lines observed on the screen beyond

5.4 Index profile determination

the angle expected for the given barrier height. They are not observable in the bright mode display as they are not true confined states. Up to this point we have seen the whole view of the reflectivity calculation method. It can be concluded that this method is ideally suitable for simulation of the complete dark mode pattern. How this can then be used to predict and characterise the waveguide refractive index profiles will be discussed in the following subsection.

5.4.4 Index profile characterisation

It appears that the reflectivity calculation method can be used to determine the index profile shape of a waveguide. This may involve the following steps: first, choose an analytic function for the index profile characterised by several parameters; second, calculate the mode indices of this hypothetical profile by using the reflectivity method described above; third, compare these theoretical mode index values with the experimental values obtained from dark mode measurements; and finally, keep changing the parameters to alter the index profile shape until the theoretical mode indices match the experimental ones within a satisfactory error. This final profile is therefore assumed to be the optimum shape for the given mode index data.

In this subsection, a typical analytical profile will be illustrated in detail; the 'mode index curve' will be introduced to relate the mode spacing to the index profile, and the dependence of profile shape on this curve will be explained. Finally, two specific examples of index profiles will be given to illustrate the performance of the reflectivity method.

5.4.4.1 Analytic profile for a barrier waveguide

Figure 5.14 shows a typical analytic index profile for a barrier waveguide (e.g. ion implanted quartz). The profile is constructed from a skewed flat-topped Gaussian peak, with a sloping base plateau. The shape is characterised by nine possible parameters, which are defined as the diagram shows. These parameters are presented as follows:

P_1 the distance at which the maximum barrier height occurs;
P_2 the refractive index at the surface;
P_3 the refractive index at the maximum of the barrier;
P_4 the standard deviation of the barrier rising edge;
P_5 the width of the barrier flat-top;
P_6 the standard deviation of the barrier falling edge;

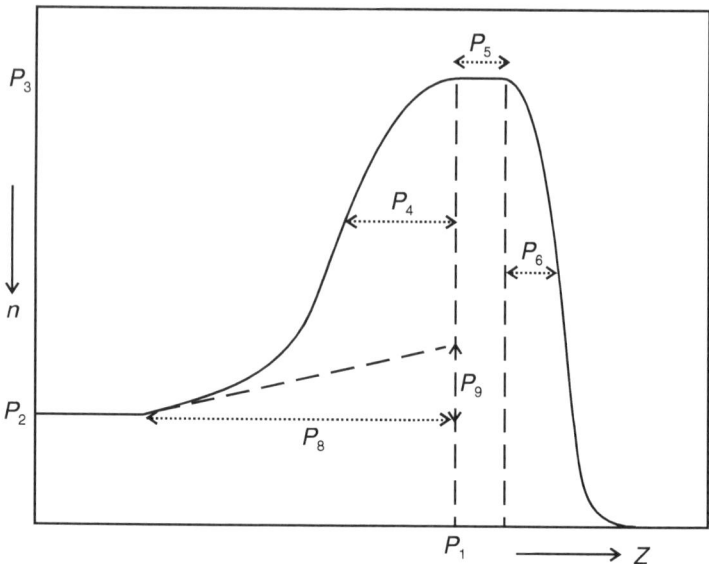

Fig. 5.14. A typical analytic index profile for an implanted barrier waveguide. The profile is constructed from a skewed flat-topped Gaussian peak with a sloping base plateau. The shape is characterised by up to nine parameters.

P_7 the parameter determining the Gaussian or exponential nature of the barrier shape;
P_8 the depth over which the sloping base occurs;
P_9 the height of the sloping base.

In practice, a more complicated analytic profile can be chosen if necessary (e.g. double-barrier or triple-barrier profiles), characterised by as many parameters as necessary. In principle, the number of parameters is only limited by the number of experimental data points, the former not being allowed to exceed the latter. In most cases, however, a much simpler profile is found to be quite adequate, characterised by only four or five parameters.

5.4.4.2 Mode index curve

Recalling quantum mechanics basic theory, it is known that an infinitely deep square potential well corresponds to a straight line when the energy levels are plotted against n^2 (Figure 5.15) – n being the order of the energy level.

An ideal high Δn index slab waveguide is equivalent to an infinite

5.4 Index profile determination

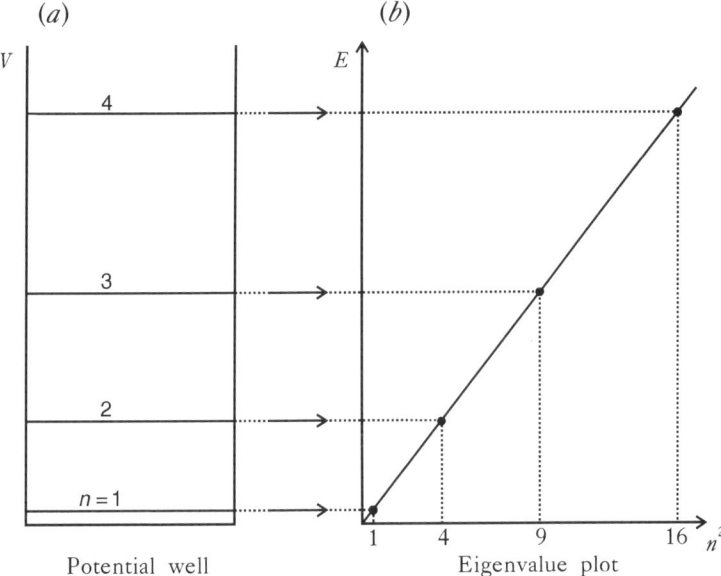

Fig. 5.15. An infinitely deep square potential well (a) results in a straight line (b) when the energy levels are plotted against n^2, n being the order of the energy level.

potential well. In this case, the phase changes at the two interfaces (air–guide, guide–substrate) become π, and therefore the mode index eigenvalue equation becomes:

$$2k_z d = m2\pi + \pi + \pi = 2(m+1)\pi \qquad (5.20)$$

Substituting $k_z = kn_1 \cos\theta$, where $k = 2\pi/\lambda$ into Equation (5.20), and using $n_m = n_1 \sin\theta_m$ we obtain:

$$n_m^2 = n_1^2 - (m+1)^2 \left(\frac{\lambda}{2d}\right)^2 \qquad (5.21)$$

Equation (5.21) indicates that a very high Δn index slab waveguide has a similar behaviour to an infinite potential square well – a straight line (in Figure 5.15) represents the linear relation between the energy level (square of mode index) and the square of the energy level quantum number (mode number +1). For the sake of convenience, in the case of a waveguide, this is referred to as the guide's 'mode curve'. Equation 5.21 gives a close relationship between the waveguide index profile shape and its mode curve. It is easy to see from the equation that the slope of the mode curve $(\lambda/2d)^2$ gives the width of the waveguide, and its intercept

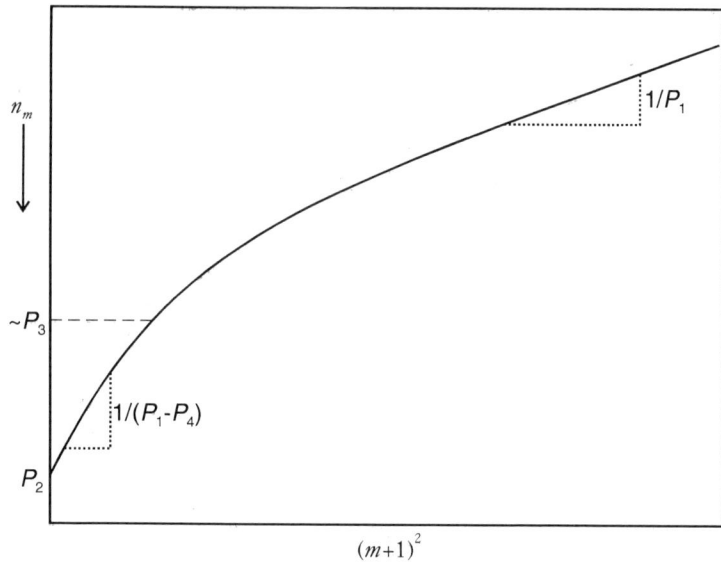

Fig. 5.16. The mode curve corresponding to an implanted barrier waveguide. The mode index values are plotted against $(m+1)^2$, m being the order of the waveguide mode. The shape of this curve is related to the parameters characterising the profile.

(n_1^2) gives the index of the waveguiding layer. Naturally, a waveguide with graded index profile can also be related to its mode curve. This leads to a simple method of employing interactive computer graphics in conjunction with the mode reflectivity calculation explained above, to obtain a preliminary estimate of a practical graded index profile.

For a chosen analytical profile of a barrier waveguide, the mode indices can be calculated by using the reflectivity calculating method explained above. The calculated (theoretical) mode indices are then plotted as a mode curve. It has been found in practice that the nature of the mode curve shape is closely related to the parameters which characterise the analytic index profile.

As shown in Figure 5.16, the mode curve is no longer a straight line. It is now the case that: the intercept of the mode curve gives the surface index (P_2) of the waveguide; the point at which the curve changes slope leads to a value of the barrier height (P_3); the slope of the upper curve is roughly inversely proportional to the waveguide width P_1; the slope of the lower curve is related to the width of the lower well, from which the standard deviation of the barrier rising edge can be estimated

($\propto 1/(P_1 - P_4)$); and the degree of curvature of the bottom curve gives an idea about the sloping base of the guide profile (P_8, P_9). By using those relationships between the mode curve and the profile parameters, the computer calculation program initially estimates the main parameters (P_1, P_2, P_3, P_4) very roughly, from the experimental mode curve (data), and hence gives a first theoretical mode curve, calculated from this first estimated profile. This theoretical mode curve and the experimental data points are plotted on the computer screen. Visual assessments of the necessary parameter changes are then made, followed eventually by a computer 'least-squares' fitting program to adjust the parameters, and hence the theoretical mode indices, until finally they match adequately within the experimental errors of the input data.

5.4.5 Examples of refractive index profile fitted by using RCM

5.4.5.1 Single-barrier profile

Figure 5.17(a) shows the dark mode data plotted as a mode curve for a typical barrier waveguide produced by a single energy He$^+$ implantation in quartz. It is plotted as the mode index value (β/k or n_m) against the mode number squared $(m + 1)^2$. The circles represent the experimental mode index data, and the various curved lines depict the 'theoretical' mode values that would be obtained for the various (approximate) profiles shown in Figure 5.17(b). The dotted profile (in (b)) and dotted mode curve (in (a)) correspond to the initial guess, and this can be improved by 'inspection' to give the dashed results. Finally, the least-squares fit produces the profile and mode curve given by the solid lines. The fact that the final mode curve goes very precisely through all the points indicates that this chosen profile represents the true profile of this waveguide. Furthermore, the degree of accuracy of this profile can be seen from Table 5.1. In the table the second column gives the experimental mode index data; the third column shows mode index data calculated from the profile; and the differences between them are given in the last column. It can be seen from the table that for the first seven modes (which are real confined modes), there is a very small difference ($\pm 6 \times 10^{-5}$) between experimental data and calculated data. The error gradually increases for the higher order modes. This is because the location of the higher modes is less precise as they get broader with increasing mode number. However, the difference still averages at less than $\pm 5 \times 10^{-4}$. Any systematic error observed in Table 5.1 can be attributed to the fact that the chosen analytic

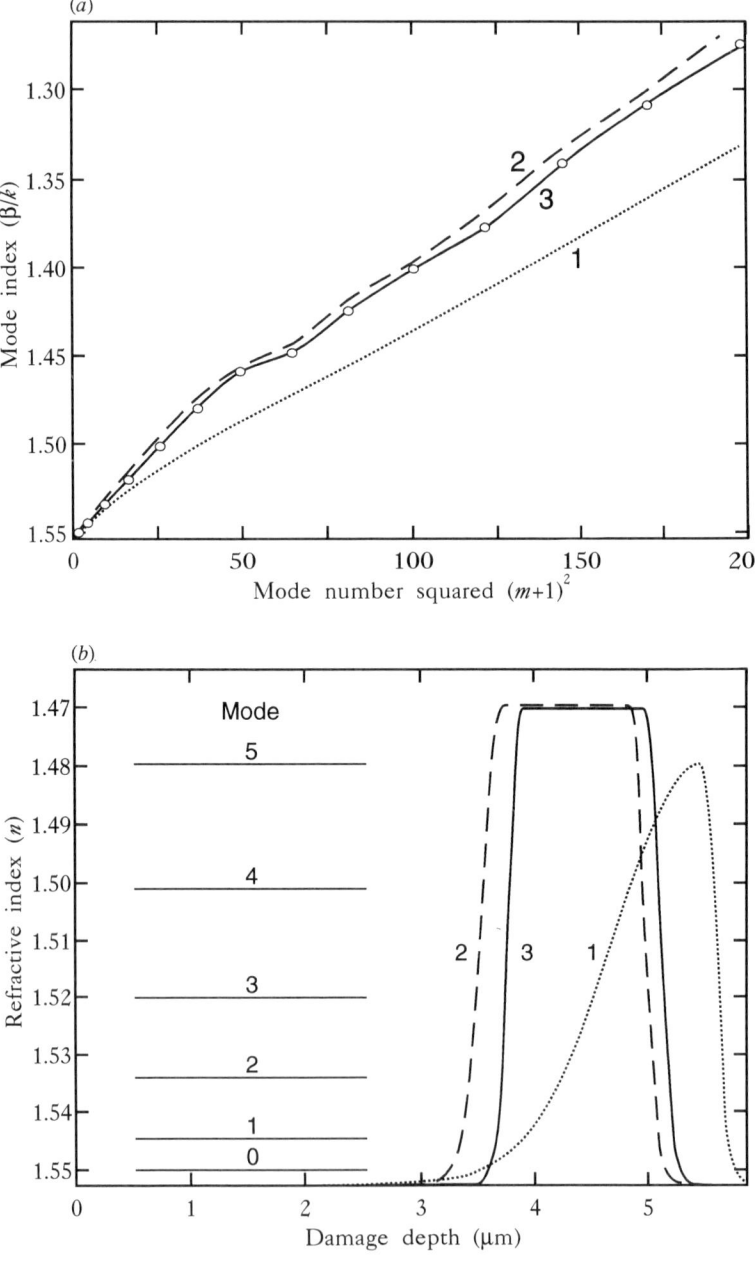

Fig. 5.17. (a) Mode curve plot for a single-barrier waveguide. The circles are the experimental mode index data. The curves represent the 'theoretical' modes calculated from the refractive index profiles given in (b). The solid curve going through all the points indicates that this chosen profile is an accurate representation of the true profile for this waveguide. The other curves are computed mode values corresponding to the other two profiles in (b).

5.4 Index profile determination

Table 5.1. *Typical mode indices for a barrier waveguide profile in quartz*

Mode number	Experimental n_m	Calculated n_m	Error Δn
0	1.550 50	1.550 47	+0.000 03
1	1.544 33	1.544 32	+0.000 01
2	1.534 03	1.534 08	−0.000 05
3	1.519 74	1.519 77	−0.000 03
4	1.501 53	1.501 51	+0.000 02
5	1.479 83	1.479 86	−0.000 03
6	1.459 01	1.458 97	+0.000 04
7	1.448 11	1.447 85	+0.000 26
8	1.423 70	1.423 48	+0.000 22
9	1.399 02	1.399 77	−0.000 75
10	1.376 18	1.376 02	+0.000 16
11	1.339 66	1.340 24	−0.000 58
12	1.306 76	1.304 95	+0.001 81
13	1.271 31	1.272 62	−0.001 31

profile is not a perfect description of the experimental profile, but this error appears small.

5.4.5.2 Double-barrier profile

Figure 5.18(a) and (b) are the mode curve plot and the refractive index profile for a double-barrier waveguide formed in LiNbO$_3$ by using two He$^+$ ion beam energies (1.1 and 2.2 MeV). It can be seen that the calculated mode curve goes through all the modes, which suggests that the mode pairing effect must be due to the existence of a double-barrier profile, and that the two optical wells characterised by this profile must have similar widths. If two wells having similar widths are closely coupled, the modes of each well appear to be 'split' into two, producing the paired effect seen here. The differences between the mode data measured experimentally and that calculated from the chosen profile are listed in Table 5.2. It can be seen from the table that there is only about ±0.0002 difference for the lower (real) modes, and on average ±0.0005 for the higher order (substrate) modes. These errors are remarkably good when it is considered that the analytic functions used to model the profiles here (exponentials) may not be the ideal ones.

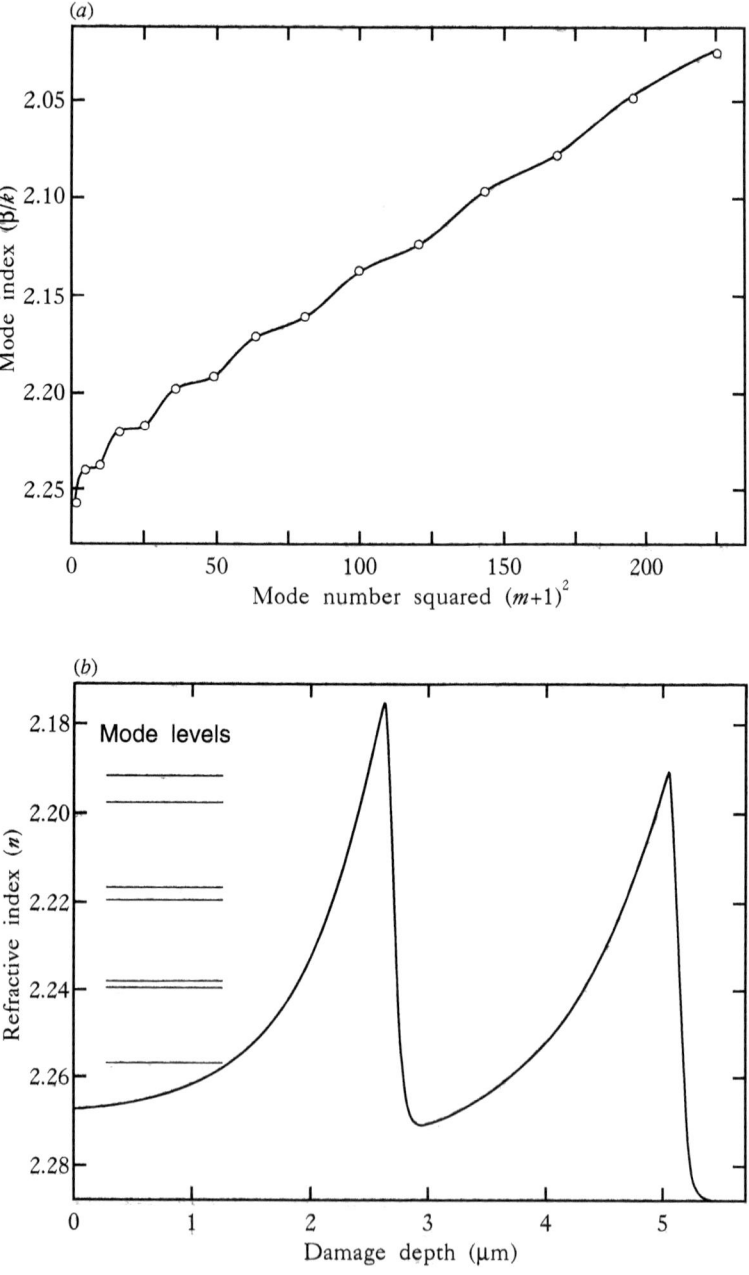

Fig. 5.18. Mode curve plot (a) and fitted profile (b) for a double-barrier implanted waveguide. The circles are the experimental mode indices, and the mode pairing effect is due to the two optical wells having similar widths. The theoretically fitted mode curve goes through all the experimental modes very impressively. This indicates that the chosen profile is correct for the double-waveguide structure.

Table 5.2. *Typical mode indices for a double-barrier waveguide in $LiNbO_3$*

Mode Number	Experimental n_m	Calculated n_m	Error Δn
0	2.256 76	2.256 74	+0.000 02
1	2.239 87	2.240 02	−0.000 15
2	2.237 66	2.237 98	−0.000 32
3	2.220 17	2.220 12	+0.000 05
4	2.217 21	2.216 75	+0.000 46
5	2.198 48	2.197 81	+0.000 67
6	2.191 81	2.192 00	−0.000 19
7	2.171 56	2.171 18	+0.000 38
8	2.160 73	2.161 18	−0.000 45
9	2.136 53	2.137 59	−0.001 06
10	2.123 00	2.123 09	−0.000 09
11	2.095 51	2.096 00	−0.000 49
12	2.077 43	2.076 85	+0.000 58
13	2.047 80	2.045 96	+0.001 84
14	2.024 90	2.022 27	+0.002 63

5.4.6 Thin film reflectivity method

The technique outlined above for the analysis of waveguide refractive index profiles from their mode line spacings, is obviously limited to multi-mode spectra, the precision of the result being dependent on the number of modes available for the analysis. However, in practical applications, mono-mode waveguides are common, or guides with very few modes, and in these cases this method of profile analysis is impossible. Furthermore, in some experimental trials (e.g. ion implantation with heavy or low energy ions) it is possible that a profile may be produced whose only index change is a decrease at the surface, thereby being unable to act as a waveguide. In both these cases, the index profile may be determined by an alternative method, such as surface reflectivity analysis, which involves measuring the reflection coefficient as a function of either wavelength or angle of incidence. The theory behind the latter technique has similarities to the waveguide mode reflectivity calculation method (RCM) developed by Chandler and Lama (Section 5.4.3) which also derives the reflection coefficient as a function of angle at the coupling prism waveguide interface.

Analysis of the optical properties of thin films has been studied for many years (e.g. Heavens, 1955). In the case of uniform thin film coatings (e.g. for anti-reflection), the essential parameters are the thickness of the film and its refractive index relative to the substrate. These values can be extracted efficiently by recording the oscillations in reflectivity, which are seen as a function of wavelength for the coated material. The oscillations merely result from multiple interference between beams reflected at the two interfaces. Similar features are apparent in the angular dependence of reflectivity data for a single wavelength.

The reflection coefficient of an uncoated substrate at normal incidence is given by the Fresnel formula $R = [(n_s - n_a)/(n_s + n_a)]^2$, where n_s is the substrate index, and n_a is the air index. A uniform film of index n causes the reflectivity to vary between this value and a higher one (if $n > n_s$) or a lower one (if $n < n_s$) depending on its thickness.

The two extremes are given by

$$R_1 = \left(\frac{n_s - n_a}{n_s + n_a}\right)^2 \quad \text{and} \quad R_2 = \left(\frac{n^2 - n_s n_a}{n^2 + n_s n_a}\right)^2$$

(NB. an anti-reflection coating has $n = (n_s n_a)^{\frac{1}{2}}$ giving $R_2 = 0$.) By virtue of index dispersion, it can be seen that these values vary slightly with wavelength.

The spacing of the turning points is at approximately equal intervals in $1/\lambda$, such that the optical phase path changes by π between adjacent maxima and minima. Thus the film thickness d is given by

$$2\pi d \left(\frac{1}{n_1 \lambda_1} - \frac{1}{n_2 \lambda_2}\right) = \pi \tag{5.22}$$

where n_i and λ_i are adjacent (max, min) turning point film indices and wavelengths. Once again, because of dispersion, the turning points are not quite equally spaced. Thus, to a first approximation, the index of a uniform thin film may be obtained from the maximum and minimum values of its reflectivity, and its thickness from the spacing of these maxima and minima with wavelength. If wavelength is kept constant (e.g. using a laser), and the reflectivity is measured as a function of incident angle, the problem of dispersion is eliminated and rigorous values may be obtained for the two parameters (n and d).

For a non-uniform index film, the analysis becomes more complicated. As with the waveguide mode analysis (Section 5.4.3), a multi-layer approximation is made and boundary conditions are applied at each interface. A matrix multiplication method is again employed to determine the reflectivity, and a least-squares fit can then be used to match an index profile to the experimental oscillations in reflectivity. The effects of ion

5.4 Index profile determination

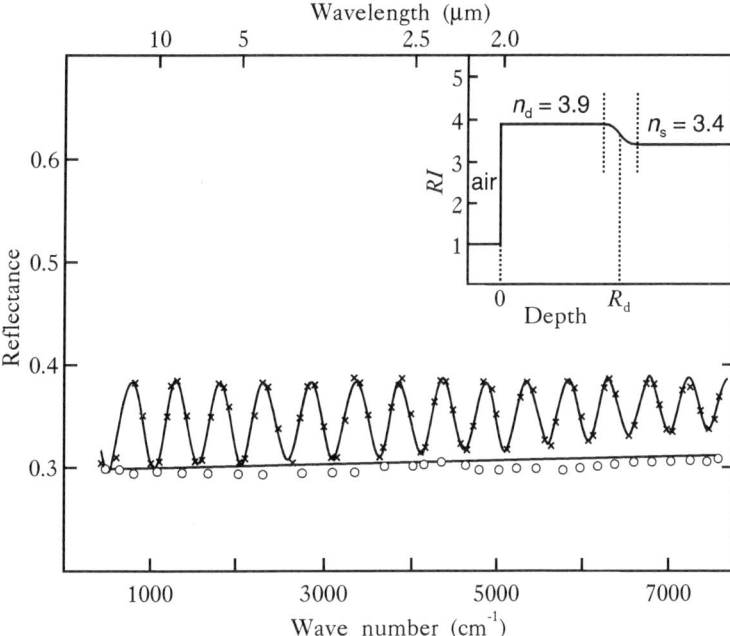

Fig. 5.19. Reflectivity (×) of silicon implanted with phosphorous as a function of wavelength. The lower curve (o) is for the unimplanted sample. The inset shows the refractive index profile predicted by the computer fit (———).

implantation mentioned earlier in this section may be analysed in this way.

For an approximately uniform layer, which is the case for the effect produced by ion beam amorphisation of silicon or germanium, then the reflectivity oscillations are of equal magnitude. Figure 5.19. shows data by Hubler *et al.* (1979a,b) for P^+ implanted silicon. In this example the reflectivity is increased and describes a series of almost equal oscillations with wavelength. The inset shows that these are attributed to a higher index layer extending throughout the ion beam range. For a non-uniform layer the oscillations vary in magnitude with wavelength. The observed pattern can be quite complex, depending on the detailed structure within the layer but, at least in principle, it should also be possible to analyse the depth profile of the refractive index from the reflectivity data. In practice, it is straightforward to model a multi-layer film and then compute the expected effects on the reflectivity functions. Indeed, such simulations can be confidently matched to experimental data to give film thickness

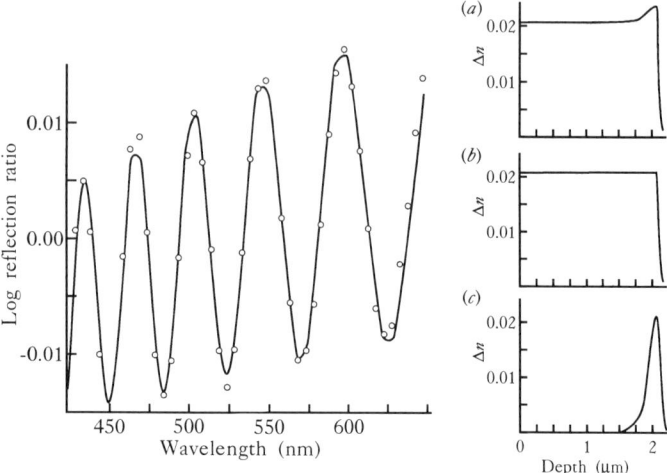

Fig. 5.20. Reflectivity data (o) as a function of wavelength for He implants in silica. The curves (a), (b) and (c) show the effects of an absolute shift in the reflectivity measurement on the consequent shape of the computer analysed refractive index profile. Curve (a) is for data as measured; the scale is shifted by +0.03 in (b) and by -0.016 in (c). Note that the data predict the same maximum index change and ion range, but different profiles shapes.

and maximum change in index. However, further progress to yield the index profile is normally unjustified as the experimental data are rarely obtained with sufficient absolute precision.

Figure 5.20 presents experimental data for a silica substrate bombarded with He^+ ions. As recorded on a spectrophotometer the reflectivity is obtained by reference to an unimplanted sample and emerges as the ratio of the two signals, typically as $\log_{10}(R_{ref}/R_{imp})$. The reflectivity only changes by a few percent, and systematic errors introduced by misalignment, dust, non-uniformity and the spectrophotometer itself produce large systematic shifts of the data which may be comparable with the signal, typically on the scale of ± 0.01 in the log reflection ratio. Therefore, different index profiles are obtained by considering different systematic errors within this range. For the same experimental data on the left of Figure 5.20, the three inset profiles on the right (a), (b) and (c) are therefore equally valid. These show that, as far as the values obtained for the basic depth and maximum index change are concerned, a systematic error is not serious. However, if the absolute value of the signal is varied within realistic limits, then the detailed shape of the fitted index profile varies significantly.

5.4 Index profile determination

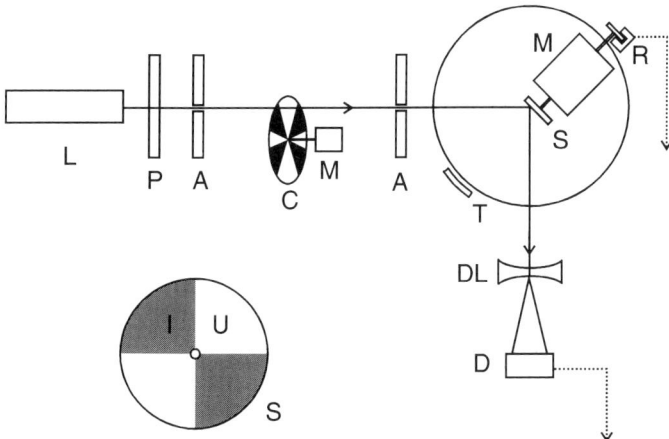

Fig. 5.21. Experimental arrangement for the measurement of angular dependence of reflectivity. L – laser, P – polariser, A – 1 mm aperture, C – chopper, S – sample, R – reference, M – motor, T – turntable, DL – diverging lens, D – detector. Inset: sample disc showing I – implanted region, U – unimplanted region.

Rather than vary the wavelength of the light it may be more convenient to select a single laser wavelength and vary the angle at which the reflectivity is monitored. Such angular changes in reflectivity were tried by Heibei and Voges (1978, 1980, 1982) and Faik *et al.* (1986) for ion implanted silica. In the first attempts the sample was implanted through a mask and then analysed by spinning it in front of a laser beam, and detecting with a phase sensitive amplifier to give the change in reflectivity, ΔR. Addition of a second beam chopper, Figure 5.21, allows an assessment of the reflectivity, R.

The spinning disc method is unsuitable for anisotropic materials and so Lax (1987) and Chandler *et al.* (1986b) used a polarised laser and a beam splitter, and switched between the implanted and reference samples. Absolute errors accrue for the same reasons as in the wavelength dependence method, but additionally, the use of a mask requires very high precision alignment to ensure equal dwell time in each segment. Such systematic errors have similar effects on the computed profile as mentioned above. Further, Chandler *et al.* noted reflectivity anomalies at the edges of the implant regions caused by both sputtered mask material and small steps resulting from density changes in the implanted layer. Figure 5.22 shows an interesting example of one set of angular reflectivity

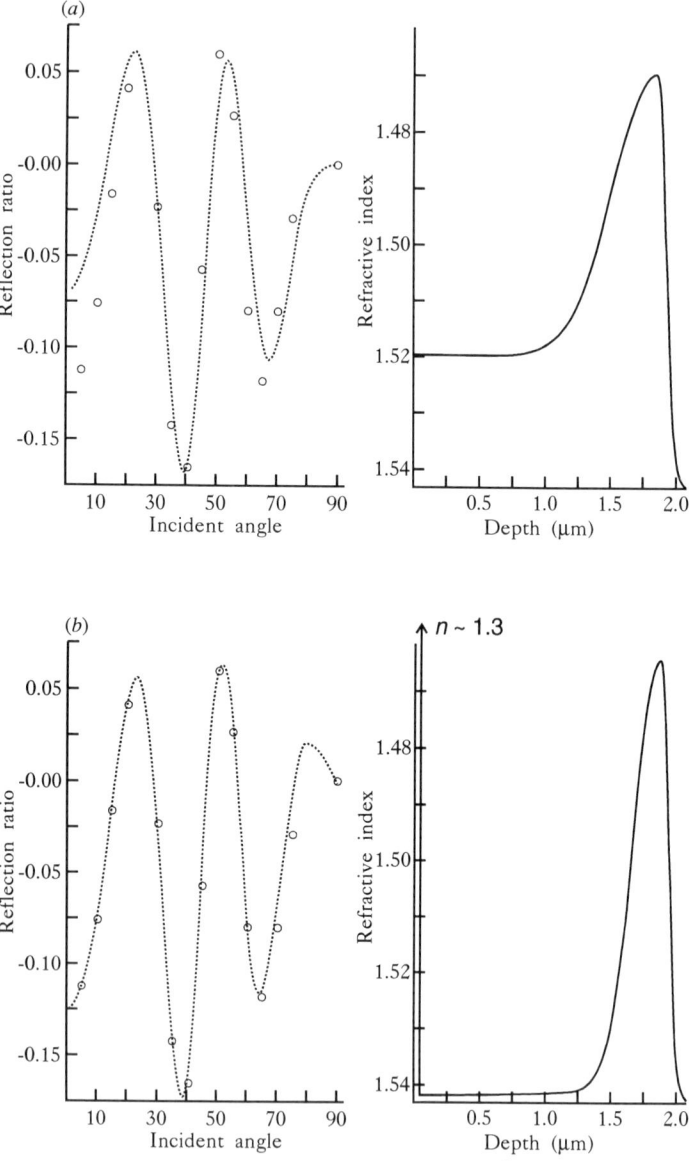

Fig. 5.22. (a) Angular reflectivity data for N$^+$ implanted quartz (o). The solid line shows the index profile giving the best, but rather poor, fit to the data as measured(......). (b) The data have been shifted to give a better fit. This can be interpreted in terms of a low index surface barrier in the index profile.

data and two possible resulting profiles. If the data are fitted to a simple Gaussian-type profile as in Figure 5.22(a), it can be seen that the error of the reflectivity curve (dotted) is not very satisfactory. However, Figure 5.22(b) shows that by the inclusion of a very thin low index surface layer, a much better fit is obtained to the same experimental data (dotted line). Also the profile is now remarkably similar to that obtained by waveguide mode measurements. It may thus be that the angular reflectivity approach has revealed a layer of surface contaminant, such as moisture, to which the waveguide mode analysis is insensitive. Both methods can thus be complementary in obtaining a detailed refractive index profile of an ion implanted surface layer.

5.5 Planar waveguide attenuation

An optical waveguide is the basic component in integrated optics. Its attenuation is a significant parameter in the evaluation of its usefulness for many applications. Attenuation in planar optical waveguides is the result of three principal mechanisms: (1) absorption by the material; (2) scattering at the guide interfaces or by imperfections and intrinsic density fluctuations in the waveguide material which are comparable in size to the optical wavelength; and (3) tunnelling loss if the guide is confined by a narrow low index optical barrier. The absorption in the waveguide can be an inherent effect resulting from impurities in the original materials, which are often present causing several absorption bands. Alternatively, it can be an induced absorption resulting from the process of fabrication of the waveguides – for instance the colour centres produced by irradiation. The latter can usually be annealed out by thermal treatment. Other causes of absorption can be from strain due to polishing or dislocation loops, especially at the surface. These losses in the surface (guiding) region are far greater than those measured in the bulk. The second loss mechanism can also be an inherent scattering from defects or any surface roughness originally existing in the material, but for planar optical waveguides, large scattering occurs at the two interfaces, the smoothness of which depends on the fabrication techniques. For example, a high dose implant can increase surface roughness. It is therefore nearly always advisable to use as small an ion dose as possible to form waveguides. An exception to this may be when micro-domains are produced, and it is then possible to remove this by high dose amorphisation and subsequent recrystallisation on annealing. Since light can be thought of as travelling in a waveguide in a zig-zag manner, then the surface or interface scattering is proportional to the number of zig-zags, and so for a real application, to minimise this loss the lowest order mode should be selected for excitation. The tunnelling loss mechanism only exists in a barrier optical waveguide, which is mainly produced by ion implantation. A very narrow barrier lets light leak through it, but the thickness of the barrier increases with ion dose. Therefore, in one way, it may be advantageous to use a high dose in order to form a thick barrier to stop tunnelling loss, but on the other hand the high dose creates more absorption and scattering. It is important to consider an ion dose comprehensively for all loss mechanisms. It is encouraging to discover that multi-energy implants can produce sufficiently thick barriers whilst using relatively small ion doses by comparison with single energy implants. The tunnelling light is an evanescent wave, which decays exponentially in the optical barrier in terms of $E(z) = A\exp(-k_2 z)$, and therefore the tunnelling loss increases with the wavelength.

The most widely used method for determining the attenuation of a waveguide is to measure the transmitted power as a function of waveguide length. This can be written as

$$I(x) = I_0 10^{-x\alpha/10}$$

giving (5.23)

$$\log I(x) = \log I_0 - (\alpha/10)x$$

where I_0 is the initial power, $I(x)$ is the transmitted power through the waveguide at a distance x (cm), and α is defined as the attenuation of the waveguide, measured in dB/cm. The attenuation is obtained from the slope of this log/linear plot of intensity against guide length.

5.5.1 Prism methods

In principle, the attenuation of a waveguide can be measured by cutting it in a series of stages with reduction in length at each stage. Obviously, this may be possible for fibres where the losses are small, and long lengths are involved, but it is not a feasible method for normal waveguides. It needs a well polished end face at every stage, which is rather difficult, and it is imprecise in reproducibility of coupling efficiency. To avoid this problem one has to use an output coupling prism to extract the energy from the guide. Based on this principle, several techniques have been devised. They can be categorised as either two-prism methods or three-prism methods.

5.5 Planar waveguide attenuation

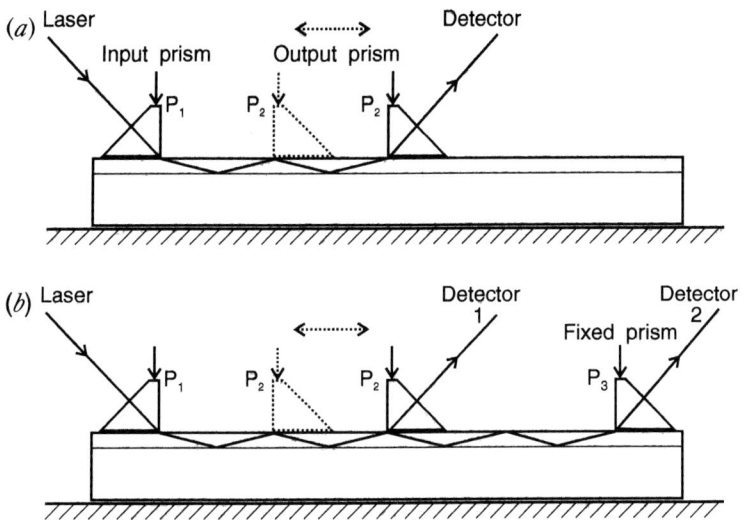

Fig. 5.23. (a) Arrangement of the two-prism method for loss measurements. (b) An improved setup for loss measurements – using three prisms.

Figure 5.23a shows an arrangement of the two-prism method for loss measurement. Such a measurement employs two right-angle prisms to couple light into and out from the waveguide. By moving the second prism along the waveguide surface, the measured output intensity $I'_2(x)$ can be related to the transmitted intensity $I_2(x)$ by

$$I'_2(x) = I_2(x)\gamma_2 \qquad\qquad I'_2 < I_2 \qquad (5.24)$$

where γ_2 is the coupling efficiency of the second prism. The attenuation is still given by the slope of the log plot, provided γ_2 is constant. This method is comparatively simple, but it may be carried out accurately only if there is a sufficiently high degree of reproducibility in the coupling efficiency (γ_2) of the movable prism. Constant coupling efficiency may be obtained more easily using high contact pressure, but this should really be avoided in order not to vary the efficiency of the input coupling prism, and also not to perturb the mode position if any piezo-optic effect is present. In practice, the simply arranged two-prism method cannot be used to measure low loss waveguides accurately. An improved two-prism method has been proposed by Weber et al. (1973). In their experimental arrangement an output coupling prism moving along a liquid coupling film is used to couple-out the light from the waveguides easily. This method is convenient to use, but there are considerable difficulties if the losses are to be measured in waveguides with high refractive indices ($n \simeq 2$), and an accuracy improvement can only be made under conditions where the liquid has a matching refractive index.

To overcome the disadvantages of the two-prism method, a three-prism technique has been designed and used (Won, 1980). As shown in Figure 5.23(b), one prism is used to excite the waveguide, and the other two are output couplers. The extra third prism may be used in either of two ways. The simplest approach is to maintain a constant detected signal at the third prism (I_3'), thereby ensuring a constant coupling efficiency at the second prism (γ_2). The loss calculation is then as above (Equation (5.24)). Alternatively, the second prism coupling efficiency (γ_2) need not be kept constant, but its value can be derived using the two output signals from the third prism, firstly with the second prism absent (giving I_3) and then with the second prism in contact (giving I_3'). The efficiency γ_2 is given by $1 - I_3'/I_3$, and its value may again be used in the above formula (5.24) to derive the attenuation:

$$I_2'(x) = I_2(x)[1 - I_3'/I_3] \qquad (5.25)$$

However, laser fluctuation can still cause inaccuracy in the use of this method.

Several other methods which are used to measure waveguide loss have been reported. They have essentially evolved from the two-prism and the three-prism methods. One of them is the fibre probe technique (Nourshargh et al., 1985), which uses one prism to excite the waveguide, and a fibre probe moving along the waveguide surface to detect the intensity in different positions, via the evanescent wave at the surface. Another is the four-prism method, which can even overcome the laser fluctuation problem (Arutunyan and Galoyan, 1986). All these methods have their disadvantages as well as advantages. The selection of which of these methods is best to use depends on the surface properties of the particular waveguide and the different loss value ranges encountered.

5.5.2 Insertion loss

For a mono-mode waveguide, it is simple to apply the end firing method to measure modal loss. But for a multi-mode guide, this gives a value averaged over whichever modes are efficiently coupled in the guide. A necessary stage for this method is end-polishing of the sample, but it is very difficult to achieve a good finish to the ends without chips in the outer few microns. A pair of microscope objective lenses are used

5.5 Planar waveguide attenuation

to couple light into and out from the waveguide mode, as discussed in Section 5.3.1 (Figure 5.6(a)). To achieve high coupling efficiency, it is necessary to consider the match of laser spot size and the guide dimension. In general, a waveguide has a width in the range of $\sim$2–5 μm, and so short focal length microscope objective lenses (with magnification $\times$20 or $\times$40) are chosen to focus the laser beam down to this size. But the smaller the focal length the larger the numerical aperture, in which case the numerical aperture may be larger than the acceptance angle of the waveguide, which is limited by the index change of the barrier material. In this case all the light will again not be coupled in to the guide. The optimised coupling condition is therefore obtained by considering Gaussian beam optics (Self, 1983) and matching the laser beam waist diameter to the waveguide size.

For a multi-mode guide, if the prism coupling efficiency can be estimated, the individual mode loss can also be measured by using the prism coupling-in and lens coupling-out method (Figure 5.6(b)). Apart from considering coupling efficiency, to estimate the mode loss value, the Fresnel reflection of the lenses and of each surface of the sample has to be taken into account, especially for high refractive index optical materials. To summarise all these effects, the mode loss can be calculated from the overall transmission factor α given by:

$$I = I_0 c f_l^2 f_s^2 \alpha \tag{5.26}$$

where c is the lens (or prism) coupling efficiency, I_0 and I are the incident and detected light intensity, respectively, and f_l and f_s are the Fresnel reflection factors for lens and sample. Then

$$\text{Loss (dB/cm)} = \frac{10}{x} \log_{10}\left(\frac{1}{\alpha}\right) \tag{5.27}$$

where x is the guide length in cm.

In practice, loss measurements are a necessary experimental stage in the characterisation and development of optical waveguide structures. Despite this, the current techniques are difficult and subject to considerable error.

References

Arutunyan, E.A. and Galoyan, S.Kh. (1986). *Opt. Comm.*, **57**, 391.
Chandler, P.J. and Lama, F.L. (1986a). *Optica Acta.*, **33**, 127.
Chandler, P.J., Glavas, E., Lama, F.L., Lax, S.E. and Townsend, P.D. (1986b). *Rad. Effects*, **98**, 211.
Chandler, P.J., Lama, F.L., Townsend, P.D. and Zhang, L. (1988). *Appl. Phys. Lett.*, **53**, 89.
Chandler, P.J., Zhang, L., Cabrera, J.M. and Townsend, P.D. (1989a). *Appl. Phys. Lett.*, **54**, 1287.
Chandler, P.J., Zhang, L. and Townsend, P.D. (1989b). *Appl. Phys. Lett.*, **55**, 1710.
Chandler, P.J., Lama, F.L., Townsend, P.D. and Zhang, L. (1990). *J. Lightwave Tech.*, **8**, 917.
Faik, A.B., Chandler, P.J., Townsend, P.D. and Webb, R. (1986). *Rad. Effects*, **98**, 399.
Harris, J.H., Shubert, R. and Polky, J.N. (1970). *J. Opt. Soc. Am.*, **60**, 1007.
Heavens, O. (1955). *Optical Properties of Thin Solid Films* (Dover, New York).
Heibei, J. and Voges, E. (1978). *IEEE Trans. Quant. Elect.*, **QE-14**, 501.
Heibei, J. and Voges, E. (1980). *Phys. Stat. Sol.*, **57**, 609.
Heibei, J. and Voges, E. (1982). *IEEE J. Quant. Elect.*, **18**, 820.
Hocker, G.B. and Burns, W.K. (1975). *IEEE J. Quant. Elect.*, **11**, 270.
Hubler, G.K., Malmberg, P.R., Carosella, C.A., Smith, T.P., Spitzer, W.G., Waddell, C.N. and Phillippi, C.N. (1979a). *Rad. Effects*, **48**, 81.
Hubler, G.K., Waddell, C.N., Spitzer, W.G., Fredrickson, J.E., Prussin, S. and Wilson, R.G. (1979b). *J. Appl. Phys.*, **50**, 3294.
Lama, F.L. and Chandler, P.J. (1988). *J. Mod. Opt.*, **35**, 1565.
Mahdavi, S.M. and Townsend, P.D. (1990). *J. Chem. Soc. Faraday Trans.*, **86**, 1287.
Lax, S.E. (1987). D.Phil. Thesis, Sussex, (unpublished).
Nourshargh, N., Starr, E.M., Fox, N.I. and Jones, S.G. (1985). *Elect. Lett.*, **21**, 818.
Self, S.A., (1983). *Appl. Opt.*, **22**, 5.
Tien, P.K., Ulrich, R. and Martin, R.J. (1969). *Appl. Phys. Lett.*, **14**, 291.
Tien, P.K., Ulrich, R. and Martin, R.J. (1970). *J. Opt. Soc. Am.*, **60**, 1325.
Tien, P.K., Smolinsky, G. and Martin, R.J. (1972). *Appl. Opt.*, **11**, 637.
Tien, P.K. (1977). *Rev. Mod. Phys.*, **49**, 361.
Ulrich, R. (1971). *J. Opt. Soc. Am.*, **61**, 1467.

Weber, H.P., Dunn, F.A. and Leibolt, W.N. (1973). *Appl. Opt.*, **12**, 755.
White, J.M. and Heidrich, P.F. (1976). *Appl. Opt.*, **15**, 151.
Won, Y.H., Jaussaud, P.C. and Chartier, G.H. (1980). *Appl. Phys. Lett.*, **37**, 269.
Zhang, L. (1990). D.Phil. Thesis, Sussex, (unpublished).
Zhang, L., Chandler, P.J. and Townsend, P.D. (1991). *Nucl. Inst. Methods*, **B59/60**, 1147.

6
Ion implanted optical waveguides

Ion implantation may be used to change the optical properties of insulators, either because of the chemical presence of the dopant ions, or more generally because of the radiation damage caused during their implantation. The latter effect produces a significant change in the refractive indices of most materials, and consequently He^+ implantation has been used to define optical waveguides in a wide variety of substrates. These include electro-optic, non-linear and laser host materials, with key successes in quartz, $LiNbO_3$, $KNbO_3$, $KTiOPO_4$ (KTP), $Bi_4Ge_3O_{12}$ (BGO), garnets such as $Y_3Al_5O_{12}$ (YAG), and amorphous glasses such as silica and lead germanate.

Although this technique has wide applicability, the refractive index profiles vary considerably between materials, and even between different indices of the same material. The index change may vary in degree, and even in sign, for both the nuclear collision and the electronic ionisation regions. These effects are discussed in this chapter, together with their applicability in the formation of optical waveguides, and more complex structures. Of particular interest are the three detailed examples of quartz, $LiNbO_3$ and $Bi_4Ge_3O_{12}$ since between them they embody most of the features so far observed in ion implanted waveguides in insulating materials. The performance of the implanted waveguides is considered in terms of their thermal stability and their attenuation due to absorption, scattering and tunnelling losses. The He^+ guides are first compared with those produced by conventional chemical diffusion methods. At the end of the chapter, waveguides formed by implantation of chemically active components are discussed.

6.1 Practical waveguide structures

The basic theory of optical waveguides was presented in the previous chapter and it is now necessary to compare alternative fabrication routes to the production of optical waveguides. In particular we shall examine how ion implantation can be used, either in isolation, or in combination with other processing methods.

6.1.1 Conventional fabrication methods

Several alternative methods may be employed for the fabrication of planar waveguides. The simplest technique is that of epitaxial layer growth. This method is capable of use in a very wide range of optical materials (both crystals and glasses). The basic principle of this method is to grow a higher index thin film epitaxially on a lower index substrate. In practice, it includes many different processes, such as liquid-phase epitaxy (LPE), molecular beam epitaxy (MBE), melt-phase epitaxy (ME), vapour-phase epitaxy (VPE) or chemical vapour deposition (CVD), and sputtering. Epitaxial growth produces an index profile which is closely approximated by a step function. These methods are not always reliable, as the film properties do not correspond to those of the bulk material, and are inconsistent, being very dependent on growth conditions. The common problem of epitaxial growth, especially for crystals, is the relatively high scattering loss at the interface between the optical film and substrate due to their in-plane lattice mismatch. In general, an amorphous higher index layer will be produced by this method, therefore it is not really an ideal technique for 'active' materials which must be deposited as a poled single-crystal layer.

Techniques of diffusion or ion exchange give more reproducible profiles, and to a large extent are suitable for active materials. The principle is to increase the index above that of the substrate by chemical means, but this implies that the guiding region must to some extent be 'contaminated'. The presence of impurities (in the diffusion case) or the change in composition (ion exchange case) would be expected to affect the crystalline properties of the substrate, and to produce possible absorption bands. However, after sufficient research, these effects may be controlled and satisfactory devices produced. One major drawback is that research time must be spent to ascertain the optimum dopant and procedure, and the results are not easily transferable to new substrates. A second big problem is that for some dopants negligible diffusion may be possible in the temperature range set by the useable phases of the crystal.

The index profile produced in LiNbO$_3$ by proton ion exchange is close to a step-function (Jackel et al., 1981, 1982; Wong et al., 1989), and quite satisfactory for most applications. Titanium diffusion, however, gives a profile which is typically half-Gaussian or error-function in shape (Korotky et al., 1987), with a maximum change at the surface. The profiles can be reproducible provided care is taken with the environmental conditions. In some cases field assisted diffusion may be able to drive the guide slightly beneath the surface, but in general there is no simple means of producing a buried profile, other than by the secondary deposition of an extra surface layer with low index.

6.1.2 Fabrication by ion implantation structural effects

The dominant effect of ion implantation on refractive index is usually due to the partial lattice disorder produced by nuclear damage processes (Townsend, 1976). This invariably leads to a decrease in physical density and hence to a reduced refractive index. This decrease in index below that of the substrate is in the opposite sense to that of conventional chemical techniques. Thus for an energetic light ion (e.g. 1 MeV He$^+$), the nuclear damage peak at the end of the ion track generally produces a low index 'optical barrier' (Townsend, 1976). The region between this barrier and the surface is therefore surrounded by regions of low index and is able to act as a waveguide. The guiding region itself is subject mainly to ionization processes which have little effect on lattice order for most types of insulator, and any colour centres produced are easily annealed. It is this type of 'implantation damage' guide which is of considerable interest because its production mechanism is applicable to a very wide range of materials. The detailed index profile of such a guide varies greatly with substrate, ion species, and implantation/annealing conditions, but all cases have certain general properties. The main characteristic is that it is a 'barrier confined' guide, having essentially a 'pure' substrate structure for its guiding medium, with its preserved active crystalline properties (electro-optic, non-linear, etc.), unlike the conventional enhanced-index chemically produced guides.

The height of the optical barrier may be controlled by the ion dose. For saturation damage (at a dose $\sim 10^{17}$ ions/cm^2) which may or may not correspond to amorphisation, it may reach as much as 15% (decrease in index) but in general it is about 5%. In most materials a 2 or 3%

6.1 Practical waveguide structures

barrier is produced by a moderate dose of $\sim 10^{16}$ ions/cm^2 and in some cases a dose of 10^{15} ions/cm^2 is sufficient to produce this change (e.g. KNbO$_3$, KTaO$_3$). The thickness of the guide is controlled by the ion energy, and 1 MeV He$^+$ generally has a range of $\sim 2\,\mu$m, sufficient for several modes if the barrier has $\Delta n \sim -2\%$. For mono-energetic ions the barrier width varies with substrate material, depending on the stability of the lower-complexity defect structures in the nuclear damage tail region. If optical attenuation due to narrow barrier tunnelling is a problem (e.g. barrier width$< 1\,\mu$m), then multiple energy implants may be used to broaden it.

The versatility of this barrier confinement method facilitates several processes which are not available for conventional methods. For instance, guides can be buried away from the surface using very low energy implants, and multi-layer structures can be fabricated. These will be discussed in more detail later.

Although the general profile shape, as expected from the nuclear stopping profile, is that of a negative Δn optical barrier which confines a guiding region of practically unchanged index, the actual result is usually much more complex. Firstly, the profile of retained lattice damage may not correspond directly to the nuclear energy deposition profile. This is because point defects in the nuclear damage tail may be annealed during implantation (either thermally or by ionisation enhancement); or, conversely, defect production may be increased in the nuclear tail region by synergistic effects due to the electronic stopping energy. On top of this, low dose threshold effects and high dose saturation effects may give a typical S-shape to the dose dependence of the barrier growth curve. The second important thing to realise is that the refractive index is not only dependent on the volume expansion due to the defect content, but should be more realistically related to several defect dependent parameters which include changes in not only volume (ΔV), but also atomic bond polarisability ($\Delta \alpha$) and structure factors (F). The Wei adaptation (Wei *et al.*, 1974; Wenzlik *et al.*, 1980) of the Lorentz–Lorenz equation gives

$$\frac{\Delta n}{n} = \frac{(n^2 - 1)(n^2 + 2)}{6n^2} \left[-\frac{\Delta V}{V} + \frac{\Delta \alpha}{\alpha} + F \right]$$

In general the $-$ve ΔV term dominates as described above, giving the typical $-$ve Δn optical barrier which may have a tail extending to the surface due to isolated point defects. However, it is important to realise that lattice restructuring also results in chemical bond changes and subsequent polarisability effects ($\Delta \alpha$). These and other contributions, such as stress in piezo-optic materials, lead to the possibility of small index changes of either sign, even in the guiding region. Furthermore, defect and atomic diffusion may occur. Such cases will be considered when actual examples are discussed. In the special case of +ve Δn in the guiding region, it must be realised that the profile would then be similar to that of a 'conventional' waveguide, and it would then be possible for the lower modes to be confined without the need for an optical barrier (i.e. existence of a non-tunnelling regime). This can lead to low dose, low loss guides in some materials.

To summarise, ion implantation damage (from MeV He$^+$) generally produces optical barrier confined waveguides due to volume expansion in the nuclear damage region. However, a variety of effects may also lead to index changes in the surface region and then a non-tunnelling guide may be feasible. In general, the ΔV effect is widely applicable and waveguides are possible in some form or other in nearly all materials. Usually, absorption losses are due to the presence of activated colour centres and may be reduced by moderate temperature annealing. The tailoring of profiles by choice of dose and multiple energies makes ion implanted waveguides very versatile, e.g. the production of double-barrier structures, as described later.

6.1.3 Chemically formed ion implanted waveguides

Apart from the radiation damage effects described above which normally produce barrier confined waveguides in most optical materials (and sometimes non-tunnelling +ve Δn waveguides), implantation of chemically active ions can be employed to produce chemically doped waveguides with similar properties (+ve Δn) to those produced by, for example, diffusion. Here the same chemical limitations apply as in the conventional guides, requiring dedicated recipes for each substrate, but there can be advantages over the conventional techniques. For instance, some limitations which apply in the conventional diffused guides may be relaxed as a high diffusivity is not required. The ability to position the dopant below the surface prior to diffusion facilitates the production of a buried guide and also avoids precipitation due to surface chemistry effects. Such devices will be mentioned later.

6.2 Summary of the effects of ion implantation on refractive index

Before detailing results for specific materials, we will summarize the general effects contributing towards ion implanted waveguide refractive index profiles and their annealing characteristics.

(a) The structural changes which might possibly give rise to refractive index effects are predominantly caused by nuclear collision damage (+ve $\Delta V \rightarrow$ –ve Δn). This generates point defects along the ion track and more complex clusters (or amorphisation) near the end of the range.

(b) In a few materials, ionisation or electronic damage (at the start of the ion track) may produce point defects, but it is more likely that the ionisation acts to enhance the efficiency of the nuclear damage (synergistic effect). Ionisation damage may have the opposite effect by enhancing the self-annealing of point defect pairs. Ionisation primarily activates colour centres.

(c) The defects produced by electronic and nuclear damage may or may not be retained after implantation.

(d) Defect retention is more effective for the more disordered part of the damage track (at the end of the range).

(e) Defect retention may be more effective once lattice disorder exists – making the change super-linear with dose. Eventual saturation of the damage results in a typical S-shaped dose curve.

(f) Defect retention is better for low temperature implants – where self-annealing is inhibited.

(g) The non-thermodynamic nature of the excited lattice may result in the stabilisation of new phases.

(h) Changes in stoichiometry or phase may result in –ve or +ve changes in index, and these may be opposite for the different indices of anisotropic materials. In the more stable ionic materials, amorphisation may only be reached at ion doses which correspond to impurity levels where segregation (bubbles) and resulting surface shearing may occur.

(i) At high doses, stress effects may dominate the index change. In some materials the effects of stress are sufficient to inhibit total amorphisation.

(j) Colour centres activated by the ionisation, which may produce optical absorption in the guide, are often annealed out at moderately low temperatures (200 °C).

(k) Annealing at higher temperature ($\sim 400\,°C$) tends to remove point defects (i.e. in the guide region) but preserves complex defects (the barrier), but these may be annealed at much higher temperature ($> 600\,°C$).

(l) The existence of stable structures for both crystalline and amorphous phases of a particular material may lead to a profile which is extremely stable even at very high temperature.

(m) Implantation may result in radiation enhanced migration (of species or defects). This effect should be dose-rate dependent.

(n) Small concentrations of implanted impurity ions may act as nucleation sites to stabilise larger volumes of disorder, or new phases.

(o) Preferential sputtering will occur as a result of mass differences and, more importantly for insulators, electronically driven sputtering mechanisms may remove one species dramatically.

6.3 Materials exhibiting index changes

Optical waveguides have now been produced in a large number of materials (more than 40) by the effects of ion implantation damage. In nearly all substrates tested, some degree of guiding has been achieved. Whether or not a guide can be produced with sufficiently low attenuation for general applications ($\sim 1\,dB/cm$) depends on several factors. A suitable index profile can usually be achieved, either with a broad enough barrier, or with Δn +ve in the guiding region. It is then usually necessary to anneal the sample ($200\,°C$) to remove colour centre absorption produced by electronic damage in the guiding region. Problems arise if this annealing is not possible because of low temperature distructive phase transitions, or if constituent diffusion can take place at these temperatures. In the following sections, various representative materials will be considered, and the diversity of their waveguiding properties after ion implantation will be discussed. Table 6.1 comprises a list (circa 1992) of substrate materials which have been successfully implanted with He^+ to produce waveguides, and included are some of the main parameters which are available at present.

6.4 Crystalline quartz

During ion implantation, quartz is gradually converted into amorphous silica (Primak, 1958; Hines and Arndt, 1960), producing an index change of $\Delta n \sim -5\%$. It does not quite reach the state of fused silica ($\Delta n \sim$

6.4 Crystalline quartz

−6%), but, rather, a phase equivalent to that produced by the saturated radiation damage of fused silica. The growth curve in the nuclear damage region is a pronounced S-shape due to strong initial collaborative mechanisms giving a 'threshold' effect, followed by eventual saturation. When energetic light ions are used, the depth profile of the refractive index shows a nuclear damage barrier whose characteristics are typical of quartz (Red'ko et al., 1981; Zhang et al., 1989; Chandler et al., 1990a). Characteristic index profiles are shown in Figure 6.1. For low doses the barrier is almost Gaussian in shape and quite narrow, showing practically no point defect tail extending into the surface region. This suggests that ionisation induced annealing of these point defects is occurring, or alternatively this may possibly be due to a high 'threshold' effect for the initial damage mechanism as described above (Primak, 1958; Hines and Arndt, 1960). After a dose of 10^{16} ions/cm^2 a distinct saturation occurs and a flat-topped, almost square, barrier results. This barrier broadens asymmetrically with further increasing dose, which indirectly supports the damage threshold model just mentioned. Up to very high doses ($\sim 10^{17}$ ions/cm^2) the surface region shows practically no index change. As explained earlier, low temperature implants result in more efficient damage retention. Figure 6.2 compares profiles for 300 K and 77 K implants. The latter produces a barrier which has a width equivalent to a high dose implant at 300 K – clearly demonstrating the enhanced damage efficiency at low temperature. However, the difference in barrier height simply implies that the degree of amorphisation attainable increases with temperature. This can also be seen in the annealing results described below.

On annealing it is found that the regions with lower damage levels ($\Delta n \sim -2\%$) recover back towards crystalline quartz (by $\sim 600\,°C$) but that the semiamorphised material achieves an even lower index and becomes more like fused silica. This results in an almost square edged barrier with high thermal stability (Chandler et al., 1990a). In the guiding region, colour centres and any possible point defects may be annealed out by $\sim 400\,°C$, leaving a guide with very low absorption (Zhang et al., 1989).

The close approach of the refractive index profile in quartz to that of a square optical well has simplified both data analysis and the prediction of more complex structures. For example, it is possible to use multi-energy implants to form a wide optical barrier layer, to form coupled wells, or to make perturbations to a simple well profile. However, in a multi-energy example there is the possibility of differences between ion doses delivered

Table 6.1. Summary of substrates in which He^+ implanted waveguides have been fabricated

Material	Substrate index (633 nm)		Max. % Δn Guide	Max. % Δn Barrier	Dose/1%Δn 10^{16} cm^{-2} 77 K	Thermal stability	Loss dB/cm
Crystals							
Al_2O_3	1.766	(n_o)		-2	>5	800 °C	
	1.758	(n_e)					
BaB_2O_4	1.669	(n_o)	-0.5	-0.7	1		~2
	1.551	(n_e)	$+0.7$	~0			
BaF_2	1.473		~0	-0.5	~5	600 °C	~7
$Ba_2NaNb_5O_{15}$	2.218	(n_a)	-1	-5	1	550 °C	
	2.320	(n_c)	$+1$	-4			
$BaTiO_3$	2.40	(n_o)		-4.5			
	2.37	(n_e)		-3			
$BeAl_2O_4$:Cr	1.747	(n_b)	$+0.1$	-0.5	>2	800 °C	~1
$Bi_4Ge_3O_{12}$	2.098		$+2.5$	-3	~2	800 °C	~0.2
$Bi_{12}GeO_{20}$	2.545		-0.5	-3	~1		~1
$CaCO_3$	2.267		~0	-5			
$CaMoO_4$	1.91		$+0.3$	-0.8	4	300 °C	
$CaWO_4$	1.92		$+0.25$	-0.3	4	300 °C	
$Gd_3Ga_5O_{12}$	1.966		$+0.2$	-2			~1
$Gd_3Sc_2Al_3O_{12}$	1.928			-1			
$Gd_3Sc_2Ga_3O_{12}$	1.967			-1			
$KNbO_3$	2.174	(n_a)	-0.5	-10	0.5		~2
	2.280	(n_c)	$+1$	-7			
$KTaO_3$	2.229		-0.5	-16	0.5	900 °C	~1
$KTiOPO_4$	1.865				0.5		
$LaNdMgAl_{11}O_{12}$	1.789	(n_e)	-0.04	-1	~6		
	1.574	(n_x)	~0	-1.5			
LiB_3O_5	1.601	(n_y)	~0	-5	1	500 °C	~2
	1.616	(n_z)	~0	-4			

Crystals—contd.

Material	n		col1	col2	col3	Temp	col5
LiCaAlF$_6$	1.388	(n_o)	+0.1		1		
	1.386	(n_e)					
LiNbO$_3$	2.287	(n_o)	−1	−5	0.5	350°C	1
	2.203	(n_e)	+1	−3			
LiTaO$_3$	2.177	(n_o)	−1	−7	1	350°C	1
	2.182	(n_e)					
LuY$_2$Al$_5$O$_{12}$	1.837		~0	−2	~3		
MgO	1.736		~0	−1	~5		
SiO$_2$	1.543	(n_o)	~0	−5	0.5	1100°C	~0.2
	1.552	(n_e)					
Sr$_{.65}$Ba$_{.35}$Nb$_2$O$_6$	2.314	(n_o)	−0.5	−3	~1		
	2.280	(n_e)	+0.2	−1.5			
TeO$_2$	2.412		~0	−2.5	~2	700°C	~1
YAlO$_3$	1.95		−0.7	−3	2	300°C	~5
Y$_3$Al$_5$O$_{12}$	1.830		+0.3	−2	~3	800°C	~1
YLiF$_4$	1.477		−0.15	−1	2		
Y$_3$Sc$_2$Al$_3$O$_{12}$	2.287		+0.1	−1	2		
Y$_2$SiO$_5$	1.785	(n_{xy})	~0	−1.7	2	900°C	1.6
ZnWO$_4$	2.19			−4	2	800°C	~1
ZrSiO$_4$	1.921		~0	−7.5	1		
Glasses							
Fused silica	1.456		+2.0	+2.5	2	500°C	0.2
Fluoride glass	1.502		−0.2	−0.3			
Phosphate glass	1.526		−0.5	−0.8		350°C	
Silicate glass	1.565		−0.4	−0.8			
Lead germanate glass	1.822		+0.4	~0		350°C	~0.15

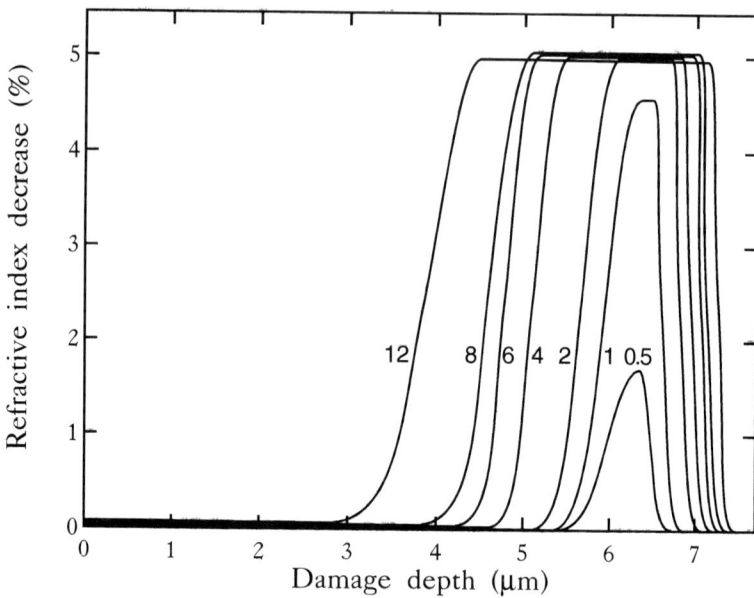

Fig. 6.1. Optical barrier effect of index profiles for quartz waveguides formed by 2.2 MeV He$^+$ with doses from 0.5 to 12×10^{16} ions/cm^2. The damage saturates when the index almost reaches that of amorphous silica.

in different energy sequences. This is possible both because the nucleation of damage is sensitive to pre-existing damage, and because there is the possibility that predominantly ionising radiation might induce relaxation and annealing of the lattice. Low loss waveguides have been formed in quartz by multi-energy implants but in these cases a decreasing energy sequence has always been used. It was found that a 1.5 μm wide barrier required a single energy implant of dose 6×10^{16} ions/cm^2, whereas by the use of four distributed energies the total dose could be reduced to only 2×10^{16} ions/cm^2, with a subsequently lower loss (0.2 dB/cm).

These annealed waveguides are ideal, because the high Δn ($\sim -6\%$) square-edged barrier confines the light to the low loss crystalline guiding region, where all properties such as non-linearity and acousto-optic effects are preserved. Second harmonic generation (SHG) has been demonstrated in such a waveguide, where modal phase matching was achieved by using ion-energy dimensional control of the profile (Babsail et al., 1991a). This has allowed the first mode ($m = 0$) of the fundamental (at 800 nm) to be matched to the third mode ($m = 2$) of the harmonic (at 400 nm). Acousto-optic interactions have also been demonstrated in ion

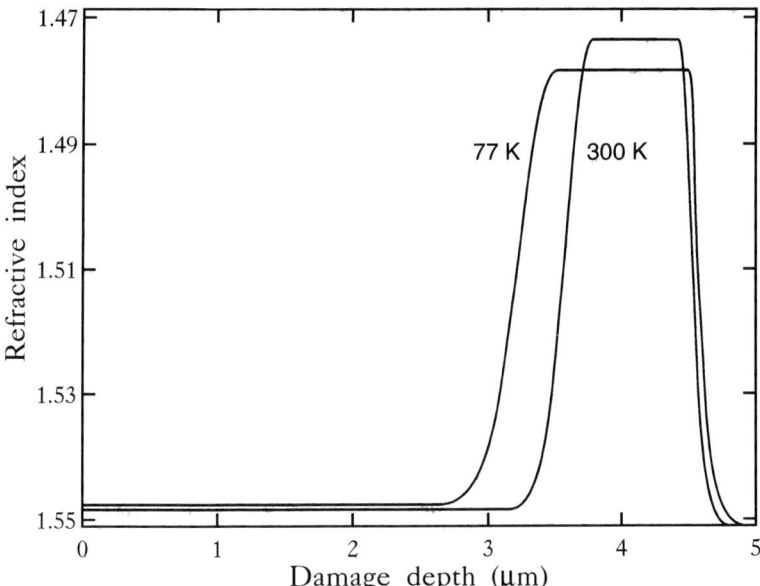

Fig. 6.2. Comparison of n_o profiles in quartz for 77 K and 300 K implants with 1.5 MeV He$^+$ to a dose of 3×10^{16} ions/cm^2.

implanted quartz waveguides (Pitt et al., 1984). This was a precursor for a Fourier transform radar frequency spectrum analyser, which efficiently used the diffraction of guided light by a surface acoustic wave (SAW) grating. These applications of ion implanted quartz waveguides will be discussed in more detail in Chapter 7.

6.5 Niobates

6.5.1 Lithium niobate

The material which has been most extensively developed for optoelectronic and non-linear applications is LiNbO$_3$ (Korotky and Alferness, 1987; Abouelleil and Leonberger, 1989). The effects of ion implantation have been studied for many years and it has long been known to exhibit a large decrease in refractive index due to implantation damage (Wei et al., 1974). It was therefore soon realised that energetic light ions could produce barrier-type waveguides (Destefanis et al., 1978, 1979), and subsequent research projects by various groups have now been able to achieve quite acceptable waveguides in this material, in both its pure and doped forms, as reviewed by several authors (Gotz and Karge, 1983; Townsend, 1987; Reed and Weiss, 1989; Zhang et al., 1990a, 1991a). Careful study has revealed, however, that the mechanisms for refractive index change are far more complex than was suspected at first sight. Despite basic similarities to the quartz model, there are major differences in the effects of the radiation damage, and in their response to annealing. The production of a nuclear damage optical barrier at the end of the ion range is the major effect, and its height is of a similar magnitude to that in quartz (eventually reaching $\Delta n \sim -5\%$) – but its production efficiency is less than quartz giving only $\sim 2\%$ for 10^{16}ions/cm^2, as opposed to $\sim 5\%$. It is caused by the destruction of long range crystallinity, but it does not appear to reach amorphisation, as the material never becomes isotropic (Chandler et al., 1986). Figure 6.3 shows that the profiles for the ordinary (n_o) and the extraordinary (n_e) indices are significantly different. As n_o appears more straightforward we will deal with this first. The profile closely resembles that of the nuclear stopping damage profile (as given by simulation programs such as TRIM), with a point defect tail extending to the surface. This implies that there is no appreciable annealing (thermal or ionisation) of isolated point defects taking place during irradiation, and also that the Δn growth curve does not have an initial subthreshold region (S-shape) due to cumulative mechanisms (as in quartz). The growth curve does begin to show saturation effects at high dose, however, although amorphisation is not reached. The rate is temperature dependent, showing an increased retention of damage at low temperature, where any self-annealing effects are minimised. More surprisingly, the rate is dependent on the ion energy (Zhang, 1990; Zhang et al., 1991a), as shown in Figure 6.4. This has been attributed to stress effects in the strongly piezo-optic material. At very high doses stress relaxation can be seen to occur, particularly for implants near to the surface, and eventually the implanted layer is delaminated from the surface, especially after heat treatment ($\sim 300\,^\circ$C).

The extraordinary index profile presents a much more complex picture. Immediately apparent (Figure 6.3) is an index increase in the guiding region (where the isolated defect damage tail occurs for n_o) such that the 'lowest' modes may have an effective mode index greater than that of the substrate. This means that they are completely confined in a non-tunnelling well, and hence low attenuation may be achievable for n_e without resort to broad barrier multi-energy implants. This refractive index enhancement is wavelength dependent, reaching $\Delta n \sim +1\%$ for blue but only $\sim +0.7\%$ for red. This increase in dispersion is seen in

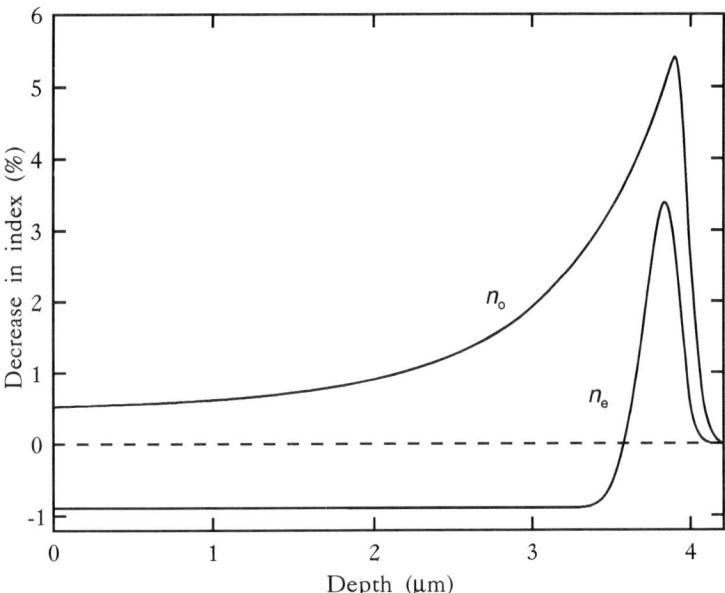

Fig. 6.3. Profiles for the two indices of LiNbO$_3$ implanted at 77 K with 1.75 MeV He$^+$ to a dose of 2×10^{16} ions/cm^2. Note the increase of n_e in the guiding region.

other materials, and is possibly due to a reduced bandgap in the new structure. This surface region has been investigated by high precision X-ray structure determinations at low incidence angles, and a reduction in the original microcrystallinity has been observed (Haycock, 1989).

The increase in refractive index of n_e has now been attributed to a depletion of lithium due to radiation enhanced diffusion. Lithium loss moves the phase composition towards niobium oxide where n_e has a higher value, but n_o remains unchanged. This is the effect used in the fabrication of n_e waveguides produced simply by the thermal out-diffusion of lithium from the crystal surface, or by proton exchange where lithium is replaced by hydrogen. Initial suggestions were that under irradiation, even at low temperatures, highly mobile lithium ions were able to diffuse to the surface where Li$_2$O was lost. However, it is now believed that the lithium in fact diffuses in the opposite direction, towards the highly disordered nuclear damage peak where ionic mobility is enhanced. This model is favoured experimentally because of the phenomenon of 'missing modes' (Chandler et al., 1989a). As shown in Figure 6.5, a profile having an optical well with its maximum index

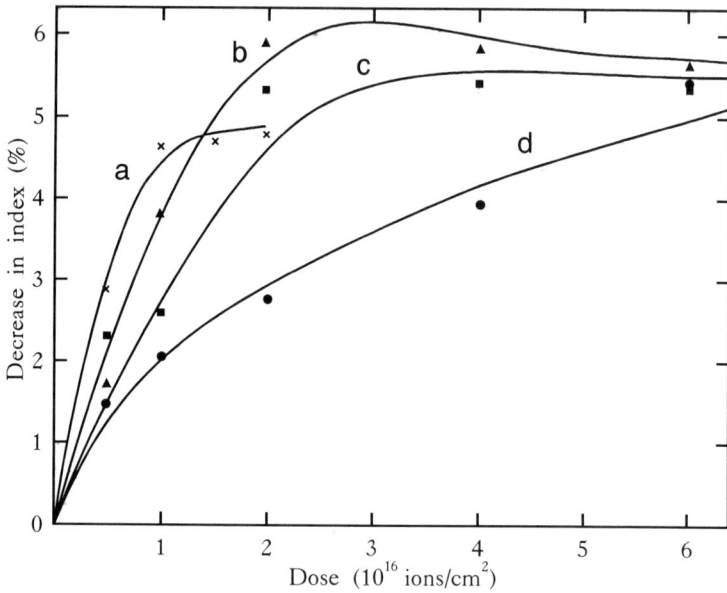

Fig. 6.4. Dependence on ion dose and energy of the peak index change in LiNbO$_3$ implanted with He$^+$ at 77 K. The energies are a: 0.75, b: 1.1, c: 1.75 and d: 2.2 MeV.

position buried at some distance beneath the surface could contain modes which are not able to be detected by a surface coupling prism, because of the large tunnelling distance between the well and the surface. In fact, for prism coupling, the bottom modes for n_e are quite often missing, especially at shorter wavelengths, and other low modes are much fainter than the upper ones. In early experiments, the misnumbering of modes led to considerable confusion, especially when comparing 'bottom mode' performances in similar structures. Their existence was first realised by the need for mode spectra at different wavelengths to give profiles with the same linear dimensions (Chandler et al., 1989a). Surface removal (polishing) experiments were then performed to confirm this effect, in which the missing modes appeared after a certain layer had been removed (Figure 6.6). Precise mode measurements between polishings were then able to determine profiles such as the one shown in Figure 6.5. It should be realised that such profiles confine modes away from the surface where scattering may be a problem, and hence they are likely to produce very low loss buried waveguides.

Subsequent to the satisfactory explanation of the phenomenon de-

6.5 Niobates

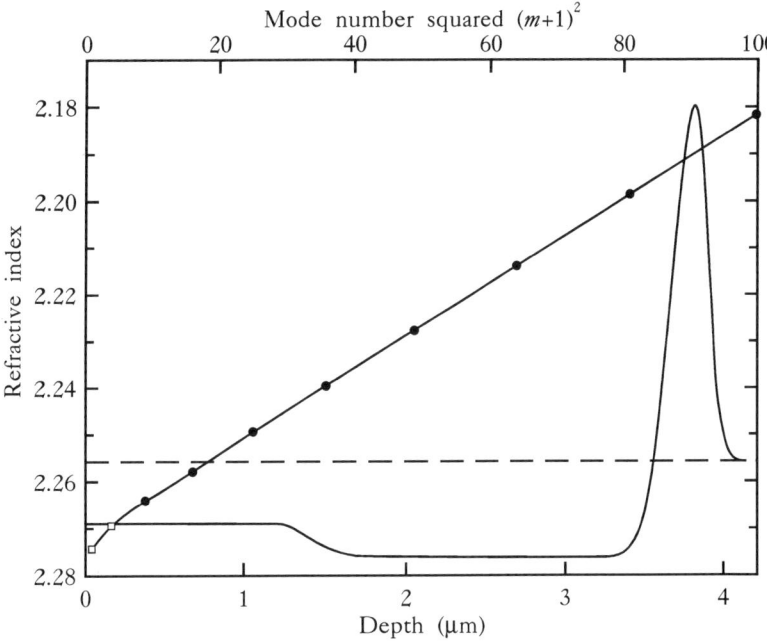

Fig. 6.5. Improved n_e profile for the bottom modes of a LiNbO$_3$ waveguide (at 488 nm). The bottom 'missing' mode cannot tunnel to the surface coupling prism, and the second mode is faint.

scribed above, it has now been realised that a second smaller optical well also appears to exist beyond the nuclear damage barrier (Zhang et al., 1991a,b). This is because if diffusion of lithium into the barrier region can occur from the surface direction, then it is possible that it also occurs from the substrate direction – but to a lesser extent because of the lower damage density in that region. For intermediate ion doses this secondary optical well is large enough to support an isolated mode, i.e. once sufficient lithium diffusion has occurred, but whilst the nuclear peak (-ve Δn) has not yet grown broad enough to cover it over. This extra isolated 'strange' mode is confined at a considerable distance from the surface and may only just be detectable by a surface prism as a very faint, but quite sharp, mode line. Its actual observation is quite fortuitous because the wavelength can be too long for the mode to exist, or too short for the tunnelling to be possible. It is therefore observed only for a very limited range of ion energy and dose and also wavelength. Examples of these n_e 'strange' modes have been seen in several samples

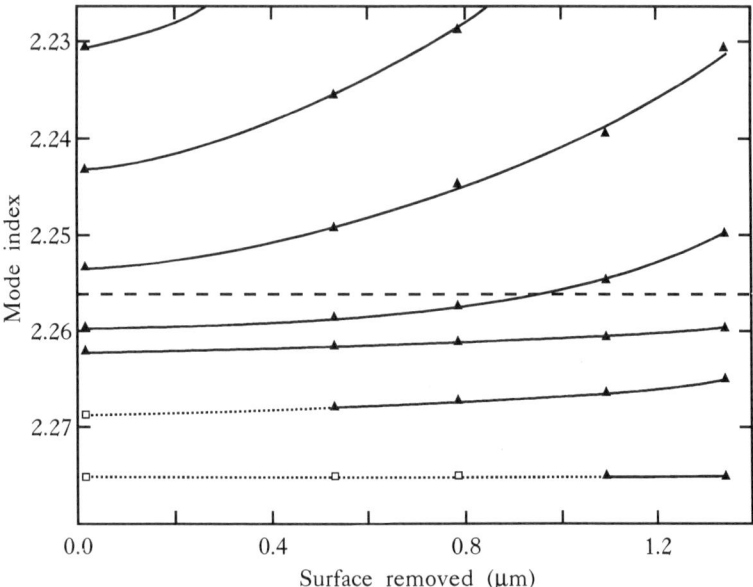

Fig. 6.6. Results of surface polishing to detect 'missing' modes. The bottom mode is coupled once ~ 1 μm of surface has been removed. The dashed line shows the substrate index.

of $LiNbO_3$. They did not fit into the pattern of the remaining mode spectrum – except possibly to slightly displace any adjacent mode. By ignoring their existence it was quite possible to fit a single-well refractive index profile of the correct linear dimensions. However, their true nature was discovered by observing the waveguide mode spectra as a function of laser wavelength, and eventually by surface polishing (Figure 6.7). The levels of the normal modes were sensitive to wavelength and polishing, whereas the strange mode was only slightly affected. Once sufficient surface had been polished off to displace all the normal modes, and most of the barrier had been removed, the strange mode became very clear and could be coupled-into as a real guiding mode. This confirmed its existence in a secondary optical well beyond the nuclear damage barrier, as depicted schematically in Figure 6.8. The annealing characteristics of $LiNbO_3$ are not like those of quartz. Because no stable amorphous phase exists, the barrier does not exhibit the same high temperature stability. In fact, the damage recovers smoothly with temperature, showing no clear annealing stages (Glavas et al., 1988). Figure 6.9 shows the isochronal annealing behaviour of n_o for the nuclear and guiding re-

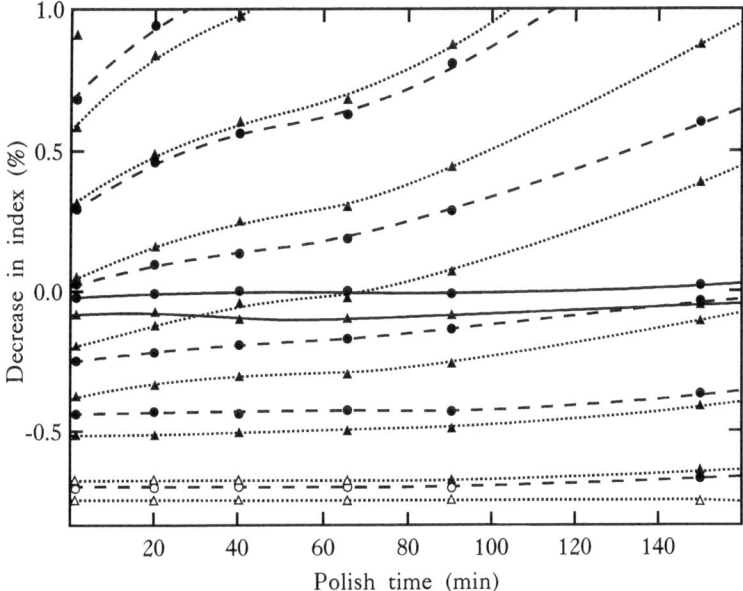

Fig. 6.7. Results of polishing to show the unusual behaviour of the 'strange' modes (solid lines) sometimes seen in LiNbO$_3$ – 633 nm (circles), 488 nm (triangles), 'missing' modes are shown as open symbols.

gions. Thirty-minute anneals were used, but isothermal tests showed that stability had more or less been reached by this time at each temperature. For practical applications a 200 °C anneal (for ~ 30 minutes) is chosen, as this removes most of the absorbing colour centres without reducing the barrier confinement appreciably. Prior to annealing, losses can be as high as 30 dB/cm (0.1% transmission) but these are now routinely reduced for the bottom modes to ~ 2 dB/cm for n_o (63% transmission) and ~1 dB/cm for n_e (80% transmission) (Zhang et al., 1991a). It has been claimed that rapid thermal annealing may be more effective than the normal furnace method (Al Chalabi et al., 1987; Weiss et al., 1987).

6.5.2 Optical damage in lithium niobate

Despite its useful combination of electro-optic, non-linear and laser host properties, LiNbO$_3$ suffers from a high susceptibility to optical damage (Carrascosa and Agullo-Lopez, 1988). This is a consequence of strong photo-refractive effects which can be produced by the presence of trace impurities. It causes light above a certain threshold power density to be

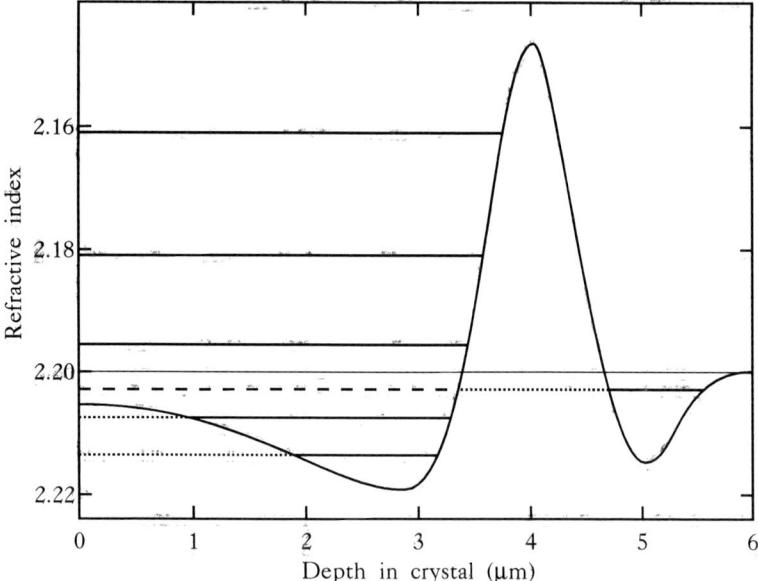

Fig. 6.8. Index profile model suggested to explain the 'strange' mode effect observed in LiNbO$_3$ n_e. A subsidiary well is situated beyond the main optical barrier.

'self-focused', and hence possibly lost due to scattering. Such impurities as iron (initially in the Fe^{3+} state) are responsible, and so the effect may be minimised by the introduction of dopants such as Mg which encourage a reduced (Fe^{2+}) state. Despite nearly two decades of study, the mechanism is not well understood and the simple view of the role of Fe is only one stage in a much more complex process. Some dopants may exacerbate the problem by charge compensation mechanisms, and unfortunately one such example is Ti. It has been found that Ti diffused waveguides in LiNbO$_3$ can suffer a 100-fold reduction in the threshold power density from the bulk value, whereas proton exchanged waveguides remain less affected. Diffused waveguides therefore have the disadvantage that they must be co-doped with high concentrations (~6 mol.%) of MgO (Feng et al., 1991), if they are to be used at even moderate power densities (only milliwatts in a typical guide cross-section). It has now been demonstrated that He$^+$ implanted waveguides, in common with proton exchange, do not suffer the same problem as Ti diffused ones (Glavas et al., 1987, 1989; Reed and Weiss, 1987). Implanted waveguides in both 'pure' and Mg doped material were tested and shown to have the same optical damage

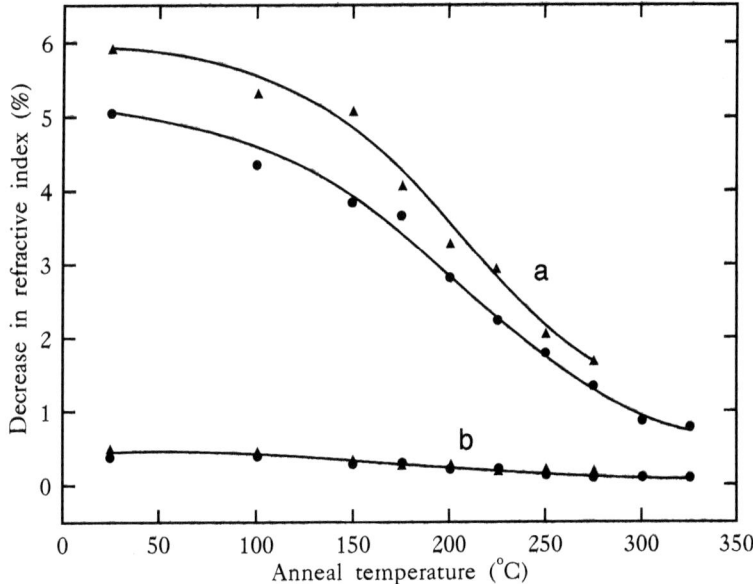

Fig. 6.9. Isochronal annealing of LiNbO$_3$ n_o for samples implanted with 2×10^{16} He$^+$ions/cm^2, using 30 min heating stages: a, nuclear peak; b, surface index. Implants were at 77 K (triangles) and 300 K (circles).

thresholds as in the respective bulk crystals. Application of the optical damage effect to holographic writing and erasing will be discussed in the next chapter.

6.5.3 Other niobates

Ion implanted waveguides have been reported in KNbO$_3$ (Bremer et al., 1988, 1989; Zhang et al., 1990; Strohkendl et al., 1991a,b; Fluck et al., 1991, 1992a,b; Fleuster et al., 1993) and in Ba$_2$NaNb$_5$O$_{15}$ (BNN) (Zhang et al., 1988). These materials are not isomorphous with LiNbO$_3$. KNbO$_3$ has a bi-axial orthorhombic perovskite structure, and exhibits extremely strong non-linear properties, and BNN has a tungsten bronze structure. However, the index profiles of their waveguides show remarkable similarity to those of ion implanted LiNbO$_3$.

KNbO$_3$ is obviously of considerable interest for non-linear applications, despite its present high cost. However, its low Curie temperature (223 °C) prevents a thermal diffusion doping method being used for waveguide production. Ion implantation has the advantage of being an ambient temperature process, and is therefore not precluded by this fact. Helium ion implanted waveguides were first reported by Bremer et al. (1988, 1989) and, subsequently, index profiles for the two major indices were measured by Zhang et al. (1990) and Fleuster et al. (1992). The refractive index changes are both large and efficient (doses of 2×10^{16} ions/cm^2 produce barriers with $\Delta n \sim -10\%$). Figure 6.10 shows how two of the indices (n_a and n_c) behave in a similar manner to those of LiNbO$_3$, the one producing an index reduced barrier with a tail extending out to the surface, and the other a non-tunnelling 'well' due to an index enhancement of up to $\sim 1.5\%$ in the region between the barrier and the surface. In the latter case, missing modes were detected at 0.488 μm, as with LiNbO$_3$ n_e. Ionic diffusion of potassium is thought to be responsible for this effect. The scarcity of samples prevented the performance of sufficient trials to detect 'strange' modes as seen in LiNbO$_3$. The great interest in this material for SHG, and the fact that good waveguides are easily produced by low dose ion implantation has led to several further publications by the Gunter and Buchal group (Fluck et al., 1992c,d; Fleuster et al., 1993), who are now able to produce low loss guides for extremely low dose implants ($< 10^{15}$ ions/cm^2). This study has been necessary because, unlike other materials, post-implant annealing is impossible in KNbO$_3$ due to its rather low Curie temperature. In order to obtain a better understanding of the influence of the ion dose on the shape of the index profile, they produced a series of waveguides with increasing dose, and determined the profiles. In Figure 6.11 the refractive index of the barrier is plotted as a function of dose, showing that all three indices decrease to a common (isotropic) value of 2.119 after a dose of $\sim 10^{16}$ ions/cm^2, explained by complete amorphisation of the crystal in the barrier. The view was supported by RBS data showing 100% damage at this dose (Fluck et al., 1992b). From this diagram they estimated an optimum dose (balancing good confinement against high radiation induced losses) of only 7.5 $\times 10^{14}$ ions/cm^2. Using this low dose and no subsequent annealing they were able to produce guides with losses $\sim 1 - 3$ dB/cm. Problems still exist with confinement into channel guides, and these will be discussed, together with the applications of these guides, in the next chapter.

Barium sodium niobate has a large electro-optic coefficient, but waveguide production in it by alternative methods had only achieved $\Delta n \sim 0.01\%$, using hydrogen in-diffusion (Hopkins and Miller, 1974). Despite its different structure, BNN exhibits very similar profiles to LiNbO$_3$ and KNbO$_3$ after helium ion implantation (Zhang et al., 1988). The extent

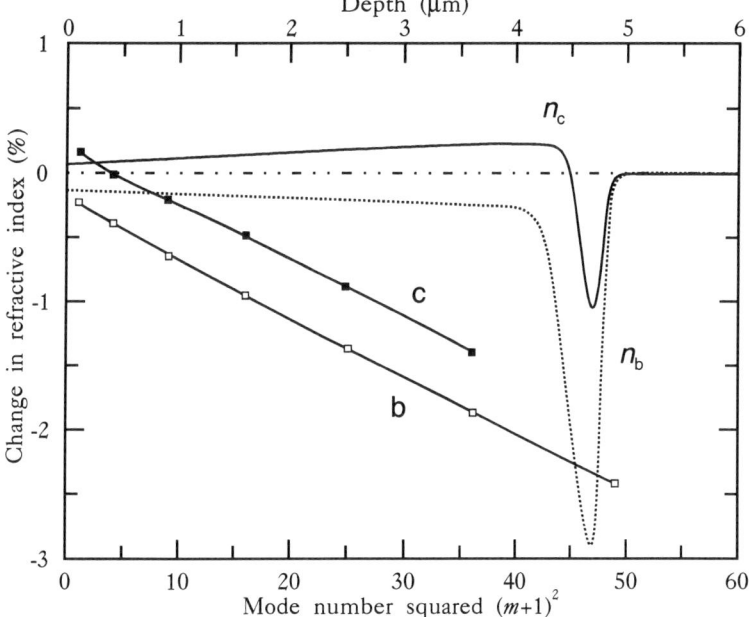

Fig. 6.10. The experimental mode indices of a 2 MeV planar waveguide in KNbO$_3$ implanted at 300 K, measured at 0.6328 μm, for n_b (- - -) and n_c (——). Also shown are the reconstructed index depth profiles.

of the index change (~ 5%) and its efficiency is comparable to that for LiNbO$_3$. A further niobate to be successfully implanted is strontium barium niobate (SBN). Like BNN, this is also a tungsten bronze structure. An alternative technique of doping with sulphur had been reported to produce an index increase of only ~ 0.1% (Bulmer et al., 1986).

6.6 Tantalates

LiTaO$_3$, which is isomorphous with LiNbO$_3$, might originally have been expected to show similar ion implantation refractive index profile effects, if these are caused by lithium diffusion. However, this has not been found to be the case in practice (Glavas et al., 1988). In fact both its indices (n_o and n_e) behave in the same way, showing only decrease in magnitude producing typical barrier profiles with only a minor tail extending towards the surface – somewhat in between the two extreme cases of quartz and LiNbO$_3$ described above – suggesting that some isolated defect annealing is occurring during implantation. This difference in behaviour between

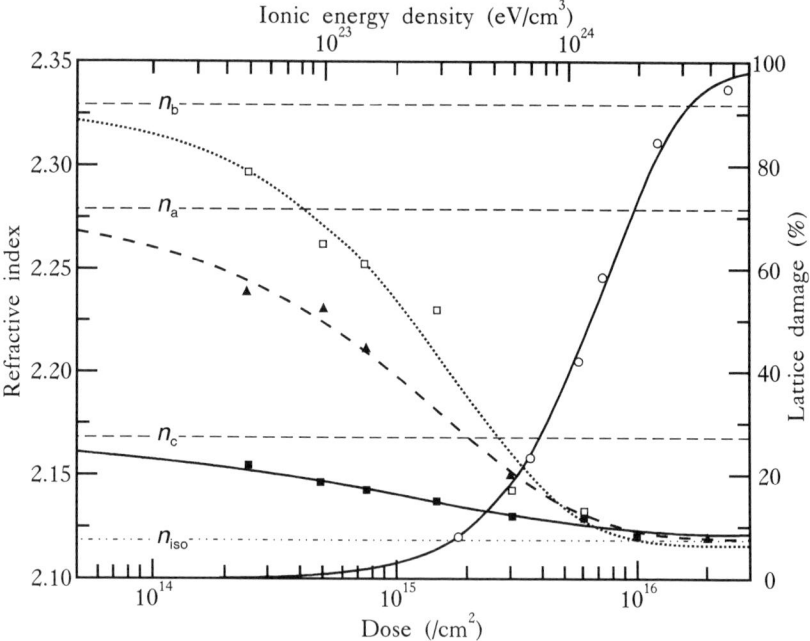

Fig. 6.11. Dose dependence of damage effects caused by 2 MeV He$^+$ implantation into KNbO$_3$ – lattice damage (circles); refractive index values of the optical barrier peak for n_a (triangles), n_b (open squares), and n_c (filled squares).

LiNbO$_3$ and LiTaO$_3$ may be justified by the higher stability of LiTaO$_3$, as indicated by its stoichiometry and melting point. The magnitudes and efficiencies of the changes are similar to those of LiNbO$_3$, also being greater at low temperature. At 300 K the barrier height reaches a saturation level of $\sim 4\%$ for a dose of 3×10^{16} ions/cm^2, but at 77 K this increases to $\sim 6\%$ for a dose of 7×10^{16} ions/cm^2. Once again, complete amorphisation (isotropy) is not reached by this dose. The lack of a stable amorphous phase leads to a steady annealing at temperatures above $\sim 200\,°$C with complete recovery by $\sim 400\,°$C (Glavas et al., 1988).

KTaO$_3$ is cubic at room temperature. Its ion implanted index profile shows a typical barrier-type decrease (Wong et al., 1992), but this occurs with very high efficiency (as with KNbO$_3$). A dose of 1×10^{16} ions/cm^2 is sufficient to produce a barrier of $\Delta n \sim -5\%$, and consequently very low loss (~ 1 dB/cm) waveguides are produced after annealing. The barrier is stable up to $\sim 400\,°$C and does not fully recover until above $\sim 900\,°$C. However, high dose implants show stress effects – the maximum barrier height with $\Delta n \sim -15\%$ occurs at a dose of $\sim 4 \times 10^{16}$ ions/cm^2, and above this a stress relaxation possibly occurs reducing the barrier considerably. Even medium dose implants are not reliable, showing surface cracking and anomalous 'range' behaviour on annealing above $\sim 400\,°C$, possibly due to bubbling effects of the implanted gas. The ideal behaviour at very low doses, however, makes this a very good candidate for device applications.

6.7 Bismuth germanate

In all of the preceding examples, even though there are considerable differences between the shapes of the index profiles, an optical barrier with decreased index is always clearly produced in the nuclear damage region. This is expected on the grounds that collision damage modifies the crystal lattice into a lower density structure. By contrast, Mahdavi *et al.* (1989) have reported that the refractive index profile of bismuth germanate, $Bi_4Ge_3O_{12}$, shows no such index decrease after He^+ implantation. Instead, as shown in Figure 6.12 (25 °C curve), there is a pronounced index enhancement throughout the entire region which peaks at a depth close to the nuclear damage maximum. The mechanism suggested which would enable the index to increase is a lattice relaxation into a more dense structure, such as a new crystalline phase. Changes in atomic arrangement which produce more polarisable bonds might also enhance the index since Bi_2O_3 and GeO_2 form a wide range of compounds including stable mixtures with the ratios 1:3, 2:3, 6:1 and 7:1 (Weber and Monchamp, 1973; Gevay, 1987). Consequently, it is suggested that the energy supplied from the ion beam, coupled with the non-thermodynamic pressure and temperature conditions within the ion track, can induce relaxation into a separation of closely related BGO phases. Mahdavi *et al.* (1989, 1990a) studied the stability of the waveguide profile during annealing and noted that the optical well steadily decreased in magnitude between 250 °C and 425 °C, consistent with normal annealing of defects. However, in the narrow interval between 425 °C and 450 °C the index profile 'switched' in character from one with index enhancement throughout the guide to one with only a narrow low index optical barrier, as shown by Figure 6.12 (450 °C curve). For helium doses greater than 4×10^{16} ions/cm^2 the new index profile was stable up to 800 °C. At lower ion doses there was complete annealing at moderate temperatures. Mahdavi *et al.* considered that this rapid conversion might reflect a further phase transition in the region of the new barrier. To test this idea, measurements of stored en-

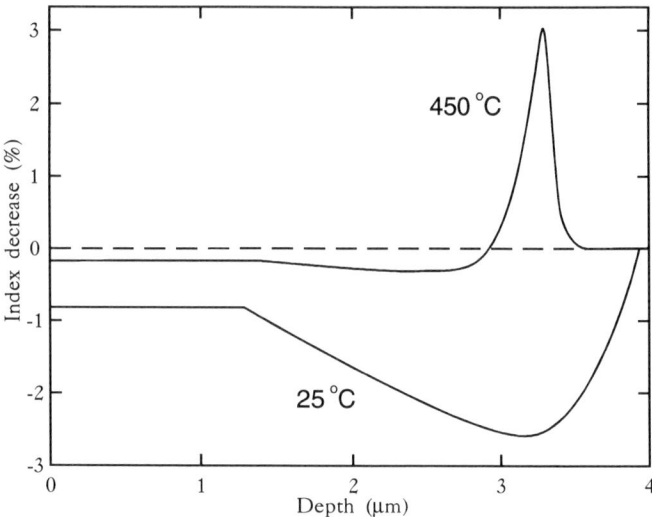

Fig. 6.12. Unusual enhanced index profile of He$^+$ implanted Bi$_4$Ge$_3$O$_{12}$ (at 25 °C). On annealing it recovers gradually and then by 450 °C it 'inverts' to give a typical barrier profile as shown here.

ergy were made (Jimenez de Castro et al., 1991). However, whilst the ion implanted material was found to have retained a considerable amount of stored energy, there was no evidence of a first-order phase transition.

As implanted, the guides were initially very lossy (~40 dB/cm), but annealing proved successful. By the choice of low temperature ion implantation, optimised dose and energy, and subsequent annealing, excellent waveguides have now been formed with losses ~0.2 dB/cm. The optimum conditions using high purity Chinese samples were reported to be 1.5 MeV He$^+$ to a dose of 4 ×10^{16}ions/cm^2, at a target temperature of 77 K, followed by annealing at 200 °C for 30 minutes. The most crucial factors were the quality of the starting material and the implant temperature.

The loss measurements using a polarised input beam had very unusual characteristics (Mahdavi et al., 1990b). The log/linear plots of intensity versus distance, instead of giving the normal straight lines, showed pronounced oscillations (Figure 6.13). These were found to result from rotation of polarisation of the beam within the waveguide. Optical activity is a well known property of some BGO compositions, and so this phenomenon was not totally unexpected, assuming that the ion beam damage had led to the formation of a modified phase. At first sight it

6.7 Bismuth germanate

might be expected that the region of optical activity would correspond to that of enhanced refractive index. However, the samples annealed above 450 °C, where no index enhancement existed, were also optically active, and this activity was extremely high with values reaching 90°/mm (at 633 nm). The magnitude of the optical activity was found to be a function of the mode being measured, and this was interpreted as occurring when the modal field profile had maximum overlap with the region in which the optically active phase had been produced. In the original enhanced index guide (annealed at < 400 °C), the maximum activity occurred for the $m = 2$ mode, but after annealing at 450 °C when the index decrease barrier was observed, the maximum activity occurred for the $m = 0$ mode. Thus it appeared that the refractive index measurements were detecting regions of certain phase structure, and the optical activity was detecting regions in which other phases dominated. In summary it was concluded that with the wide range of phases which could be formed in the Bi_2O_3/GeO_2 system, processes might be expected to occur which would lead to combinations of new crystal phases, and these might dominate over simple lattice damage effects. It was thus not surprising to observe index enhancement and optical activity in different parts of the guide profile. Nd doped $Bi_4Ge_3O_{12}$ has been found to exhibit waveguide lasing (Field et al., 1991a). An unusual observation in the behaviour of ion implanted BGO is the apparent self-focusing of a laser beam within the implanted region to produce a channel waveguide structure (Brocklesby et al.,1992). These results will be discussed in more detail in Chapter 7.

Detailed measurements of ion implantation in another BGO composition, $Bi_{12}GeO_{20}$, have been reported to show a much more conventional behaviour (Mahdavi et al., 1990a), with a typical low index damage barrier and no major peculiarities in annealing or optical properties. Low temperature implants produced a 5% decrease in index at the damage peak whereas at room temperature this was only ~ 1%. There were certain anomalies in the wavelength dependence of the profiles. It was reported that the guiding region exhibited a relative increase in index for short wavelengths, such that for room temperature implants there was an increase for blue but a decrease for red. The effect was less pronounced for low temperature implants. This relative increase in index was attributed to shifts in the narrow bandgap of this particular BGO composition (Mahdavi et al., 1990a).

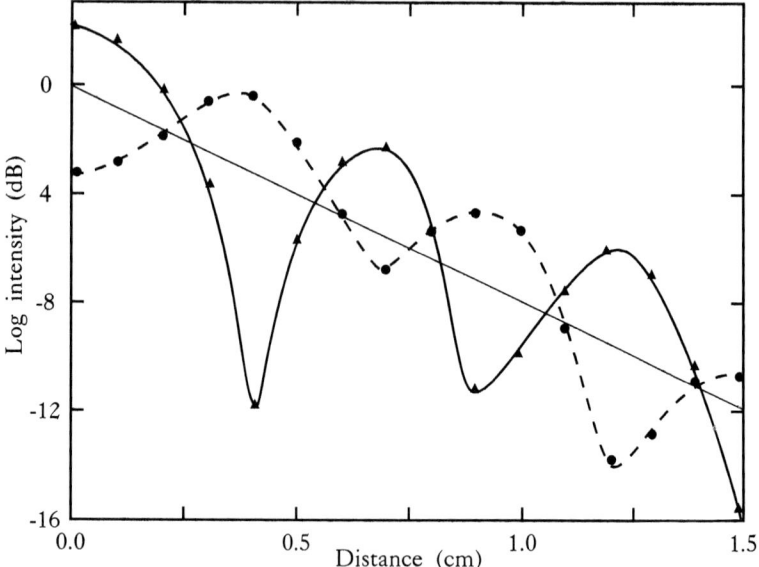

Fig. 6.13. Intensity versus propagation distance of the $m = 2$ mode in a $Bi_4Ge_3O_{12}$ waveguide formed by 4×10^{16}ions/cm^2 He$^+$ at 1.5 MeV, showing optical rotation – TE (solid) and TM (dashed).

6.8 Laser hosts

6.8.1 Garnets

An interest in the possible fabrication of waveguide lasers has prompted attempts to produce He$^+$ implanted waveguides in conventional laser systems. These include the Nd and Cr doped garnets, such as YAG ($Y_3Al_5O_{12}$), GGG ($Gd_3Ga_5O_{12}$), YSAG ($Y_3Sc_2Al_3O_{12}$), GSAG ($Gd_3Sc_2Al_3O_{12}$) and GSGG ($Gd_3Sc_2Ga_3O_{12}$) and related materials such as the perovskite YAP ($YAlO_3$). Of these, the first two and the last one have met with considerable success and planar waveguide lasers have been demonstrated in all three.

YAG waveguides were first reported by Arutyunyan et al. (1985), but a much more detailed study was made by Zhang et al. (1991c) and Chandler et al. (1991). The index profile in YAG presented some problems for infra-red (1.064 μm) mode confinement and so this will be discussed in detail here. Figure 6.14 shows the build up of the index profiles for three different doses at the same energy. The optical barrier caused by the nuclear damage is rather narrow and is not produced with great efficiency – requiring $\sim 3 \times 10^{16}$ions/cm^2 for $-\Delta n$ of $\sim 2\%$ even

6.8 Laser hosts

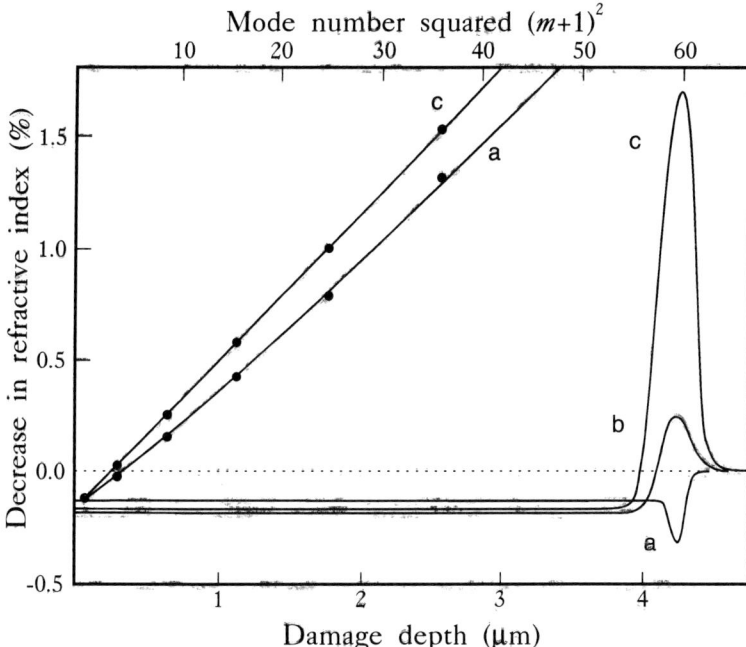

Fig. 6.14. Index profiles of Nd:YAG implanted with 2 MeV He$^+$ at 300 K for doses: a, 1.5; b, 3.0; and c, 5.0×10^{16} ions/cm^2 showing initial increase of ~ 0.3%. The slopes of the two mode curves indicate the different effective guide widths.

for a 77 K implant. Barrier confinement would therefore require a multi-energy implant with a total dose of, say, at least 9×10^{16} ions/cm^2. The undesirable consequences of such a high dose are two-fold. Firstly, even after annealing, the presence of a high concentration of He$^+$ leads to scattering losses within the guide. Secondly, its effect on the fluorescence efficiency and on the transition cross-section has inhibiting consequences for the lasing action. It is therefore desirable to keep the dose as low as possible (~ 5 × 10^{16} ions/cm^2 maximum). From this observation, a barrier confined waveguide in YAG does not therefore seem feasible.

However, there may be alternative ways to achieve a satisfactory profile. From Figure 6.14(a) it can be seen that a small index enhancement occurs in the surface region (which has not yet been fully explained). Unfortunately, this is only ~ 0.15%, and so for a 6 μm-deep guide (~3 MeV He$^+$) this can confine a non-tunnelling mode for the pump wavelength (~ 0.59 μm) but not for the lasing wavelength (1.064 μm). A 4 MeV He$^+$ implant would be required to achieve this. The solution to the problem appeared when it was discovered that the nuclear peak region also produced an initial index enhancement (for very low doses, $< 10^{16}$ions/cm^2, profile (a) in Figure 6.14), prior to the build up of the −ve Δn barrier. This inverted 'barrier' amounted to $\Delta n \sim +0.3\%$ and by the superposition of several of these low dose implants at different energies, equally spaced right up to the surface, a non-tunnelling region with Δn averaging 0.25% was produced. This was able to confine a 1.064 μm mode, especially when combined with a high energy optical barrier (dose 3×10^{16}ions/cm^2). The profile was stable up to 800 °C, and by $\sim$ 400 °C the guiding loss had been reduced to $\sim$1 dB/cm.

A further discovery was that, despite the isotropic nature of the starting material, the implanted guide had a far superior index profile for TM propagation. For TE measurements the index enhancement was only $\sim$ 0.1%, and mode confinement was insufficient to allow lasing for this polarisation. Performance was also found to be slightly dependent on the orientation of the sample with respect to the boule axis.

For GGG, initial trials of waveguide fabrication were performed on undoped samples. A guide with an ion dose of 4×10^{16}ions/cm^2 at 2.5 MeV gave a 0.25% index increase in the electronic stopping region and a 2% index decrease at the end of the ion track. These data are a theoretical fit to the dark mode spectrum observed experimentally for TM polarisation. But similar measurements for the TE polarisation proved impossible as no dark modes could be detected in this case. This suggested that there are different index profiles for the TE and TM polarisation, leading to a lower cut-off wavelength in the TE case. Guides fabricated with the same implant in Nd:GGG (3.35 at.%), while still showing good transmission in the visible region, had no mode at 1.06 μm. To increase the cut-off wavelength and allow a 1.06 μm mode to propagate, a deeper guide was fabricated using 2.9 MeV energy ions at a dose of 2×10^{16}ions/cm^2. Figure 6.15 shows the index profiles for both 2.7 at.% and 3.35 at.% Nd doped GGG. These results are the only ones so far reported which exhibit a significant dependence of the refractive index profile on the doping level. The best waveguide showed $\sim$ 80% transmission over 2.5 mm including launch losses (for TM polarisation), indicating that the guide loss must be $\leq$ 1 dB/cm (Field et al., 1991b). There is no report of attempts at waveguide fabrication in undoped GSAG and GSGG, but the Sussex group has tried ion implantation in Cr doped GSAG and GSGG. It appears that there is a common problem in all Cr doped materials so far tested (especially for high doping concentration), i.e. that a high absorption or scattering

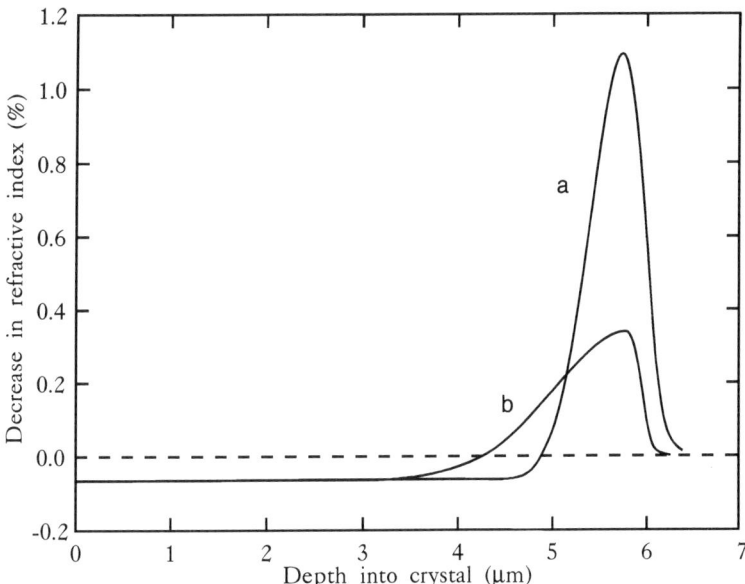

Fig. 6.15. Comparison of the refractive index profiles of two He$^+$ implanted waveguides in Nd:GGG with different dopant concentrations. The samples were implanted with 2.9 MeV He$^+$ at a dose of 2×10^{16} ions/cm^2. The index increases slightly in the guiding region (a: 2.7 at.%, b: 3.35 at.%).

mechanism is introduced after implantation. As a consequence, only dark mode spectra but no propagating modes have been observed in these guides. The losses in these guides have not yet been reduced sufficiently to permit meaningful measurement.

So far the exception to this problem for the Cr doped garnets is Cr:YSAG. A double energy implanted Cr:YSAG waveguide appears to support guiding modes after a moderate temperature annealing. The waveguide loss was estimated at about 7 dB/cm in the IR (Rodman et al., 1993). This guiding behaviour in Cr:YSAG may not be a case of exception to the high absorption in other implanted Cr doped materials. It has been found that the absorption effect is strongly related to the Cr doping level, so ion implantation can be used to form a waveguide in a low Cr concentration doped material. The higher threshold pump power required for a Cr doped laser means that the improved power density of a stripe configuration guide will probably be necessary.

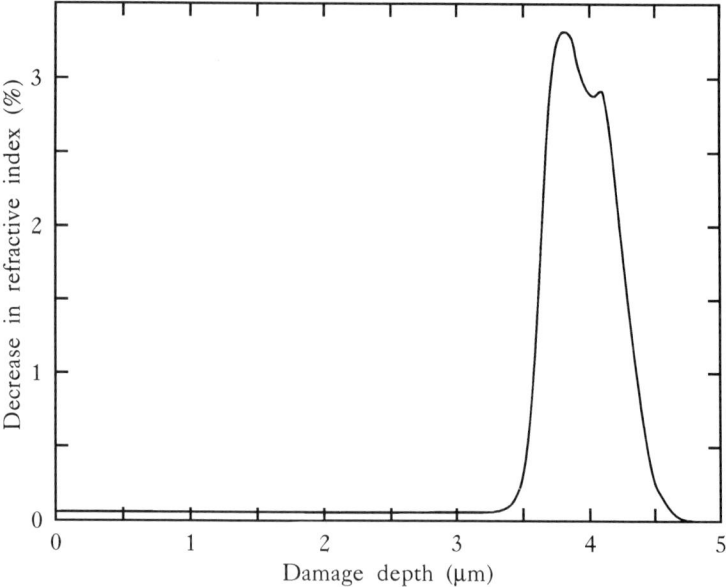

Fig. 6.16. Refractive index profile of Nd:YAP implanted (at 77 K) with He$^+$ at energies 2.3, 2.15 and 2.0 MeV to a total dose of 4.5×10^{16} ions/cm^2.

6.8.2 Other laser substrates

Because of large scale electro-optic development programmes, conventional fabrication techniques are now able to produce very good quality waveguides in undoped LiNbO$_3$ (by Ti diffusion or proton exchange). The techniques are now being extended to LiNbO$_3$, which has been grown doped with laser active elements, and these substrates are readily available for ion implanted waveguide trials. Chandler *et al.* (1990b) have used a range of stoichiometries and doped samples, and analysed the refractive index profiles of He$^+$ implanted material. They have shown that the index profiles (in percentage terms) and the irradiation induced losses of these guides are independent of the dopant content. As a consequence, waveguide lasing has been demonstrated in Nd doped LiNbO$_3$ (Field *et al.*, 1991c), and suitable low loss guides have also been produced in both Tm and Er doped samples. As stated above, the guiding behaviour is identical to that in the undoped material, described earlier (Section 6.5.1).

Neodymium doped yttrium aluminium perovskite (YAP) has also produced a planar waveguide laser (Field *et al.*, 1990). Figure 6.16 shows the index profile of the implanted Nd:YAP crystal which supported wave-

6.8 Laser hosts

guide lasing. As can be seen, there is no index enhancement, and so the guiding mechanism relies on a broad optical barrier. This is produced by a triple-energy implant which gives a barrier width of $\sim 1\,\mu$m. However, as yet the attenuation has not been reduced below $\sim 10\,$dB/cm and this results in a high threshold pumping value of $\sim 50\,$mW. A possible reason for the high loss is that this material forms a profile which is confined by a rather narrow barrier with no index enhancement in the guiding region. The nuclear peak has a higher damage efficiency than YAG and so low doses are possible, but because it is narrow, several doses at different energies must be used to produce a broad enough barrier for good confinement. At the moment, therefore, a triple-energy broad barrier has been constructed to give a moderately low loss guide.

An ideal candidate for waveguide laser fabrication, which has shown good ion implantation characteristics is $Cr:BeAl_2O_4$ (alexandrite). Unfortunately, the Be content presents potential chemical and radiation hazards when considering, respectively, the polishing and the (α, n) nuclear reaction during He^+ ion irradiation. However, despite these problems successful planar waveguides have been produced which exhibit profiles similar to those of the garnets, with an enhanced index in the guiding region. Good end-launched modes have been obtained with low loss, and prior to annealing the attenuation is $\sim 2\,$dB/cm in the planar guide. It is estimated that a stripe waveguide will achieve sufficient pump power density to reach lasing threshold.

Waveguide formation by He^+ implantation into sapphire (Al_2O_3) and ruby was attempted and reported by Townsend (1985). However, the highly ionic nature of the lattice resulted in a very low damage retention efficiency. Amorphisation is expected to require doses in excess of 5×10^{17}ions/cm^2, and at this concentration He gas can precipitate during ion implantation or annealing, and may result in the physical removal of the implanted surface layer. At these high doses a poor quality barrier-confined waveguide is produced, but at lower doses of He^+ or N^+ ($\sim 10^{16}$ions/cm^2) it was reported that a very small index increase occurred, but not sufficient to contain a guiding mode. It was suggested that oxygen loss was possibly producing a metal-rich surface layer.

An alternative approach to the production of waveguides in Al_2O_3 was suggested by the data of surface amorphisation published by White *et al.* (1988), indicating that group four ions were very effective in stablising an amorphous layer. Townsend *et al.* (1990) investigated the effects of implanting 6 MeV C^+ into Al_2O_3. They implanted at 77 K to minimise defect recombination but, more importantly, they used a *c*-cut surface orientation. Because defect diffusion is highest along the c-axis, this allowed better separation of the vacancies and interstitials, thus further inhibiting recombination of defects. They suggested that the carbon forms a carbide bond by occupying the oxygen lattice site. This stable structure produces lattice distortion which enhances defect aggregation. It is predicted that the defects are retained in the region of the carbon implant, but that the lattice nearer to the surface recovers during implantation. Initial profile measurements suggest that this appears to be the case (Townsend et al., 1990). Overall, the method seems to have met with some success, but losses have not yet been reduced to an acceptable level. However, work is continuing as a successful solution of the problem will allow fabrication of tunable waveguide lasers in ruby or titanium sapphire.

Several materials which are known to be interesting laser hosts have been implanted with He^+ and have exhibited planar waveguide properties, but these have not yet been produced with sufficiently low attenuation to achieve lasing threshold. They include $LiCaAlF_6$, YLF ($LiYF_4$), LYAG ($\sim LuY_2Al_5O_{12}$), LNA ($La_xNd_{1-x}MgAl_{11}O_{19}$, $x \sim 0.1$) and $ZnWO_4$. The topic is still at an early stage of development, and with further experimentation many of these materials may form usable laser waveguides. This confidence is justified by reference to the progress in loss reduction for BGO (Section 6.7) or the Nd:YAG waveguide lasers where the pump threshold was reduced from $\sim 70\,mW$ to $0.5\,mW$ (Field et al., 1991d).

6.9 Non-linear materials

The fabrication of optical waveguides is desirable for optical non-linear applications, such as frequency conversion, wave mixing and phase conjugation. The reason for this is that the waveguide provides compatibility with other miniaturised, integrated optical waveguide devices, and in particular laser diodes and fibre geometries. Additionally, the optical confinement of waveguides maintains a very high confined power density in contrast to the bulk for a given input power. This suits the requirement of the non-linear applications. However, a number of non-linear materials have very low phase transition temperatures. This inhibits the use of conventional techniques, which involve high temperature processes, to form waveguides. In contrast, ion implantation which may be performed at ambient or reduced temperature has been shown capable of form-

6.9 Non-linear materials

ing waveguides in a wide range of optical non-linear materials, and the non-linearity is preserved in most of these materials after implantation.

Apart from $LiNbO_3$ and $KNbO_3$, the other established non-linear optical materials in high demand are KTP ($KTiOPO_4$), LiB_3O_5, β-BaB_2O_4 and $BaTiO_3$. As discussed in Section 6.5, He^+ implantation can be used to make very low loss waveguides in the non-linear optical materials $LiNbO_3$ and $KNbO_3$. SHG has been successfully demonstrated in implanted planar and channel waveguides in $KNbO_3$ (Fluck et al., 1992c,d). KTP is another successful example of the production of optical waveguides by ion implantation (Zhang et al., 1992a,b). The index profile of implanted KTP conforms to the common pattern – barrier waveguide – of He^+ implanted optical materials. Figure 6.17 shows a comparison of n_z profiles for single and double energy implanted waveguides. The results show that for a double energy implant with a total dose of 2×10^{16}ions/cm^2, the index in the guiding region increased by $\sim 0.14\%$ for n_z and $\sim 0.05\%$ for n_{xy}, whilst in the optical barrier regions, the index decreased by $\sim 9\%$ for n_z, and only $\sim 1\%$ for n_{xy}. In general, the modes in the multiple and high energy implanted waveguides have a better confinement than those in the single and low energy implanted waveguides. The loss for the high energy implanted guides is ~ 1 dB/cm at $0.6328\,\mu m$, and ~ 2 dB/cm for IR. Frequency doubling has been achieved in these guides with a conversion efficiency up to 25% (Zhang et al., 1992c), as will be discussed in Chapter 7.

Planar optical waveguides have been fabricated in LBO (LiB_3O_5) using He^+ implantation, by Sharp Laboratories (Europe) in collaboration with Sussex (Davis et al., 1993). In contrast with most optical materials, ion implantation only produces waveguide structures in LBO at room temperature. The mechanism for this effect is not fully understood. The refractive index changes for He^+ implanted LBO have been investigated for the three principle indices n_x, n_y and n_z. A threshold ion dose of about 0.75×10^{16}ions/cm^2 is required to form an optical barrier, and ion doses higher than $\sim 2.5 \times 10^{16}$ions/cm^2 saturate the refractive index decrease. For a dose of 1.5×10^{16}ions/cm^2, the refractive index changes in the optical barrier region at $0.488\,\mu m$ are -1.5%, -5.25% and -4% for n_x, n_y and n_z respectively, whereas there is almost no change for the three indices in the guiding region. The optical loss is ~ 9 dB/cm for this single energy implant after 200 °C annealing. The high loss is mainly due to poor confinement by the optical barrier, as it is only $\sim 0.3\,\mu m$ thick. However, the crystal quality and the surface and end-polishing may also play a significant role in loss problems at this stage

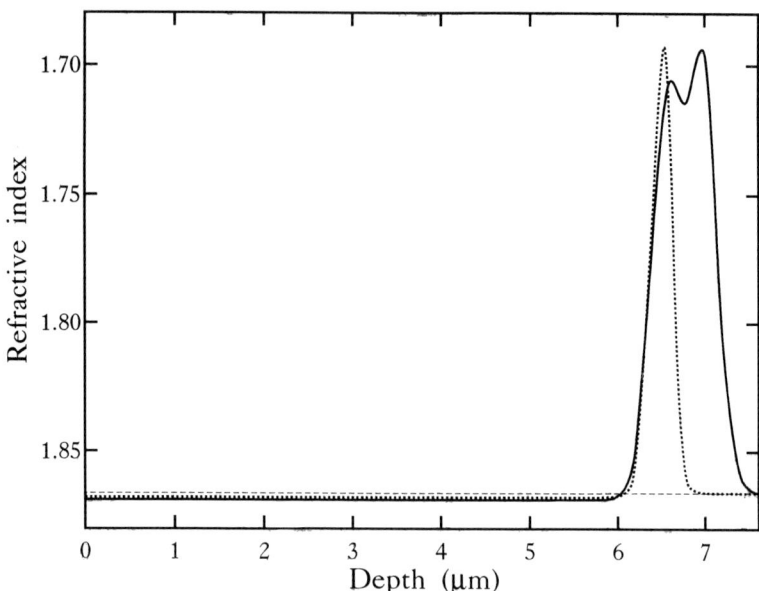

Fig. 6.17. Comparison of refractive index profiles of implanted KTP waveguides. Single energy (2.5 MeV, dotted); double energy (2.5 + 2.7 MeV, solid).

of development. Figure 6.18 shows the comparison of the index profiles between single and triple energy implants. The latter broadens the optical barrier significantly, and the loss is < 4 dB/cm after annealing treatment. An SHG signal has been detected from the waveguide. This indicates that the non-linearity is preserved after implantation (Chapter 7).

Ion implantation has also been used to form waveguides in another highly non-linear borate – BBO (β-BaB$_2$O$_4$). In this uniaxial material, the ordinary index (n_o) is practically unaffected, and hence waveguides can only be produced for the extraordinary index (n_e). The guides appear to have diffusion-type index profile shapes with an entirely enhanced index region. They are therefore expected to have very low loss. The index change for n_e after He$^+$ implantation is considerable. In fact, at saturation (dose $\sim 3 \times 10^{16}$ ions/cm^2 at 77 K) n_e reaches the value of the bulk n_o ($\Delta n \sim 8\%$). This suggests that the crystal may no longer be anisotropic, and in this case it would have lost its non-linearity.

BaTiO$_3$ is a promising non-linear material for photorefractive applications. Moretti et al. (1990, 1991) implanted He$^+$ and H$^+$ into BaTiO$_3$, for the first time, to form optical waveguides. Like KNbO$_3$, BaTiO$_3$

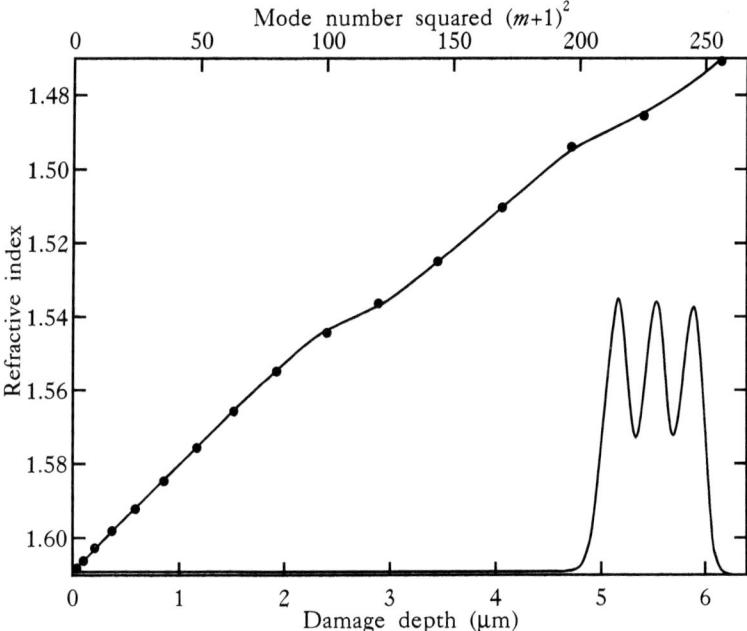

Fig. 6.18. Refractive index profile and fitted mode curve for a triple energy (2.0 + 2.12 + 2.24 MeV, 4.5×10^{16} ions/cm^2) He$^+$ implanted waveguide in LBO. The filled circles are the experimentally measured mode indices, and the curve gives the calculated mode indices for this profile.

has phase transitions close to room temperature (from tetragonal to orthorhombic at $\sim 9\,°C$ and from tetragonal to cubic at $\sim 132\,°C$). This prevents normal diffusion processing, but is quite suitable for the ambient temperatures of ion implantation. Mode measurements suggest that the index profiles are similar to those of the other titanates reported above with little index change in the surface region. The results indicate a peak change for n_o of $\sim -4\%$. There were problems for n_e but after surface polishing a peak value of $\Delta n \sim -3.5\%$ was obtained. Apparently, the attenuation is sufficiently low that, even without annealing, good mode confinement is demonstrated. For a waveguide in BaTiO$_3$ to be photorefractive it must be of the correct crystalline phase, poled, electro-optic, and the photorefractive donor/acceptor mechanism must be retained. The importance of satisfying all these criteria simultaneously has been highlighted in recent research carried out at Southampton University on He$^+$ ion implanted waveguides in SBN (Sr$_{1-x}$Ba$_x$Nb$_2$O$_6$) and BSO (Bi$_{12}$SiO$_{20}$). In the case of BSO the intensity dependent photochromic properties (Vainos et al., 1989) produced such high losses in the guide as to make characterisation impossible. For SBN, the waveguide losses were high (~25 dB/cm), but this did not preclude experiments to determine that the waveguide was still electro-optic. However, no photorefractive properties were observed for the SBN waveguide. Returning to the case of $BaTiO_3$, the Southampton group found that H^+ implants were preferable to He^+, and that all these photorefractive properties were preserved in high energy, low dose H^+ implanted waveguides. Because of the low phase transition temperature, the sample cannot be annealed after implantation to reduce the optical absorption loss. In general, this annealing process is a necessary stage for ion implanted waveguides. The guide loss is estimated to be ~14 dB/cm in H^+ implanted $BaTiO_3$. The effective photorefractive two-beam coupling response time is improved by a factor of 100 in the guide relative to that in the bulk (Youden et al., 1992).

6.10 Other crystalline materials

Various other materials have produced optical waveguides when implanted with He^+, for example TeO_2, PLZT and certain flourides (Hamelin et al., 1992). Table 6.1 gives more details of these observations. Interesting results have been reported for zircon ($ZrSiO_4$) (Babsail et al., 1991b). As a naturally occurring mineral this is known to exist as a stable phase in both the crystalline and amorphous states. Reference to the discussion for quartz indicates that this may therefore produce very stable waveguide structures. Also significant is the fact that the mineral exhibits amorphisation due to naturally occurring radioactivity. It was found that ion implantation damage is less efficient than in crystalline quartz, requiring a He^+ dose of $\sim 8 \times 10^{16}$ ions/cm^2 to produce amorphisation (having $\Delta n \sim -9\%$) – but again a lower dose is required at 77 K. As expected, the optical barrier is quite square, and isolated point defects in the guiding region were found to anneal out during implantation, giving guides with low attenuation even prior to thermal annealing. As with quartz, low dose barriers anneal out (by $\sim 700\,°C$) but high dose amorphisation is extremely stable, even at 1000 °C, producing the ideal square barrier waveguide structure. Some further annealing anomalies were reported, but in general this appears to be a promising substrate if suitable high quality crystals can be produced. It is suggested that this may be an ideal waveguide laser host since natural crystals invariably act as hosts for rare earth elements.

6.11 Non-crystalline materials

Despite the absence of 'active' properties such as non-linear or electro-optic effects, non-crystalline waveguides have many applications. For example, they may exist as overlays on active crystalline substrates or they may be used as hosts for laser dopants or matrices of organic crystals with high non-linear properties. Various rare-earth doped laser glasses are commercially available, so it is very interesting to explore possibilities of waveguide laser formation in these glasses.

The first ion implanted waveguides were produced in 1968 by proton implantation into fused silica glass (Schineller *et al.*, 1968), and the index changes for H^+, He^+ and N^+ (see Section 6.13) have since been characterised by several other groups (Webb, *et al.*, 1975,1976a,b). Unlike the majority of crystalline materials, fused silica was found to suffer compaction under the influence of both nuclear damage, and (slightly) from electronic effects. This produced solely enhanced index optical wells (non-tunnelling) with very low attenuation levels. The damage efficiency is quite high, requiring only $\sim 10^{16}$ions/cm^2 to produce a saturation change of $\Delta n \sim +2\%$ at 77 K or $\sim +1\%$ at 300 K. A comparison of fused and crystalline quartz, implantation and annealing behaviour has been made by Zhang *et al.* (1989) and Chandler *et al.* (1990a). The saturated nuclear damage region of both materials is found to have similar properties, suggesting the formation of a stable damaged state with high annealing stability.

The waveguiding effects produced by implantation of light ions into several types of glass (silicates, phosphates, fluorides, germanates) have been studied by Kakarantzas *et al.* (1992, 1993a,b,c,d). The first three produced standard optical barrier-type structures. Unfortunately, they exhibited a considerable index decrease in the surface region far in excess of most crystalline materials. This meant that broadened barriers could not be produced by multiple-energy implants without removing the optical well. A typical refractive index profile for a silicate glass is shown in Figure 6.19. As can be seen, there is a considerable decrease of refractive index in the guiding region. This is not improved by changing the implantation conditions (e.g. temperature) or by subsequent annealing. The index profile is stable up to $\sim 300\,°C$. The best quoted waveguide transmission after annealing is 15 dB/cm.

It was found that a few glasses (depending on their composition), such as ordinary window float glass, had similar (compaction) properties to fused silica, but that the addition of alkali metals increased the influence

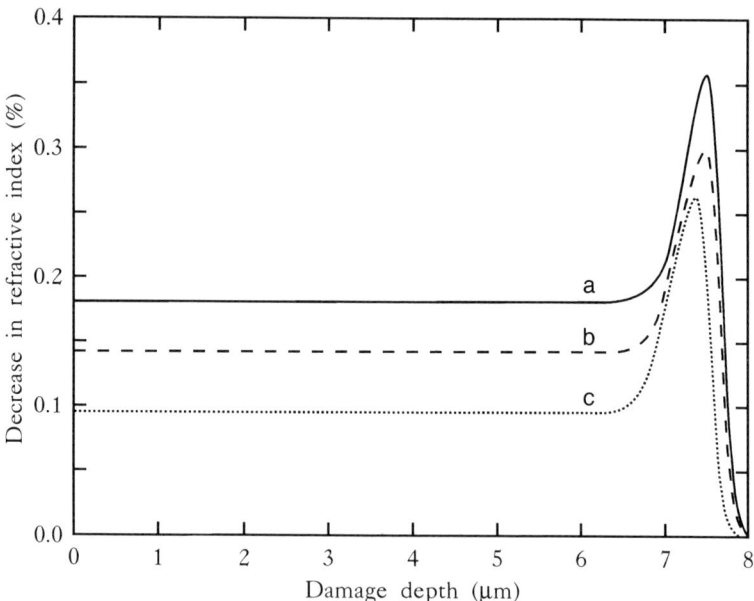

Fig. 6.19. Refractive index profiles of LG680 silicate laser glass implanted at a, 77 K; b, 300 K; and c, 625 K with 2.5 MeV He$^+$ dose 4×10^{16} ions/cm^2.

of the electronic excitation on the damage process (producing expansion) in such a way that the compaction was reversed, as shown in Figure 6.19.

Phosphate glasses showed a similar behaviour to silicates but with a larger decrease in index (1.4% peak compared to 0.6% peak). This was attributed to the relative weakness of the P—O bond compared to the Si—O bond. Thermal stability was reported up to $\sim 400\,°C$ and improved waveguide losses of $\sim 10\,dB/cm$ were obtained.

For fluoroaluminate glass, the refractive index profiles were very similar to those of silicates and phosphates, producing rather poorly confined waveguides with large index changes in the guiding region. However, a spectacular difference in behaviour was produced by implanting with hydrogen rather than helium, as shown in Figure 6.20. In this case an optical barrier was still produced, as with He$^+$, but in the guiding region there was practically no change in index. This effect has not been observed with any other glasses so far, and is possibly caused by the chemical activity of F and H, leading to H—F bond formation and refractive index change. It is unclear why H—F bond formation at the end of the ion range is able to modify the much wider electronic

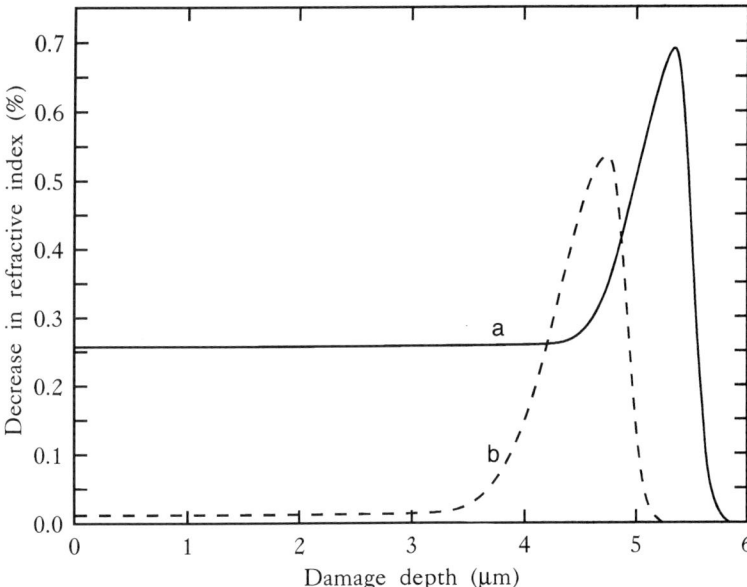

Fig. 6.20. Refractive index profiles of Nd doped fluoroaluminate glass implanted at 300 K with a, 2.0 MeV He$^+$ dose 4×10^{16} ions/cm^2 (solid); and b, 0.6 MeV H$^+$ dose 8×10^{16} ions/cm^2 (dashed).

stopping layer of the waveguides. Initial suggestions include stress effects as well as charge compensation. With this improved profile, it should now be possible to produced broad barriers using multi-energy implants, resulting in good quality, low loss waveguides.

Lead germanate glass behaves completely differently to those glasses described above. As shown in Figure 6.21, there is an optical barrier, but also a sufficient index increase in the guiding region to produce a non-tunnelling optical well. This type of profile does not require multi-energy implants to improve confinement, and is inherently capable of producing very low loss waveguides. Annealing shows two clear stages at $\sim 150\,°C$ and $\sim 350\,°C$. Losses as low as 0.2 dB/cm are reported after 300 °C annealing (Kakarantzas et al., 1993d). Lasing has been demonstrated in ion implanted doped lead germanate glass, including Tm, which at 1.9 μm gives the longest wavelength ion implanted waveguide laser yet reported (Shepherd et al., 1993).

In many heavy ion irradiated glasses, ionic motion (in particular that of the alkali ions) appears to be responsible for index changes. Mechanisms suggested include radiation enhanced diffusion, space charge fields

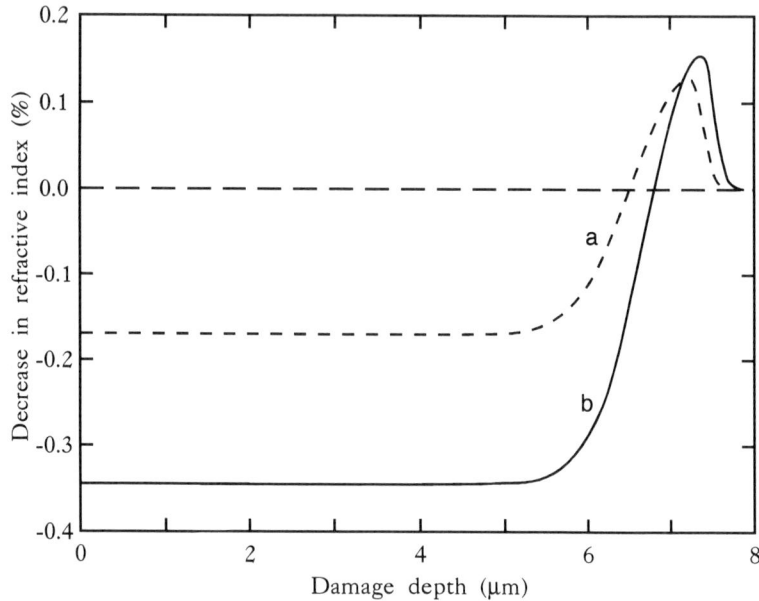

Fig. 6.21. Refractive index profiles of lead germanate glass implanted with 2.5 MeV He$^+$ dose 4×10^{16} ions/cm^2 at a, 300 K; and b, 77 K. Note the substantial non-tunnelling optical well.

and surface evaporation (Chinellato et al., 1982; Arnold, 1982; Mazzoldi, 1983; Bach et al., 1984). In soda lime glass light ion irradiation causes the sodium and calcium ions to move inwards from the surface region, but for heavier ions the sodium appears to be lost outwards from the surface. The index changes reported were a moderate $\Delta n \sim -5\%$, which is opposite to, and less extreme than, the earlier results of Bhattacharya et al. (1976), who implanted Pyrex and borosilicate glasses with gallium and argon at low doses ($\sim 10^{15}$ ions/cm^2) and found that the index increased in both materials, the latter going from 1.48 to 1.8 ($\Delta n \sim +20\%$). This was attributed to a new coordination of the silicate bonds. At first sight, it is difficult to understand the apparent contradiction of these results. The picture is obviously more complex than was thought initially. In an experiment which records the continuous dose dependence of the reflectivity across the spectrum, Hole et al. (1993) have obtained data for ion implanted pyrex which could resolve the apparent conflict. In practice, in situ measurements of colloid production using optical absorption or reflectivity changes indicate a highly complex dose/wavelength dependence, such that recording data at a single wavelength or at a se-

6.11 Non-crystalline materials

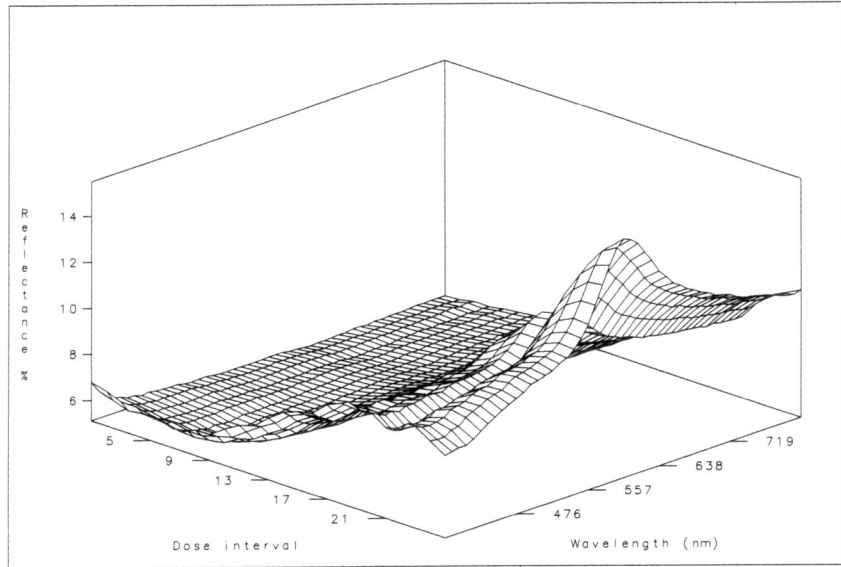

Fig. 6.22. Reflectivity of a silicate glass measured during 150 keV Cu$^+$ implantation. There are 24 scans and spectra are recorded at dose intervals of 5×10^{15} ions/cm^2. All the features were reproducible in separate implants.

lected implant dose is misleading. Examples of *in situ* plots of reflectivity versus ion dose were shown earlier in Figures 3.1 and 3.2. Figure 6.22 gives a further example, for Cu implanted into silicate glass. There is an initial dose range of minimal change followed by the growth of the Cu colloid peak, which in turn moves towards the red with increasing colloid size. Note, however, that this is not a smooth growth function but, rather, there are at least two reproducible minima. One possibility is that these represent collapse of the damaged glass to minimise stress and/or colloid precipitation. Finally, as pointed out by Arnold and Peercy (1980), some ion implanted glass layers restructure and result in the formation of crystalline material. Since this may lead to a higher refractive index, the implant can generate a waveguide. In an attempt to use this effect, Kakarantzas and Townsend (1993c) implanted a lithium silicate glass. Loss of Li is encouraged by the radiation damage, and so the implanted layer can be converted through the phase transition faster than the bulk material. In their sample, good guides were formed with losses $\sim$0.7 dB/cm at 633 nm.

6.12 Combination with conventional techniques

Examples have been reported of the advantageous employment of He^+ ion implantation into both Ti diffused and proton exchanged $LiNbO_3$ waveguides. In good quality Ti diffused guides, the small residual loss mechanisms have a high contribution factor from surface scattering and evanescent wave absorption, due to inherent roughness and contamination. It has therefore been suggested that very low energy ion implants could be used to produce an optical barrier at the surface in order to confine the mode deeper into the crystal (Heibei, 1982). After performing such an implant with He^+, it was reported that the guide losses were reduced from 0.14 to 0.07 dB/cm by this process (Drummond and Townsend, 1985).

In another example of combined technologies, ion implanted barriers have been used to 'split' a $LiNbO_3$ proton exchange waveguide into a double well structure (Glavas and Townsend, 1990). The two contributing mechanisms were found to be additive in the barrier region. More spectacular, perhaps, is the fabrication of an ion implanted optical barrier just beyond the confines of a proton exchange guide. Because the proton exchange enhances the substrate index by $\sim 7\%$ for the guiding region, and the implant reduces it by $\sim 5\%$ for the barrier, this produces an index differential of $\sim 12\%$ (Glavas and Townsend, 1990). Applications suggested for such a large Δn guide boundary are SHG phase matching at low wavelengths where high dispersion is a problem, or for the outside boundaries of sharp bends on stripe guides requiring a small radius of curvature.

6.13 Ion implanted chemical waveguides

Rather than produce waveguides by the radiation damage effects caused by light inert ions such as He^+, it is possible to implant chemically active dopants directly into the substrate, and form a new chemical compound composition in the surface layer. Silica has also been implanted with high doses of Si^+ (Heidemann, 1981, 1982) or N^+ (Webb et al., 1976a,b) to produce chemical effects due to new compound formation. In both cases the index is increased towards that of SiO_x, and SiO_zN_y respectively. The latter case has been investigated extensively (Naik, 1983; Faik et al., 1986) and good quality waveguides have been produced. Annealing to 450 °C removes all absorbing colour centres and leaves a stable chemical layer of the new silicon oxy-nitride compound.

However, the advantage of this method over the normal thermal diffusion or ion exchange methods is not immediately obvious. For instance the heavy mass of any such dopant ion (other than, possibly, C^+, N^+ or O^+) means that normal implantation energies (< 3 MeV) cannot achieve the ranges needed for optical dimensions (~ 1 wavelength). Therefore, subsequent to implantation, thermal diffusion is still required in order to obtain the necessary depth (Buchal et al., 1987). The real advantages of this implantation technique are due to the complexities of surface chemistry. For example, even in the highly developed technology of Ti diffusion into $LiNbO_3$ (Korotky and Alferness, 1987), problems can still arise from the formation of intermediate compounds of $Ti/LiNbO_3$ or because of diffusion inhibition due to TiO_2 layer growth at the surface. Attempts have therefore been made to overcome these surface problems by using a buried layer of implanted Ti as the source for the diffusion process (Ashley et al., 1989; Poker and Thomas, 1989; Bremer et al., 1990). Typically, 400 keV energies have been used with doses $\sim 5 \times 10^{16}$ ions/cm^2. One consequence of the implantation is to amorphise the surface layer of the $LiNbO_3$ and this must subsequently be regrown (Poker and Thomas, 1989). The final diffusion process is similar to that for conventional Ti-diffused waveguides. Titanium implantation has recently been performed through lithographic masks to produce stripe waveguides (Niehof et al., 1991).

An alternative technique has been to increase the index of a conventionally Ti-diffused guide by implanting excess Ti into the surface layer (Fouchet et al., 1987). In this case, the Ti was only located over a depth extending $\sim 0.3\,\mu$m from the surface, but moderate annealing temperatures were sufficient, and an excess index enhancement of more than 1% was achieved.

Although Ti implantation into $LiNbO_3$ followed by diffusion leads to a waveguide which is similar to that formed by Ti in-diffusion, it appears that the considerable radiation damage probably results in Li rearrangement. Rodman et al. (1993) explored this situation by first making a Ti doped guide in $LiNbO_3$ by implantation and annealing, and then adding a He^+ low index barrier beneath the Ti guide. The n_o index behaved as expected but the n_e index enhancement (Section 6.5) was almost 50% greater than for the undoped sample.

Erbium implantation has also been performed with the intention of producing localised waveguide laser doping (Brinkmann et al., 1990; Polman et al., 1990, 1991a,b), but with limited success. Er waveguide lasers are particularly attractive for use with optical fibre communication systems as the 1.54 emission matches a low loss optical window of germano-silicate fibres. Consequently, Er implants have been tried in a range of materials including silica, glass, quartz, Si, Si_3N_4, $LiNbO_3$ and ZBLAN glass. For example, $LiNbO_3$ was implanted with 10^{16}ions/cm^2 of Er at 200 keV (Brinkmann et al., 1990), and the sample was annealed at 1050 °C for four days to achieve epitaxial regrowth of the surface layer and to diffuse the Er into the substrate ($\sim 1.8\,\mu m$). However, a satisfactory waveguide was only produced by the subsequent in-diffusion of Ti, as with the conventional $LiNbO_3$ waveguide process. The example offers a useful approach in which the difficult diffusion problem of a reactive species, such as Er in an oxide, can be overcome by a co-diffusion approach.

The major problems with the implantation of Er in forming a waveguide laser, in all the attempts so far, appear to be:

(a) increasing the refractive index to define a guide;

(b) reducing absorption and scatter losses;

(c) regenerating the conventional Er photoluminescence spectrum;

(d) regaining the long fluorescence lifetime of the Er ($\sim$10ms);

(e) removing lattice damage;

(f) avoiding loss of Er into non-radiative imperfections such as colloids.

Problems (a)–(d) are generally solvable by normal furnace anneals, but after heating to the temperatures required to remove intrinsic defects, in, for instance, silica, the Er ions cease to give strong luminescence signals. The TEM images (e.g. Polman et al., 1991b) indicate that one reason for this is the formation of Er colloids. Additionally, not all the intrinsic damage is removed.

Whilst furnace annealing has not been successful in the case of Er implants, one may be encouraged by a recent attempt to modify the size distribution of Ag colloids by alternative treatments of furnace, rapid flame annealing and excimer laser pulses (Wood et al., 1993). The furnace and flame treatments merely converted the large Ag colloids into numerous small ones, but the laser pulses apparently totally destroyed optical evidence of colloids. If the same approach can be applied to Er colloids, then Er implanted laser waveguides may be feasible.

6.14 Summary of progress so far

The number of publications on ion implanted waveguides is as yet rather few, because the technique is only now begining to gain the interest which it justifies. Not least is the fact that the first example of an ion implanted waveguide laser in an insulator was published in 1989 (Chandler *et al.*, 1989b) and the first SHG implanted guide was reported in 1991 (Babsail *et al.*, 1991a). Since these first demonstrations the Nd:YAG threshold power has been reduced 100-fold and SHG efficiency increased by almost 10^6 in later examples. With the increasing demand for waveguide lasers and non-linear devices, more research into fabrication methods must follow. From the above summary of successes, it is apparent that ion implantation has a strong potential in this field, as its applicablity extends over a very wide range of materials, and it is able to produce device geometries which are not possible using alternative fabrication techniques.

References

Abouelleil, M.M. and Leonberger, F.J. (1989). *J. Amer. Ceram. Soc.*, **72**, 1311.
Al Chalabi, S.A.M., Weiss, B.L. and Homewood, K.P. (1987). *Nucl. Inst. Methods*, **B28**, 255.
Arnold, G.W. (1982). *Rad. Effects*, **65**, 17.
Arnold, G.W. and Peercy, P.S. (1980). *J. Non-Cryst. Solids*, **41**, 359
Arutyunyan, E.A., Galoyan, S.Kh., Il'in, V.V., Lebedev, L.S., Morozov, V.V. and Nazarova, V.Ya. (1985). *Sov. J. Quantum. Electron.*, **15**, 1168.
Ashley, P.R., Chang, W.S.C., Buchal, Ch. and Thomas, D.K. (1989). *IEEE J. Lightwave Tech.*, **7**, 855.
Babsail, L., Lifante, G. and Townsend, P.D. (1991a). *Appl. Phys. Lett.*, **59**, 384.
Babsail, L., Hamelin, N. and Townsend, P.D. (1991b). *Nucl. Inst. Methods*, **B59/60**, 1219.
Bach, H. and Hallwig, D. (1984). *Rad. Effects*, **81**, 129.
Bhattacharya, P.K., Sarma, N. and Wagh, A.G. (1976). *Pramana*, **6**, 102.
Bremer, T., Heiland, W., Hellermann, B., Hertel, P., Kratzig, E. and Kollewe, D. (1988). *Ferroelect. Lett.*, **9**, 11.
Bremer, T. (1989). *Mat. Res. Soc. Symp. Proc.*, **152**, 251.
Bremer, T., Heiland, W., Buchal, Ch., Irmischer, R. and Stritzker, B. (1990). *J. Appl. Phys.*, **67**, 1183.
Brinkmann, R., Buchal, Ch., Mohr, St., Sohler, W. and Suche, H. (1990). *Proc. Integr. Phot. Res. Conf.*, Hiltonhead, USA.
Brockelsby, W.S., Field, S.J., Hanna, D.C., Large, A.C., Lincoln, J.R., Shepherd, D.P., Tropper, A.C., Chandler, P.J., Townsend, P.D., Zhang, L., Feng, X. and Hu, Q. (1992). *J. Opt. Materials*, **1**, 177.
Buchal, Ch., Ashley, P.R. and Appleton, B.R. (1987). *J. Mater. Res.*, **2**, 222.
Bulmer, C.H., Eknoyan, O., Taylor, H.F., Greenblatt, A.S., Beech, L.A. and Neurgaonkar, R.R. (1986). *SPIE*, **704**, 208.
Carrascosa, M. and Agullo-Lopez, F. (1988). *Appl. Opt.*, **27**, 2851.
Chandler, P.J., Lama, F.L., Townsend, P.D. and Zhang, L. (1988). *Appl. Phys. Lett.*, **53**, 89.
Chandler, P.J., Glavas, E., Lama, F.L., Lax, S.E. and Townsend, P.D. (1986). *Rad. Effects*, **98**, 211.
Chandler, P.J., Zhang, L., Cabrera, J.M. and Townsend, P.D. (1989a). *Appl. Phys. Lett.*, **54**, 1287.
Chandler, P.J., Field, S.J., Hanna, D.C., Shepherd, D.P., Townsend, P.D., Tropper, A.C. and Zhang, L. (1989b). *Elect. Lett.*, **25**, 985.

References

Chandler, P.J., Zhang, L. and Townsend, P.D. (1990a). *Nucl. Inst. Methods*, **B46**, 69.
Chandler, P.J., Zhang, L. and Townsend, P.D. (1990b). *Elect. Lett.*, **26**, 332.
Chandler, P.J., Zhang, L. and Townsend, P.D. (1991). *Nucl. Inst. Methods*, **B59/60**, 1223.
Chinellato, V., Gottardi, V., LoRusso, S., Mazzoldi, P., Nicoletti, F. and Polato, P. (1982). *Rad. Effects*, **65**, 31.
Davis, G.M., Zhang, L., Chandler, P.J. and Townsend, P.D. (1993) *IEEE Photonics Technical Lett.*, **5**, 430.
Destefanis, G.L., Townsend, P.D. and Gailliard, J.P. (1978). *Appl. Phys. Lett.*, **32**, 293.
Destefanis, G.L., Gailliard, J.P., Ligeon, E., Townsend, P.D., Valette, S., Perez, A. and Farmery, B.W. (1979). *J. Appl. Phys.*, **50**, 7898.
Drummond, E. and Townsend, P.D. (1985). Plessey Internal Report (unpublished).
Faik, A.B., Chandler, P.J., Townsend, P.D. and Webb, R. (1986). *Rad. Effects*, **98**, 233.
Feng, X.Q., Zhu, Q.B., Chandler, P.J., Zhang, L. and Townsend, P.D. (1991). *Elect. Lett.*, **27**, 1504.
Field, S.J., Hanna, D.C., Shepherd, D.P., Tropper, A.C., Chandler, P.J., Townsend, P.D. and Zhang, L. (1990). *Elect. Lett.*, **26**, 1826.
Field, S.J., Hanna, D.C., Large, A.C., Shepherd, D.P., Tropper, A.C., Chandler, P.J., Townsend,P.D. and Zhang, L. (1991a). *OSA Proceedings of ASSL*, **10**, 353.
Field, S.J., Hanna, D.C., Large, A.C., Shepherd, D.P., Tropper, A.C., Chandler, P.J., Townsend, P.D. and Zhang, L. (1991b). *Opt. Comm.*, **86**, 161.
Field, S.J. Hanna, D.C., Shepherd, D.P., Tropper, A.C., Chandler, P.J., Townsend, P.D. and Zhang, L. (1991c). *Opt. Lett.*, **16**, 481.
Field, S.J., Hanna, D.C., Large, A.C., Shepherd, D.P., Tropper, A.C., Chandler, P.J., Townsend, P.D. and Zhang, L. (1991d). *Elect. Lett.*, **27**, 2375.
Fleuster, M., Buchal, Ch., Fluck, D. and Gunter, P. (1993). *Nucl. Inst. Methods*, **B80/81**, 1150.
Fluck, D., Irmscher, R., Buchal, Ch. and Gunter, P. (1991). *Appl. Phys. Lett.*, **59**, 3213.
Fluck, D., Gunter, P., Fleuster, M. and Buchal, Ch. (1992a). *J. Appl. Phys.*, **72**, 1671.
Fluck, D., Irmscher, R., Buchal, Ch. and Gunter, P. (1992b). *Ferroelectrics*, **79**, 128.
Fluck, D., Binder, B., Kupfer, M., Looser, H., Buchal, Ch. and Gunter, P. (1992c). *Opt. Comm.*, **90**, 304.
Fluck, D., Moll, J., Gunter, P., Fluster, M. and Buchal, Ch. (1992d). *Elect. Lett.*, **28**, 1092.
Fouchet, S., Carenco, A. and Duhamel, N. (1987). *Proc. 4th European Conf. on Integrated Optics*, Glasgow May 1987, p.54.
Gevay, G. (1987). *Prog. Cryst. Growth and Characterisation*, **15**, 145.
Glavas, E., Townsend, P.D., Droungas, G., Dorey, M., Wong, K.K. and Allen, L. (1987). *Elect. Lett.*, **23**, 73.
Glavas, E., Zhang, L., Chandler, P.J. and Townsend, P.D. (1988). *Nucl. Inst. Methods*, **B32**, 45.
Glavas, E., Cabrera, J.M. and Townsend, P.D. (1989). *J. Phys.*, **D22**, 611.
Glavas, E. and Townsend, P.D. (1990). *Nucl. Inst. Methods*, **B46**, 156.

Gotz, G. and Karge, H. (1983). *Nucl. Inst. Methods*, **209/210**, 1079.
Hamelin, N., Chandler, P.J. and Townsend, P.D. (1992). *Phys. Stat. Sol.*, **(b)134**, 557.
Haycock, P.W. (1989) (private communication).
Heibei, J. and Voges, E. (1982). *IEEE J. Quant. Elect.*, **18**, 820.
Heidemann, K.F. (1981). *Phil. Mag.*, **B44**, 465.
Heidemann, K.F. (1982). *Rad. Effects*, **61**, 235.
Hertel, P. and Menzler, H.P. (1987). *Appl. Phys.*, **B44**, 75.
Hines, R.L. and Arndt, R. (1960). *Phys. Rev.*, **119**, 623.
Hole, D.E., Barton, J., Townsend, P.D. and Wood, R.A. (1993), to be published.
Hopkins, M.M. and Miller, A. (1974). *Appl. Phys. Lett.*, **25**, 47.
Jackel, J.L. and Rice, C.E. (1981). *Ferroelectrics*, **38**, 801.
Jackel, J.L., Rice, C.E. and Veselka, J.J. (1982). *Appl. Phys. Lett.*, **41**, 607.
Jimenez de Castro, M., Mahdavi, S.M., Townsend, P.D. and Alvarez Rivas, J.L. (1991). *Rad. Effects*, **118**, 137.
Kakarantzas, G., Zhang, L. and Townsend, P.D. (1992). *E-MRS Symp. Proc.*, **29**, 73.
Kakarantzas, G. and Townsend, P.D. (1993a). Proc. ICDIM 92, *Nucl. Inst. Methods*, to be published.
Kakarantzas, G. and Townsend, P.D. (1993b). *Topical Issues in Glass*, **1**, 79.
Kakarantzas, G. and Townsend, P.D. (1993c). to be published.
Kakarantzas, G., Townsend, P.D. and Wang, J. (1993d). *Elect. Lett.*, **29**, 489.
Korotky, S.K. and Alferness, R.C. (1987). *Integrated Optical Circuits and Components*. Chap. 6. L.D. Hutcheson (ed.) (Marcel Dekker, New York).
Lama, F.L. and Chandler, P.J. (1988). *J. Mod. Opt.*, **35**, 1565.
Lifante, G. and Townsend, P.D. (1992). *J. Mod. Opt.*, **39**, 1353.
Lifante, G. (1991) (private communication).
Mahdavi, S.M., Chandler, P.J. and Townsend, P.D. (1989). *J. Phys.*, **D22**, 1354.
Mahdavi, S.M. and Townsend, P.D. (1990a). *J. Chem. Soc. Faraday Trans.*, **86**, 1287.
Mahdavi, S.M. and Townsend, P.D. (1990b). *Elect. Lett.*, **26**, 371.
Mahdavi, S.M. (1991). D. Phil. Thesis, Sussex (unpublished).
Mazzoldi, P. (1983). *Nucl. Inst. Methods*, **209/210**, 1089.
Moretti, P., Thevenard, P., Godefroy, G., Sommerfeldt, R., Hertel, P. and Kratzig, E. (1990). *Phys. Stat. Sol.*, **(a)117**, K85.
Moretti, P., Thevenard, P., Sommerfeldt, R. and Godefroy, G. (1991). *Nucl. Inst. Methods*, **B59/60**, 1228.
Naik, I.K. (1983). *Appl. Phys. Lett.*, **43**, 519.
Niehof, A., Renner, S., Buchal, Ch. and Heiland, W. (1991). *Nucl. Inst. Methods*, **B59/60**, 1355.
Pitt, C.W., Skinner, J.D. and Townsend, P.D. (1984). *Elect. Lett.*, **20**, 4.
Poker, D.B. and Thomas, D.K. (1989). *J. Mater. Res.*, **4**, 412.
Polman, A., Lidgard, A., Jacobson, D.C., Becker, P.C., Kistler, R.C., Blonder, G.E. and Poate, J.M. (1990). *Appl. Phys. Lett.*, **57**, 2859.
Polman, A., Jacobson, D.C., Lidgard, A., Poate, J.M. and Arnold, G.W. (1991a). *Nucl. Inst. Methods*, **B59/60**, 1313.
Polman, A,. Jacobson, D.C., Eaglesham, D.J., Kistler, R.C. and Poate, J.M. (1991b). *J. Appl. Phys.*, **70**, 3778.
Primak, W. (1958). *Phys. Rev.*, **B110**, 1240.
Properties of $LiNbO_3$ (1989). INSPEC Data Review Series 5 (IEE, London).

References

Red'ko, V., Shteingart, L.M.,Soroka, V.I., Artsimovich, M.V. and Mal'ko, A.I. (1981). *Sov. Tech. Phys. Lett.*, **17**, 399.
Reed, G.T. and Weiss, B.L. (1987). *Nucl. Inst. Methods*, **B19/20**, 907.
Reed, G.T. and Weiss, B.L. (1989). In *Properties of LiNbO$_3$*, Chap. 8, INSPEC Data Review Series 5 (IEE, London).
Rodman, M., Chandler, P.J., Zhang, L., Sharp, J.H., Townsend, P.D. and Ferguson, A.I. (1993). *Vacuum*, **44**, 281.
Schineller, E.R., Flam, R.P. and Wilmot, D.W. (1968). *J. Opt. Soc. Am.*, **58**, 1171.
Shepherd, D.P., Brinck, D.J.B., Hanna, D.C., Payne, D.N., Wang, J., Tropper, A.C., Kakarantzas, G. and Townsend, P.D. (1993). *Proc. QE-11*, in press.
Strohkendl, F.P., Gunter, P., Buchal, Ch. and Irmscher, R. (1991a). *J. Appl. Phys.*, **69**, 84.
Strohkendl, F. P., Fluck, D., Buchal, Ch., Irmscher, R. and Gunter, P. (1991b). *Appl. Phys. Lett.*, **59**, 3354.
Townsend, P.D. (1976). *Inst. of Physics, Conf. Series*, **28**, Chap.3.
Townsend, P.D. (1985). *Induced Defects in Insulators*, 207. MRS (Les Editions de Physique, Les Ulis).
Townsend, P.D. (1987). *Rept. Prog. in Phys.*, **50**, 501.
Townsend, P.D., Chandler, P.J., Wood, R.A., Zhang, L., McCallum, J. and McHargue, C.W. (1990). *Elect. Lett.*, **26**, 1193.
Vainos, N.A., Clapham, S.L. and Eason, R.W. (1989). *Appl. Opt.*, **28**, 4381.
Webb, A.P., Allen, L., Edgar, B.R., Houghton, A.J., Townsend, P.D. and Pitt, C.W. (1975). *J. Phys.*, **D8**, 1567.
Webb, A.P., Houghton, A.J. and Townsend, P.D. (1976a). *Rad. Effects*, **30**, 177.
Webb, A.P. and Townsend, P.D. (1976b). *J. Phys.*, **D9**, 1343.
Weber, M.J. and Monchamp, R.R. (1973). *J. Appl. Phys.*, **44**, 5495.
Wei, D.T.Y., Lee, W.W. and Bloom,L.R. (1974). *Appl. Phys. Lett.*, **25**, 329.
Weiss, B.L., Ahmad, C.N. and Gwilliam, R.M. (1987). *Materials Lett.*, **5**, 193.
Wenzlik, K., Heibei, J. and Voges, E. (1980). *Phys. Stat. Sol.*, **(a)61**, K207.
White, C.W., McHargue, C.J., Sklad, P.S., Boatner, L.A. and Farlow, G.C. (1988). *Mater. Science Rept.*, **4**, 41.
Wong, K.K. (1989). In *Properties of LiNbO$_3$*, Chap. 8, INSPEC Data Review Series 5 (IEE, London).
Wong, J.Y.C., Zhang, L., Kakarantzas, G., Townsend, P.D., Chandler, P.J. and Boatner, L.A. (1992). *J. Appl. Phys.*, **71**, 49.
Wood, R.A., Townsend, P.D., Skelland, N.D., Hole, D.E., Barton, J. and Afonso, C.N. (1993). *J. Appl. Phys.*, **74**, 5754.
Youden, K.E., James, S.W., Eason, R.W., Chandler, P.J., Zhang, L., Townsend, P. D. and Anderson, D.Z. (1992). *Opt. Lett.*, **17**, 1509.
Zhang, L., Chandler, P.J. and Townsend, P.D. (1988). *Appl. Phys. Lett.*, **53**, 544.
Zhang, L., Chandler, P.J., Townsend, P.D. and Lama, F.L. (1989). *NATO-ASI Series E*, **170**, 371.
Zhang, L. (1990). D. Phil. Thesis, Sussex, (unpublished).
Zhang, L., Chandler, P.J. and Townsend, P.D. (1990). *Ferroelect. Lett.*, **11**, 89.
Zhang, L., Chandler, P.J., and Townsend, P.D. (1991a). *Nucl. Inst. Methods*, **B59/60**, 1147.
Zhang, L., Chandler, P.J. and Townsend, P.D. (1991b). *Appl. Phys. Lett.*, **70**, 1185.
Zhang, L., Chandler, P.J., Townsend, P.D., Field, S.J., Hanna, D.C., Shepherd, D.P. and Tropper, A.C. (1991c). *J. Appl. Phys.*, **69**, 3440.

Zhang, L., Chandler, P.J., Townsend, P.D. and Thomas, P.A. (1992a). *Elect. Lett.*, **28**, 650.
Zhang, L., Chandler, P.J., Townsend, P.D., Alwahabi, Z.T. and McCaffery, A.J. (1992b). *Elect. Lett.*, **28**, 1478.
Zhang, L., Chandler, P.J., Townsend, P.D., Alwahabi, Z.T., Pityana, S.L. and McCaffery, A.J. (1993). *J. Appl. Phys.*, **73**, 2695.

7
Applications of ion implanted waveguides

Over the past decade ion implantation has been demonstrated as a suitable technique for the fabrication of waveguide structures in an ever increasing number of optical materials. The stage has now been reached where actual device-oriented structures are being designed and implemented using this technology. This chapter will deal with the progress which is being made in the development of useful devices, and how ion implantation is being used to achieve these ends.

Many authors have recognised the advantages of waveguide structures for signal processing, for coupling to optical fibres or for frequency conversion. Optical circuitry has been proposed which, at least initially, was purely hypothetical, but has advanced to include ideas of entirely solid state waveguide structures (sometimes termed holosteric systems). Forseeable objectives include multi-frequency, or tunable, compact lasers in which the pump power is provided by a semiconductor diode. This inherently waveguide power source could then be matched into other waveguides for a combination of frequency conversion, SHG, pumping a tunable laser waveguide such as alexandrite or Ti:sapphire, or driving an optical parametric oscillator (OPO). The conversion efficiencies in the various stages are often as high as 30%, hence even in a several-stage process a 100 W semiconductor array could result in a few watts of tunable laser power. Such systems could of course supersede the highly inefficient Ar or Kr gas laser sources for many applications. The concepts are clear, and this final chapter will indicate how close the components are to realisation when using ion implantation fabrication routes.

7.1 Waveguide construction techniques

Planar waveguides may have some practical applications (e.g. photorefractive holography) but in general it is necessary to produce them in the form of channels, either isolated or interacting with electric fields or adjacent waveguides. This may be achieved by a masking technique (similar to that developed in the integrated electronics industry), or, as will be explained later, by the implementation of effects inherent in the ion implantation process.

7.1.1 Channel waveguides

For most integrated optical devices, a channel structure is necessary for the waveguides. Ion implantation can be used in conjunction with a mask-making stage to produce channel waveguides. The mask is made by first depositing a thin protective and/or conductive layer of photoresist, metal, or oxide on to the substrate surface. Using standard photolithographic procedures, the channel region of the waveguide can be defined by a metal mask which is negative or positive, respectively, for an optical barrier type or non-tunnelling well type of waveguide. In the former case, the guiding regions (2–20 μm wide) are protected by the metal mask while the side-walls ($\sim 50\,\mu$m) are being built up using a series of successive low energy implants (Figure 7.1(a)). The channel waveguides can then finally be produced after removing the metal masks with chemical etching or a surface polish, by subsequently implanting the whole surface using the standard conditions for a planar waveguide. In the second type of waveguide, only one implantation stage is needed. The side-wall regions are protected by the metal mask while the sample surface is being implanted (Figure 7.1(b)). The refractive index increases in these implanted regions, which will therefore be surrounded by the relatively low index of the substrate.

Ion implanted channel waveguides were first reported by Kersten and Boroffka (1976) for a seven-channel distribution line in silica. For an active substrate the first example was in $LiNbO_3$ (Reed and Weiss, 1986). This was a barrier confined guide and so the first method described above was employed. Firstly, planar waveguides were produced in Y-cut X-propagating $LiNbO_3$ using 1.5 MeV He^+ at a dose of 10^{16}ions/cm^2. A thin metal layer of titanium (for adhesion) and gold was then evaporated on to the sample surface to form a conducting film. This was followed by the deposition of a thick photoresist layer (several microns), which was

7.1 Waveguide construction techniques

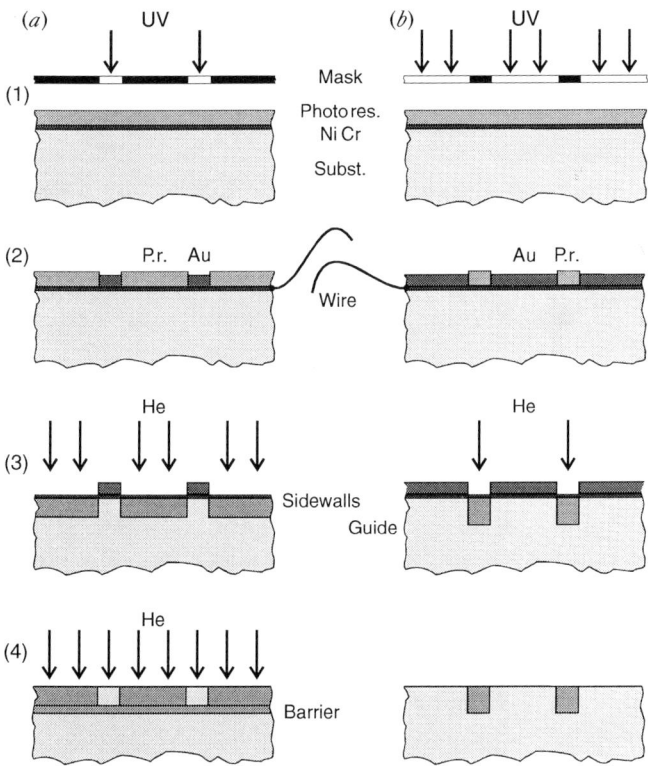

Fig. 7.1. (a) Negative masking for an optical barrier waveguide; (b) positive masking for a non-tunnelling waveguide. (1) Exposure of photoresist to ultraviolet through mask template; (2) gold plating into etched regions; (3) removal of photoresist and implantation of sidewalls (type a) or guiding regions (type b); (4) removal of metal to leave channel guides (after final implantation of barrier layer for type a).

patterned by contact photolithography using a mask containing 5-μm-wide stripes. The gold film which was used to protect the guiding regions from the ion implantation was electroplated on to the sample surface through the windows in the photoresist, using the thin evaporated Ti/Au metal layer as an electrode. The photoresist was then removed, and the remaining structure was subjected to ion beam milling to remove the thin metal layers covering the $LiNbO_3$ substrate (side-wall regions). This process produced a series of 5-μm-wide gold stripes on the $LiNbO_3$ surface, which acted as masks to protect the guiding regions of the channels during the implantation to form the side walls. Multi-energy implantation was carried out at a variety of energies (850 keV and 500 keV; 850 keV, 500 keV and 200 keV; 1.2 MeV, 850 keV and 500 keV), in order to assess the suitability of different 'side-wall' configurations. Finally, the samples were annealed in flowing oxygen at 200 °C to reduce absorption loss. The insertion loss results showed that the choice of side-wall implants had a considerable effect on the propagation loss of the resulting waveguides. To a certain extent, the side walls can introduce additional loss due to scattering from the irregular interfaces. The best conditions were found to be achieved by using two side-wall implant energies of 850 keV and 500 keV, each with a dose of 5×10^{15} ions/cm^2. The propagation loss for these guides was ~ 1 dB/cm at 0.6328 μm (Reed and Weiss, 1987).

Recently, channel waveguide lasers have been successfully formed by ion implantation in Nd doped YAG and GGG (Field et al., 1991a,1992). Unlike LiNbO$_3$, these are not barrier-confined guides, but have an index enhancement in the guiding region. The second type of masking is therefore applicable, requiring a positive mask which produces protective gold stripes covering the side-wall regions rather than the guide itself. Only one implantation stage is then required. The channel waveguide masks were made in a similar way to that described above for the LiNbO$_3$ waveguides by Reed and Weiss, except that the initial adhesive metal layer was thin enough ($<$ 100 nm) to allow implantation through it without the need for its prior removable by ion beam milling. The lasing performance in these channel guides showed a great degree of improvement over the planar waveguides. The detailed results will be discussed in Section 7.2. However, these achievements clearly show that ion implantation is a useful technique for producing channel waveguide structures in integrated optical devices. It should be realised that ion implantation can also be used in combination with other planar waveguide fabrication methods to form channel waveguides. This could be achieved either by implanting side-walls in the 'conventional' planar guide, as explained above, or by using the enhanced etch rate produced in most materials by ion implantation in order to construct ridge channels.

A quick method of forming masks for ion implanted channel waveguides was reported by Gunter's group in KNbO$_3$, for their frequency doubling experiments (Fluck et al., 1992b). Sets of parallel SiO$_2$ fibres (8 μm) or tungsten wires (13 μm) were employed as a simple mask having sufficient thickness to shield the channels of the planar waveguide from the implantation used to form the side-walls (Figure 7.2). This method avoids time consuming lithography, liftoff and additional electroplating processes, but is only suitable for the first type of waveguide profile

7.1 Waveguide construction techniques

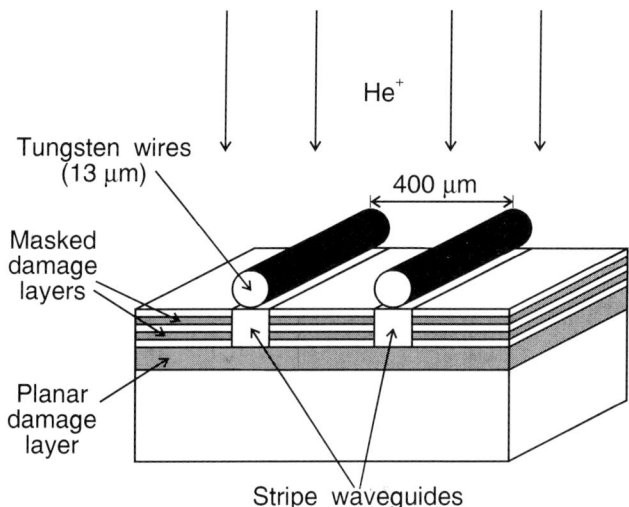

Fig. 7.2. Creation of optical channel waveguides by side-wall implantation using a tungsten wire masking technique.

(barrier-confined). The side-walls appear to be adequately smooth and vertical, but the performance of these channel waveguides has not been discussed quantitatively in comparison with other channel waveguides. Nevertheless, this could be a viable alternative way to produce channel waveguides for some non-critical applications.

7.1.2 Optical writing

An optically written channel waveguide was observed rather unexpectedly in ion implanted $Bi_4Ge_3O_{12}$ (BGO). This novel effect first occurred while lasing action was being assessed in a He^+ implanted Nd:BGO planar waveguide (Brocklesby et al., 1992). This effect was observed and repeated many times in both doped and undoped ion implanted BGO crystals. Although this is a very preliminary result, the ability to optically write channel waveguides may have a significant impact on the construction of complex waveguide circuits.

The Nd:BGO crystals which first showed this effect had been implanted at 77 K with 4×10^{16} He^+ ions/cm^2 at an energy of 1.5 MeV, and subsequently annealed for 30 minutes at 200 °C. These implant conditions were selected to be appropriate for 1.064 µm operation of a Nd:BGO waveguide laser. The formation of a channel waveguide was observed,

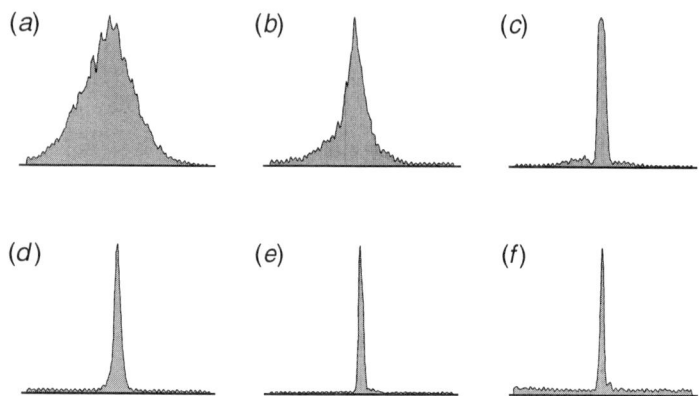

Fig. 7.3. Optically written channel waveguide output profiles in $Bi_4Ge_3O_{12}$ as a function of pump power: (a) 30 mW, (b) 55 mW, (c) 75 mW, (d) 95 mW, (e) 130 mW, (f) 350 mW.

for the first time, when an argon laser was end-coupled into one such planar waveguide. At low pumping powers (≤ 30 mW) the output spot profile appears as in Figure 7.3(a), showing the expected wide Gaussian shape resulting from the beam expansion in the unguided horizontal plane of the planar waveguide. Figure 7.3(b)–(f) shows how the output profile varies with increasing power. Between 30 and 75 mW there is a very dramatic narrowing of the profile in the horizontal plane. Beyond this point it shows a much slower narrowing up to powers of 350 mW. It seems that some self-focusing effect has taken place, effectively forming a channel waveguide with an output spot width of $\sim 10\,\mu$m. The spot size in the vertical plane is unchanged. The index change causing this self-focusing appears to be permanent. Similar effects have also been observed in undoped ion implanted BGO waveguides, but no effect has occurred in the bulk crystal (doped or undoped) up to pump powers of 3 W.

The observed effects were not consistent with photorefractive damage (which can occur for certain phases of the Bi_2O_3—GeO_2 system). To confirm this the channel waveguide was illuminated with a strong white light source for one hour. There was no observed bleaching effect upon the output profile, indicating that photorefractive damage is not the cause of the channel formation. As the argon pump light is strongly absorbed by the doped BGO crystal it is possible that the thermal load causes the index change. This is consistent with the observation that there was a much higher threshold for channel formation when pumping

7.1 Waveguide construction techniques

with an R6G dye laser tuned just off the strong 590 nm Nd absorption ($\sim$ 200 mW) than on it ($\sim$ 50 mW). However, this does not explain the occurrence of the effect in the undoped crystal.

The transmission of the channel waveguide was measured and compared with that of the planar guide. At low powers the ordinary planar waveguide transmission was 60%. At powers high enough to form channels the transmission was 50% and virtually all the power was in the narrowed profile. Launching low powers into the already formed channel guide still gave overall transmission of 50% but now only half the power was contained within the narrowed profile, the rest being contained in the side wings.

Some initial trials were also carried out to write channels with a beam focused on the top surface of the guide. By tracking the planar waveguide across the path of an argon-ion beam it is possible to write 'lines' which are visible under a microscope. However, end-launching of light into these top written lines has not yet been achieved. An investigation is needed to establish the appropriate pump power and spot size to create guides capable of supporting waveguide modes at the required wavelengths. It is encouraging that, if this is achievable, it may lead to a relatively simple method for writing complicated waveguide circuits. This may also have applications in integrated optics, especially as BGO is an electro-optic crystal. This method of channel waveguide fabrication may also lead to enhanced performance of rare-earth doped ion implanted BGO waveguide lasers and amplifiers.

7.1.3 Double barrier implants

A major advantage of ion implanted waveguides is that by a suitable choice of ion energy an optical barrier may be positioned at any depth beneath the surface. It follows that it should be possible for two barriers to be constructed at different depths. This process may be used to improve the isolation of a single guide by burying it away from the surface, or, alternatively, two or more guides may be superposed. Such devices may be used simply to carry one guide beneath another (i.e. to act as an interconnect) or else to form guides which are optically coupled in the vertical direction. In combination with lateral proximity by means of masking, a three-dimensional matrix of guides might thus be produced.

An example of an ion implanted double well structure has been demonstrated in quartz (Chandler *et al.*, 1988). The index profile of the double-barrier implant was determined by analysing its complex dark mode

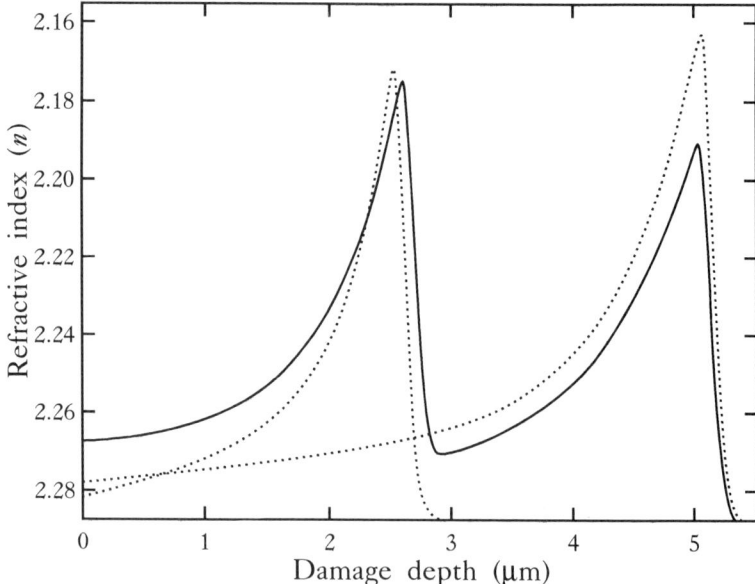

Fig. 7.4. Individual barrier profiles due to low and high energy implants in LiNbO$_3$, compared with that of the composite implant, measured at 633 nm.

spectrum using the reflectivity calculation described earlier (Chandler *et al.*, 1986). This profile was found within experimental error to match with the summation of those of the two individual implants. It should therefore be possible to design composite profiles, which might, for example, produce suitable mode matching to enable tuned coupling between the two guides.

Double well structures have also been produced in LiNbO$_3$ (Chandler *et al.*, 1989). The n_o profiles for the individual and the superposed barriers are shown in Figure 7.4. Here both implants were at room temperature and the deeper one (5 μm) was performed first. It appears that a direct summation of the damage effects has been obtained except for two minor discrepancies. Firstly, the 2.5 μm peak height does not increase above 5%, but this is to be expected because of saturation effects. Secondly, this peak in the double structure shows a small increase in ion range, because it is superposed on an initial high energy implant profile which has reduced the effective density of the material. Thirdly, the 5 μm peak which is initially 5%, is reduced to 4.5% after the second implant. This might suggest that the 2.5 μm peak has an extended influence up to a depth of 5 μm, possibly due to radiation enhanced diffusion of, for instance, lithium. However, a simple thermal annealing mechanism during the second implant may be all that is responsible. It was discovered that this effect was removed when the second implant was performed at 77 K. Despite these minor complications, it can be seen that the superposition of waveguides in LiNbO$_3$ is quite possible.

Potential applications of these coupled waveguides include optoelectronic modulators and switches. However, their use has also been considered for the improvement of second harmonic generation efficiency. In SHG the two criteria which need to be satisfied for the fundamental and harmonic waves are, firstly, their phase-matching, and, secondly, the maximization of their modal field overlap integral. It has been suggested that in a double structure such as the one described here, the longer wavelength fundamental is controlled by the overall dimensions of the double well, whereas the shorter wavelength harmonic is more tightly confined to the individual wells. Therefore, by the correct choice of dimensions for the wells and the barrier, it is possible to obtain phase-matching without unnecessary cancellation of the overlap integral. An improvement of 40% in the efficiency has been predicted (Lifante and Townsend, 1992).

A superior structure with an improvement factor of 36 has been proposed (Lifante, 1992, private communication). In this case one-half of the waveguide profile is constructed of a non-linear material, and the other half is 'modified' to lose its non-linearity, but to retain the same refractive index. In principle this could be achieved in crystalline quartz by the implantation of nitrogen. The decrease in index due to the amorphisation effect of the damage can be compensated for by the index enhancement due to the chemical effect of the nitrogen in forming silicon oxy-nitride (Faik *et al.*, 1982; Naik, 1983). It is proposed that this nitrogen implant would produce a linear section of the guide profile which has the same index as the undamaged non-linear crystalline quartz section. In such a double structure the overlap integral would only be effective within the non-linear section and it can be shown to have a theoretically large improvement over that of the single-well value. The method has been attempted, but not yet published.

7.2 Ion implanted waveguide lasers

The fabrication of lasers in the form of waveguides potentially gives a very high gain and a very low threshold, because of the high sustained pump and lasing power densities attainable along the entire cavity (guide) length. This has been demonstrated very successfully for glass lasers in the form of doped optical fibres. However, crystal laser hosts possess many useful properties which are not present in glasses. These include higher electro-optic and non-linear coefficients, a greater selection of possible laser transitions, and a very wide range of linewidths (from the extremes of Nd:YAG to Ti:Al$_2$O$_3$). The possibility can be envisaged of integrating a low threshold diode-pumped waveguide laser with broad band tunability, on to an electro-optic single chip. The multitude of applications which can be anticipated for such a device provide high motivation for investigating the prospects of laser waveguide device fabrication in these crystalline hosts.

In the development of a successful lasing device, two objectives must be achieved – firstly population inversion, and secondly an overall net loop gain which is greater than unity. If the technique used for waveguide fabrication has a degrading influence on the substrate, it is possible that one or both of these requirements may not be satisfied. Population inversion depends on the power density of the pump signal, and this is why a waveguide geometry is advantageous, because the light is confined to a focused beam over the entire cavity length. A channel waveguide is clearly superior in this respect as the pump is confined in two dimensions, and provided that no further degradation is caused by the processing (e.g. side-wall scattering or absorption of the evanescent field), a channel structure can improve the threshold over that of a planar guide by one or two orders of magnitude (depending on the guide length). Population inversion may suffer, however, if the fabrication technique introduces alternative decay routes (e.g. by stress). This would reduce the excited state lifetime and thereby require a higher pump threshold to achieve inversion. It is therefore important to maintain the lattice integrity, ideally by not introducing impurities or distortion. For this reason low dose implantation is preferable.

Even when inversion has been achieved, it is still necessary for the gain produced by stimulated emission per optical loop to exceed the total loss of the system. This includes cavity losses such as mirror reflection, coupling, etc., but the guide loss itself (absorption, scattering and tunnelling) may dominate, especially if long crystals are used. For this reason, laser waveguides must be produced with sufficiently low attenuation at the lasing wavelength. For these longer wavelengths, poor confinement and even barrier tunnelling can dominate the attenuation effects.

Only when the two criteria discussed above are satisfied, can lasing

7.2 Ion implanted waveguide lasers

threshold be achieved. Waveguides with bad crystal lattice degradation or high optical losses will have poor lasing thresholds and low conversion (slope) efficiencies. An idea of the expected pump threshold power is given (Clarkson and Hanna, 1989) by the equation:

$$P_{th} = \frac{\pi h v_p A L}{2\sigma_e \eta_p \tau_f} \qquad (7.1)$$

where P_{th} is the absorbed pump power threshold, v_p the pump frequency, σ_e the stimulated emission cross-section, η_p the pump quantum efficiency (number of ions excited to the upper laser level per absorbed pump photon), τ_f the fluorescence lifetime, L the single-pass waveguide loss, and A is a function which increases with the pump and lasing mean spot areas, given by:

$$A^2 = (X_p^2 + X_l^2)(Y_p^2 + Y_l^2) \qquad (7.2)$$

where X and Y are the $1/e^2$ half-widths of the beams in the two dimensions of the guide. As can be seen from Equation (7.1), the confinement of a waveguide is an important factor in reducing the threshold, but any reduction in the emission cross-section or fluorescence lifetime will have the opposite effect.

Once lasing has been achieved for a particular device, the threshold and slope efficiencies can be measured experimentally, to give an indication of performance. Prior to this, however, a comparison of the waveguide fluorescence spectrum with that of the bulk will show any reduction in emission cross-section, and similarly the fluorescence lifetime can be determined from a pulsed excitation measurement.

Table 7.1 presents a summary of laser materials in which the Sussex/Southampton group have so far succesfully produced waveguide lasers, together with their waveguiding and lasing performances. This represents the state of the art up to April 1993. As can be seen, success has been achieved with both barrier-confined and also index-enhanced profiles. The former generally need multi-energy implants for optimum barrier width, but the increased dose does not seem to present a major problem. The longer wavelength examples have required high energy implants and confinement has sometimes become a problem. For example, Tm:MgO:LiNbO$_3$ offers the prospect of lasing at 1.853 μm, thus requiring a high depth (energy) implant in order to have a long cut-off wavelength. This has been successfully achieved with end-launched light showing good transmissions up to $\lambda \sim 2\,\mu$m. Tm doped lead germanate glass, however, did not have sufficient index enhancement to confine

Table 7.1. Summary of ion implanted waveguide lasers

Material	Loss (dB/cm)	Length (mm)	Lasing λ (μm)	Pump λ (nm)	Threshold (mW)	Slope effic. (%)
PLANAR:						
Nd:YAG	1.2	4	1.062,1.064	590	10.5	21
				807	8.5	—
Nd:YAP	10	0.66	1.080,1.073	590	~50	—
Nd:LN	1	8	1.085	814	17	—
Nd:GGG	<1	2.5	1.062	588	12.5	30
				805	17	—
Nd:BGO	1		1.064	590	~50	—
				810	—	—
Yb:YAG	~2	1.7	1.03	941	30	—
Tm:LG/Glass	0.2		1.85	790	70	—
CHANNEL:						
Nd:YAG	1.5	2.5	1.064	807	0.5	29
Nd:GGG	~1.6	2.5	1.062	805	1.9	27

7.2 Ion implanted waveguide lasers

a 1.9 μm mode when implanted with 3 MeV ^{4}He$^+$. This problem was overcome by the use of ^{3}He$^+$ which gave a 30% increased ion range (Shepherd et al., 1993a). In most of the substrates investigated, the presence of the laser dopant appears to have had negligible effect on the waveguide characteristics (Δn and loss), but this may be due to their low concentration (< 1 at.%). A clear exception was found in the case of GGG where pure and 2.7% Nd doping gave low loss guides at 1.06 μm, but 3.35% Nd with an identical implant had too low a cut-off for mode confinement at that wavelength (Field et al., 1991b). It is important to realise that Table 7.1 merely records early examples from one research team, and does not imply that these numbers represent the optimum potential performance.

7.2.1 Spectroscopic effects

Given the disruptive nature of the ion implantation process it is important to establish whether the spectroscopy of the active ions in the waveguide region is significantly altered from that of the bulk, since this could have a deleterious effect on the lasing threshold, or even the actual lasing transition. In all the Nd doped samples it was found that both the absorption and emission spectra were broadened to some extent by ion implantation. No effect has been observed on the fluorescence lifetimes of the final (annealed) waveguides, but prior to annealing, while defects are still present in the guiding region, a slight reduction in the lifetimes has been noticed. In the case of Nd:YAG a more extensive study was performed to investigate the dependence of broadening on ion dose (Field et al., 1991c). Figure 7.5 shows how the 1.06 μm spectral region deteriorates for He$^+$ doses up to 7×10^{16} ions/cm^2, such that by the maximum dose, lasing occurs at either of two possible peak wavelengths. For the index profile with optimum mode confinement and attenuation (highest dose) the emission cross-section is reduced by 50%. However, it was shown that the lowest dose profile with 89% of the bulk cross-section value was just able to produce an adequately confined mode to give lasing action. When broader emission spectra have been examined (e.g. Tm:MgO:LiNbO$_3$), it has been found that they show negligible broadening compared with the bulk.

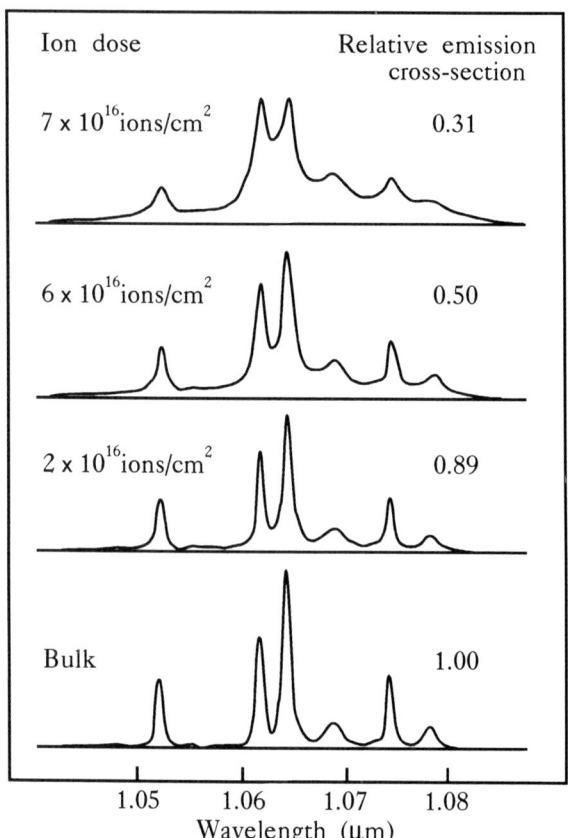

Fig. 7.5. Nd:YAG waveguide fluorescence spectra as a function of ion dose.

7.2.2 Planar waveguide laser performance

Figure 7.6 shows the experimental arrangement normally used to measure waveguide lasing threshold. The mirrors were initially held in micropositioners close to the waveguide end-faces, but it was subsequently found more satisfactory to use light, thin mirrors held on the ends of the sample by the surface tension of a thin layer of liquid. The use of non-permanent mirrors allows them to be changed for output coupling measurements, but optimum laser performance is finally obtained by the application of dielectric coatings directly on to the crystal end-faces.

The first ion implanted waveguide laser was demonstrated in a 10 mm long Nd:YAG unannealed planar waveguide with a reported lasing threshold of 50 mW and estimated slope efficiency of only 1.7% (Chan-

7.2 Ion implanted waveguide lasers

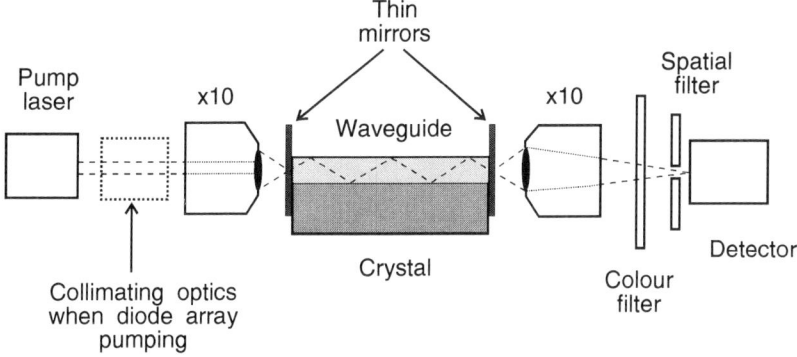

Fig. 7.6. Experimental arrangement used to test waveguide laser performance.

dler *et al.*, 1989). By using a shorter waveguide (4 mm) which was annealed and had better mirror alignment, the threshold was improved to 10.5 mW for 590 nm dye laser pumping and 8.5 mW for 807 nm diode array pumping (Field *et al.*, 1991c,d). These values compared favourably with estimates from Equation (7.1), which gave ~ 6 mW for the dye laser pumping. The slope efficiency in the diode pumped case was 21% with respect to absorbed power (using a 16% output coupler).

Yb doped YAG has recently been reported as demonstrating quasi-three-level lasing in a planar waveguide (Hanna *et al.*, 1993a,b; Shepherd *et al.*, 1993b). Room temperature continuous operation was achieved at a threshold of 30 mW, which is comparable with bulk laser performance despite a propagation loss of ~ 2 dB/cm. Being quasi-three-level, the waveguide loss is less important when compared with reabsorption of the laser light by population in the lower level. However, the enhanced confinement of a waveguide geometry over that of the bulk is extremely advantageous in reducing the threshold. A slope efficiency of 19% was reported, compared with a calculated value of 30%, predicted from the results of Risk (1988). The discrepancy was attributed to the poor waveguide loss. Channel waveguide fabrication in this material is expected to achieve an order of magnitude improvement in threshold.

A planar Nd:GGG waveguide was pumped at 588 nm with a threshold of 12.5 mW, and at 805 nm with a threshold of 17 mW. The slope efficiency was 30% when using a 17% output coupler (Field *et al.*, 1991b). Figure 7.7 shows the fluorescence spectrum of the GGG waveguide laser in the 1.06 μm region, compared to that of the bulk.

$LiNbO_3$ substrates doped with Nd, Er and Tm have been implanted

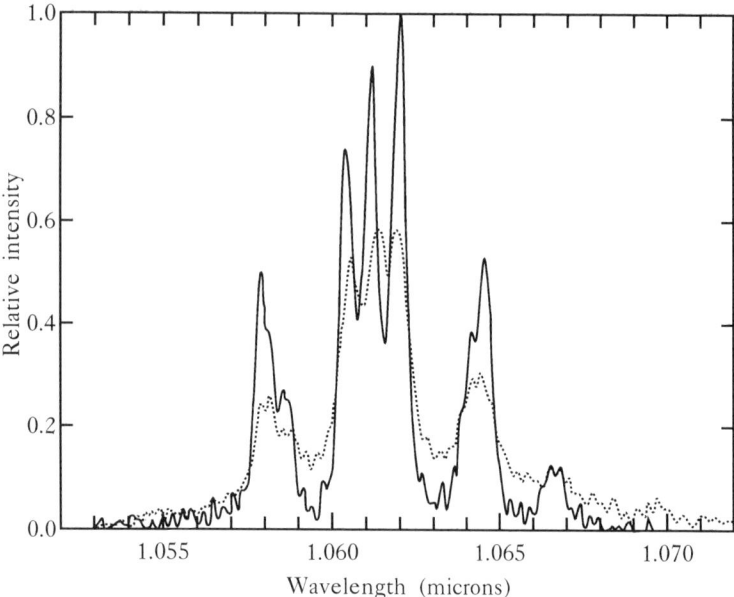

Fig. 7.7. GGG planar waveguide (TM) and bulk crystal fluorescence in the 1.06 μm region.

to produce good waveguides. Their spectroscopic properties have been assessed and their lasing potential is now being investigated. $Nd:LiNbO_3$ has lased using its n_e polarisation, which gives the best emission cross-section (Field et al., 1991e). CW operation was possible at room temperature using a diode array pump, but an elevated temperature (90 °C) was needed for it to lase continuously using a $Ti:Al_2O_3$ pump. It was suggested that the tighter beam profile of the $Ti:Al_2O_3$ laser caused more significant photorefractive damage. The threshold was ~ 17 mW. One of the interesting properties of $Nd:LiNbO_3$ is that it should exhibit self-frequency doubling. A simple and efficient device might be produced in a waveguide using temperature tuned 90° phase-matching. This would require both n_o and n_e to be guided. Here, ion implantation has a distinct advantage because the confinement of both indices is possible, whereas this is not so for proton exchange guides, and Ti-diffused guides are more prone to optical damage. Such a device would need to be forced to lase for n_o, which could be done by choice of suitable mirrors.

$Bi_4Ge_3O_{12}$ is a material which has been shown to produce low loss ion implanted waveguides, having a considerably enhanced index in the

7.2 Ion implanted waveguide lasers

guiding region, and it also displays electro-optic properties. A Nd:BGO planar waveguide is reported to have exhibited lasing for short periods of $\sim 200\,\mu s$ (Field et al., 1991d). The mechanism for this is not yet understood, but the quality of the starting material was not very good (growth problems caused by the high dopant concentration), and it is hoped that further trials will show CW operation. The reported threshold was ~ 50 to $100\,mW$ for $590\,nm$ dye laser pumping, but no lasing was obtainable with Ti:Al$_2$O$_3$ pumping at $810\,nm$. A strange phenomenon was observed in this ion implanted BGO. As reported above in Section 7.1.2, it was found that channel waveguides could be produced within the planar ion implanted region by 'optical writing'.

Nd:YAP is a typical barrier-confined waveguide which has shown lasing action (Field et al., 1990), but multi-energy, high dose implants were needed to produce a sufficiently wide barrier, resulting in rather high losses, even after annealing ($\sim 15\,dB/cm$). Using a waveguide only $0.7\,mm$ long still gave a high lasing threshold ($\sim 50\,mW$). The most recently reported ion implanted waveguide laser has been produced in Tm doped lead germanate glass (Shepherd et al., 1993a).

7.2.3 Channel waveguide lasers

The first reported ion implanted channel waveguide lasers were produced in Nd:YAG (Field et al., 1991a) and Nd:GGG (Field et al., 1992). Both materials exhibit an index increase in the guiding region (0.2% for Nd:YAG and 0.06% for Nd:GGG) which is sufficient to confine light in the two dimensions, and therefore the second type of masking was used, as explained in Section 7.1.1. Fabrication of the channels depended on the use of a 3-μm-thick gold mask, and in both cases a range of channels were formed from 4 to 20 μm wide. Both samples were cut and polished to $2.5\,mm$, and pumped with a $100\,mW$ diode laser coupled via a microscope objective lens. Nd:YAG gave a threshold of $0.5\,mW$ and slope efficiency 29%, and Nd:GGG similarly gave $1.9\,mW$ and 27% (both with 17% output couplers). Figure 7.8 shows the output power of the Nd:YAG laser as a function of absorbed pump power. The threshold here has been increased by the use of the 17% output coupler as opposed to a high reflectivity mirror. A $9\,mm$ long Nd:GGG waveguide was also employed in an extended cavity configuration, using a microscope objective lens within the cavity. In this case the threshold was $8\,mW$.

Future work on reducing losses, and more especially on improving

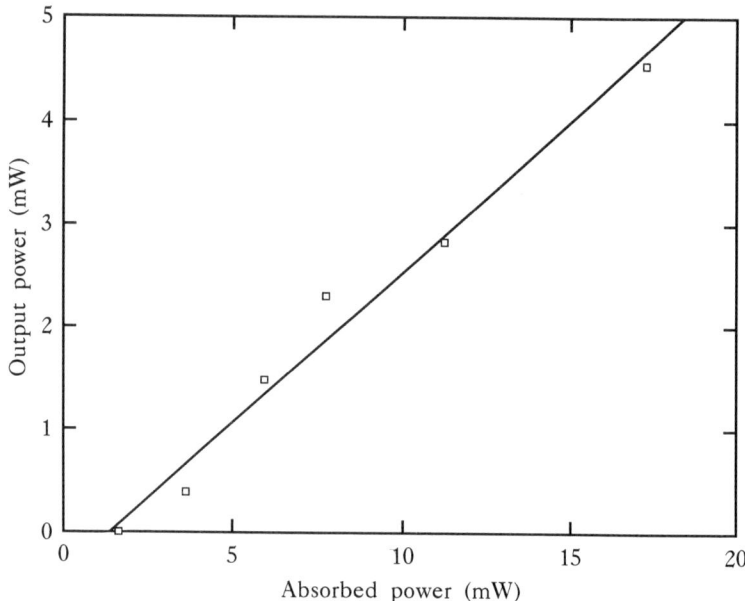

Fig. 7.8. Output power of Nd:YAG channel waveguide laser as a function of pump power, using a 17% output coupler.

lateral mode confinement by optimising side-wall fabrication methods, is expected to improve these results significantly.

7.3 Frequency doubling

Non-linear frequency conversion in a waveguide structure has the advantage of high power intensity confinement over a long interaction length. This enables the achievement of a very high conversion efficiency, even for a low power level (compared with that normally associated with non-linear effects), such as from laser diodes which are available at the present time. By selecting suitable modes, waveguide structures also provide the possibility for phase-matching wavelength ranges which are different from those of the bulk. At the present time, conventional methods of waveguide fabrication, which involve high temperature processing, can only apply to a limited range of non-linear materials, but not to those having relatively low phase transition temperatures, e.g. $KNbO_3$ and $BaTiO_3$. Therefore, ion implantation is a competitive technique in this field. This method has only been applied very recently to make

7.3 Frequency doubling   265

waveguides in a number of non-linear materials, but the early results are very encouraging.

So far, optical waveguides have been produced by ion implantation in a range of non-linear materials including quartz, $LiNbO_3$, $LiTaO_3$, $KNbO_3$, $BaTiO_3$, KTP, SBN, BNN, LBO and BBO (Glavas et al., 1988; Bremer et al., 1988; Zhang et al., 1990,1991,1992a; Davis et al., 1992; Strohkendl et al., 1991a,b). Second harmonic generation has been demonstrated in He^+ ion implanted planar waveguides for quartz (Babsail et al., 1991), $KNbO_3$ (Fluck et al., 1992a), KTP (Zhang et al., 1992b,1993), and LBO (Davis, 1992, private communication). Channel waveguide SHG has been achieved in quartz (Hamelin, 1993, private communication), $KNbO_3$ (Fluck et al., 1992b) and LBO.

7.3.1 Quartz

Quartz was used to demonstrate the first example of SHG in a He^+ implanted planar waveguide (Babsail et al., 1991). This was selected not for its non-linear properties, which are very low, but because of its ideal waveguide structure, in order to demonstrate frequency doubling by mode matching. As described in Chapter 6, He^+ implanted quartz waveguides have a profile close to a square optical well; and the wide low index optical barrier confines light in the surface region with little tunnelling probability. Post-implantation thermal annealing can reduce the absorption and scattering in the guide to give a very low loss ($\sim 0.2\,dB/cm$).

SHG was achieved in a planar quartz waveguide designed to have a width $\sim 1.8\,\mu m$ to phase-match at a wavelength of $\sim 0.8\,\mu m$, from the fundamental $m = 0$ mode to the secondary $m = 2$ mode. The guide was obtained by implanting He^+ at four different energies (0.75, 0.82, 0.89, 0.96 MeV) to give the required guide width and an optical barrier $> 1\,\mu m$ wide. After annealing, the waveguide loss was reduced to less than $1\,dB/cm$.

A typical spectrum of the output SHG intensity as a function of the input wavelength for the TE_w^0 to TE_{2w}^2 conversion is shown in Figure 7.9, where the position of the maximum is at 782.5 nm. In addition to the strong centre line there are a number of sidebands, but these are not symmetric about the centre line. This effect has also been detected by other authors, e.g. in a deposited ZnS film (Ito and Inaba, 1975). The spectrum of Figure 7.9 can be fitted using a $sinc^2$ function, $\sin^2(\Delta kl/2)/(\Delta kl/2)^2$ – to estimate the interaction length for the SHG

266   Applications of ion implanted waveguides

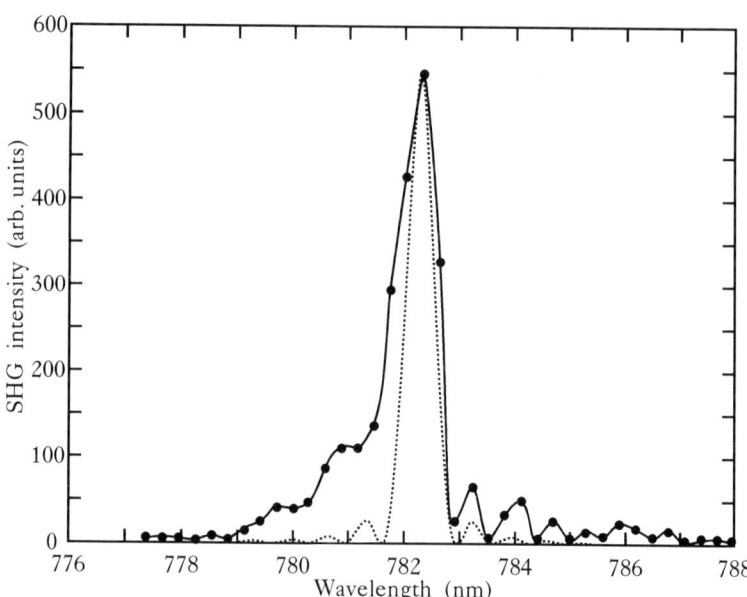

Fig. 7.9. Experimental spectral output (circles) of the SHG in an implanted quartz waveguide. The dashed line has been calculated using a sinc^2 function.

performance. The best fit from this result gives a value of $l = 15$ mm for the interaction length, which is the same as the length of the sample. This result suggests that phase-matching occurs throughout the length of the waveguide. The maximum fundamental power coupled into the waveguide is estimated to be less than 25 mW, and the efficiency for this SHG conversion process is calculated from the measured output power to be $\sim 10^{-9}$. The absolute value is very low because of the poor non-linearity of quartz, unfavourable coupling geometry and the use of a planar, rather than a channel, waveguide. An improvement by a factor of 100 in conversion efficiency has been reported from a channel quartz waveguide (Hamelin et al., 1993, private communication).

7.3.2 Potassium niobate

$KNbO_3$ has been shown to possess excellent electro-optic (Gunter, 1974) and non-linear properties (Uematsu, 1974; Gunter, 1979). The large non-linear optical coefficients (e.g. $d_{32} = 20.3$ pm/V at $\lambda = 860$ nm), very high threshold to optical damage (~ 1.3 GW/cm^2) and the combination of non-critical phase-matchability between 840 nm and 1100 nm and

7.3 Frequency doubling

transparency down to 390 nm make $KNbO_3$ a particularly attractive material for frequency conversion into the blue and green region (Gunter, 1979; Baumert et al., 1983; Biaggio et al., 1992). In addition to all these excellent properties of $KNbO_3$, a waveguide structure enables high conversion efficiency even at the power levels of presently available laser diodes. Conventional methods for fabrication of waveguides (such as ion indiffusion or ion exchange, successful for $LiNbO_3$) are not applicable to $KNbO_3$ because of its structural phase transition at 223 °C which prohibits high temperature treatment. So far, ion implantation is the only method to produce permanent optical waveguides in $KNbO_3$ (Bremer et al., 1988; Zhang et al., 1990; Strohkendl et al., 1991a,b).

SHG in He^+ implanted $KNbO_3$ planar waveguides was reported for the first time by the Gunter group (Fluck et al., 1992a). A $8.5 \times 2.0 \times 7.5 \, mm^3$ crystal of $KNbO_3$ was implanted at room temperature with 2.9 MeV He^+ to a dose of $7.5 \times 10^{14} ions/cm^2$. The optical losses in this guide were 3 dB at 488 nm and 4.5 dB at 860 nm. In order to determine which mode conversion gives the highest conversion efficiency, the field distributions of the fundamental and the secondary modes were calculated (Figure 7.10). The pronounced asymmetry of the fields of the TM_0 and TE_0 modes is due to the anisotropic refractive index profiles obtained by ion implantation in $KNbO_3$. Because the Gaussian profile of the incident laser beam ensures highest end-coupling efficiency to the TM_0 mode of the waveguide, Fluck et al. considered the TM_0 fundamental mode only for determining overlap integrals. The overlap integrals for this particular guide, they calculated, are $\gamma_{00} = 0.04/\mu m$, $\gamma_{01} = 0.08/\mu m$ and $\gamma_{02} = 0.05/\mu m$. Therefore, the TM_0 to TE_1 mode conversion was expected to be the most efficient one and to take place at a calculated phase-matching wavelength of 865.5 nm.

By scanning the $Ti:Al_2O_3$ pulsed laser from 855 nm to 875 nm, the most intense blue light was measured from the waveguide at a wavelength $\lambda = 867.7 \pm 0.2$ nm. From the far field pattern of light on the screen the SHG wave was identified to be the TE_1 mode of the planar waveguide. The SHG peak power was measured as a function of the fundamental power at 867.7 nm. The results gave an approximately quadratic relationship between the fundamental and the secondary, as Figure 7.11 shows. A maximum conversion efficiency of 28.7% was achieved for a fundamental peak power of 1.3 kW in the waveguide, giving a blue light output of 385 W peak power. The conversion efficiency is relatively lower than the theoretical value. There are three facts which could explain this. The propagation loss of the waveguide is the first obvious reason – by

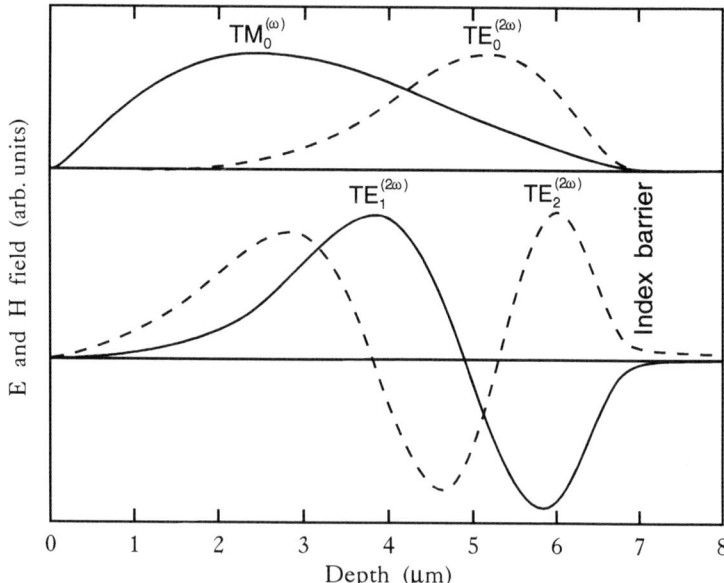

Fig. 7.10. Field amplitudes in a $KNbO_3$ waveguide for the fundamental TM_0 and the harmonic TE_0, TE_1 and TE_2 modes at a fundamental wavelength of $\lambda = 860$ nm.

reducing the mode losses to about 1 dB/cm, the SHG output power would increase by a factor of 2. The second reason is that the linewidth of the Ti:Al_2O_3 laser ($\Delta\lambda_{868} = 3.3$ Å) was larger than the acceptance bandwidth ($\Delta\lambda = 1.1$ Å) of the sample, so that only part of the total input power was phase-matched over the full length of the waveguide and thus only a fraction of the input power contributed to the SHG process. The poor overlap integral is the third reason – the overlap integral of the TM_0 fundamental and the TE_1 second harmonic modes of the planar waveguide used here was only about 30% of the value for perfectly overlapping modes. Therefore, optimisation of the implantation process has the potential to increase the conversion efficiency remarkably, as shown in Figure 7.11.

The same group has also demonstrated the Cerenkov frequency doubling of a diode laser in an implanted $KNbO_3$ channel waveguide (Fluck et al., 1992b). The channel waveguides were fabricated by first producing a planar waveguide by homogeneously irradiating the samples with high energy ions, and then applying a shielding mask of SiO_2 fibres to define channel guides, as explained previously in Section 7.1.1. The planar wave-

7.3 Frequency doubling

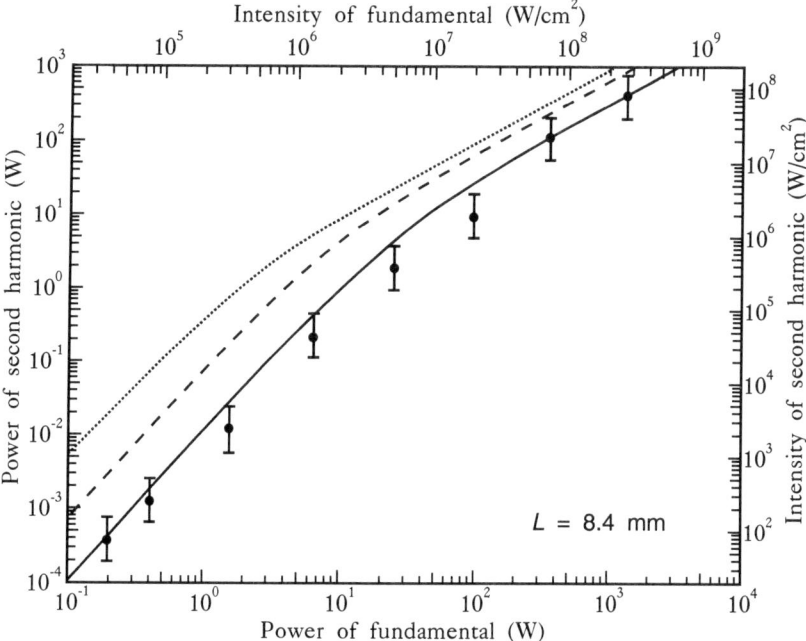

Fig. 7.11. SHG power as a function of fundamental power for a planar KNbO$_3$ waveguide. Experimental data (circles). Theoretical curves for: present guide using broad band Ti:Al$_2$O$_3$ laser (solid); improved lower loss guide and using laser with matching bandwidth (dashed); ideal guide with fully optimised confinement and overlap integral (dotted).

guide was formed by implanting the sample at room temperature with 2.3 MeV He$^+$ to a dose of 7.5×10^{14} ions/cm^2. The vertical side-walls were produced by implanting He$^+$ to 5×10^{14} ions/cm^2 at energies of 0.8, 1.8 and 2.1 MeV. The non-linear optical coefficient d_{32} was chosen to ensure that the birefringence of KNbO$_3$ could be fully exploited to achieve Cerenkov SHG at a small radiation angle. For the SHG experiments, either a CW AlGaAs diode laser or a CW Ti:Al$_2$O$_3$ laser was used with a spectral linewidth of ~ 50 and 5 GHz respectively. The mode coupling insertion loss was ~ 5 dB for the 5.6-mm-long channel guide. Using a low power diode laser, up to 13 mW of fundamental power was transmitted through the waveguide, giving 20 μW of 430 nm second harmonic radiation. With the CW Ti:Al$_2$O$_3$ laser, up to 97 mW of fundamental power was coupled into the same channel waveguides providing a blue light output power of 1.1 mW at 430 nm. The SHG radiation generated by the TE_{00} fundamental mode was emitted at a Cerenkov angle of 2.4° into the bulk crystal. The normalised conversion efficiency for this process is about 40%/Wcm, which is in good agreement with theory. It was estimated that further optimisation of the implantation process and a reduction of the cross-section of the channel guides should increase the conversion efficiency by a factor of four. In that case, up to 60 mW Cerenkov SHG blue light could be generated from 200 mW of fundamental power in a 1-cm-long $KNbO_3$ channel waveguide.

7.3.3 Potassium titanyl phosphate

KTP ($KTiOPO_4$) is another established, highly non-linear material for second harmonic generation (SHG) and frequency upconversion of IR lasers. The Sussex group has reported the achievement of SHG for the first time in He^+ implanted KTP planar waveguides, with a conversion efficiency up to 25% (Zhang et al., 1992b, 1993). KTP is a biaxial crystal. The high effective non-linear coefficient only exists for type-2 phase-matching $\{[n_z(\omega) + n_{xy}(\omega)]/2 = n_{xy}(2\omega)\}$. He^+ implantation produces a barrier waveguide in KTP, and so in order to reduce the tunnelling loss, double energy (2.5 + 2.7 MeV) implants were used to broaden this optical barrier. The loss was measured to be ~ 0.5 dB/cm at 0.633 μm and ~ 1 dB/cm in the IR.

It was expected that the phase-matching wavelength for the frequency doubled waveguide mode would be different from that of the bulk. The mode dispersion curves for the lower modes of the implanted waveguide were calculated based on a square well approximation. Their crossing points predicted the phase-matched wavelengths for the different mode matches between the fundamental and the secondary. Figure 7.12 presents the experimental wavelength scanning results for the implanted waveguide. The main peak (mode 0 to 0) appeared at $\lambda = 1.0741$ μm, which was shifted from the bulk by 0.011 μm. A small second peak was also observed at $\lambda = 1.0482$ μm, and this was attributed to the phase-matching between the $m = 0$ fundamental and $m = 1$ harmonic modes. These results are in good agreement with predictions from the mode dispersion curves. Wavelength scanning was also carried out in low single energy and triple energy implanted waveguides. As they have slightly narrower guide widths, the frequency doubling was observed at longer wavelengths.

A quadratic relationship of $P^{2\omega}$ to P^{ω} was measured, giving a logarithmic slope of 1.9 ± 0.1, as expected. The waveguide conversion efficiency

7.3 Frequency doubling

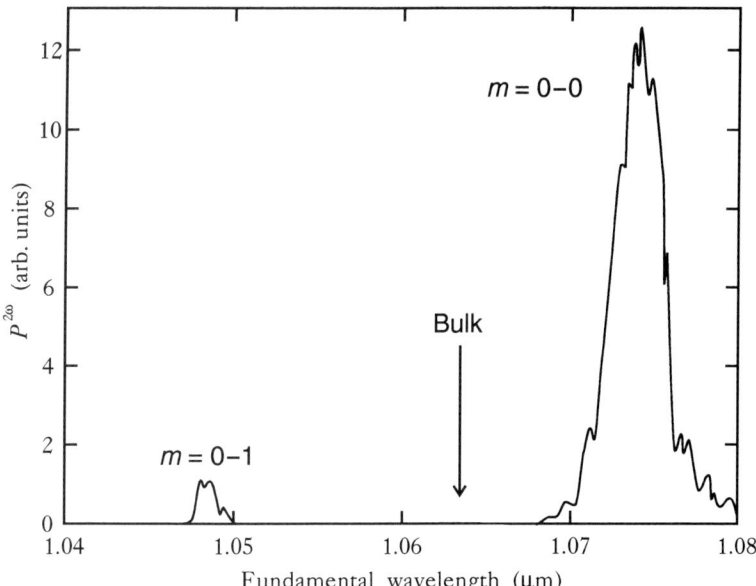

Fig. 7.12. SHG intensity as a function of the fundamental wavelength. The main peak is the fundamental mode 0 phase-matching with the harmonic mode 0. The small peak is due to phase-matching between fundamental mode 0 and harmonic mode 1.

was estimated using a photodiode detector. The efficiency for the double energy implanted waveguide was about 25% for incident pulses of $\sim 1\,\mu\text{J}$ and length 20 ns. A higher conversion efficiency is expected to be achieved in channel waveguides implanted under optimised conditions.

Phase-matching is the first important condition to achieve high conversion efficiency of SHG. It is apparent that high efficiency from a waveguide requires low loss and high non-linearity, but, in addition, the overlap integral between the fundamental and secondary electric field distributions should be high. Figure 7.13 shows the electric field profiles for the fundamental 0 modes (n_z and n_{xy}), and the secondary 0 and 1 modes (n_{xy}). The overlap integral coefficients which were calculated for this guide are $OV^{00} = 0.453$ for the 0 to 0 conversion, and only $OV^{01} = 0.017$ for the 0 to 1 conversion, which is 20 times smaller. The asymmetry of the mode 1 electric field profile results in the small value of the overlap integral for the 0 to 1 conversion. However, the small peak in Figure 7.12 is not 20 times smaller than the main peak. The experimental reason for this is that the small peak is very near to the

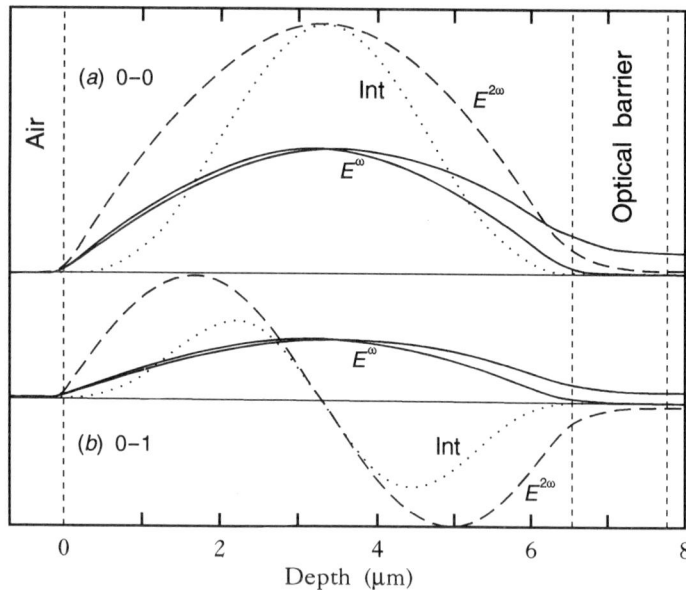

Fig. 7.13. Electric field distributions calculated from the KTP index profiles, at the two phase-matching wavelengths (1.0741 and 1.0482 μm), for fundamental E^ω $m = 0$ (n_z and n_{xy} – solid) and secondary $E^{2\omega}$ $m = 0$ and 1 (n_{xy} – dashed). The dotted curves show the respective overlap integral functions (Int).

input laser's maximum performance, while the SHG main peak is near to its falling edge.

Although, so far, SHG has only been demonstrated in ion implanted optical waveguides in five non-linear optical materials, it is clear that non-linearity can be preserved after implantation.

7.4 Photorefractive effects

Photorefractive, non-linear optical crystals, such as $BaTiO_3$ and $Sr_xBa_{1-x}Nb_2O_6$(SBN) are currently used for a wide range of applications in optical phase conjugation and two- and four- wave mixing techniques (Gunter and Huignard, 1988). The optical confinement of a waveguide structure allows a high intensity to be maintained within the crystal waveguide, and this increased intensity–length product leads to an appreciable decrease in the effective response time for a given input power. $BaTiO_3$ is one of the most interesting materials for non-linear photorefractive applications, because of the very large value of its

7.4 Photorefractive effects 273

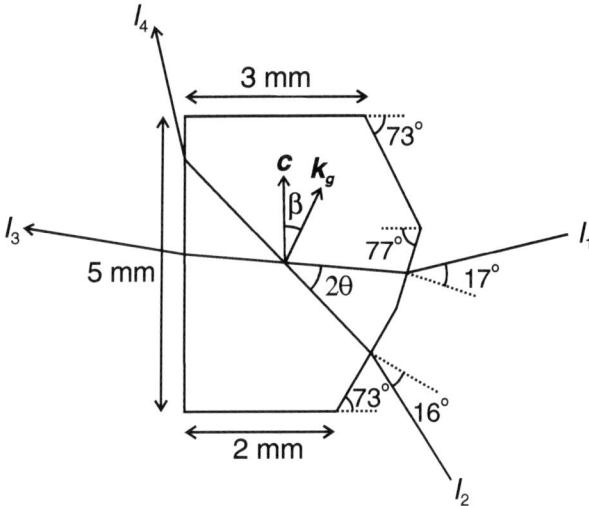

Fig. 7.14. Details of the crystal dimensions and angles of precise polished faces (k_g is the grating wave vector).

electro-optic coefficient, $r_{42} = r_{51} = 1640\,\text{pm/V}$, but its low phase transition temperatures ($\sim 9\,°\text{C}$ and $\sim 130\,°\text{C}$) limit waveguide formation by conventional methods, which involve high temperature processes. Ion implanted BaTiO$_3$ waveguides have recently been used to demonstrate the advantages of photorefractive applications in waveguide structures (Youden et al., 1992). The photorefractive waveguide in BaTiO$_3$ was produced by implantation of 1.5 MeV H$^+$ to a dose of 10^{16} ions/cm^2 at room temperature. The guide supported ~ 20 modes, and gave a propagating loss of $\sim 14\,\text{dB/cm}$. The low Curie temperature ($\sim 130\,°\text{C}$) prevented annealing treatment after implantation to reduce the loss. However, high loss is not significant because only a few mm length of crystal is required to perform the two-beam coupling experiment.

In the experiment for the characterisation of the photorefractive properties of the waveguide, extraordinary-polarised light of three wavelengths (633 nm, 568 nm and 488 nm) was used to access TE waveguide modes and hence the large r_{42} electro-optic coefficient of the crystal. Two mutually coherent beams I_1 and I_2 were launched into the waveguide at angles shown in Figure 7.14. The input crystal faces had previously been cut and polished to facilitate end-launching for the two-beam coupling. Outputs from the waveguide, I_3 and I_4, were selectively imaged on to a power meter. It was observed that the direction of the two-beam coupling

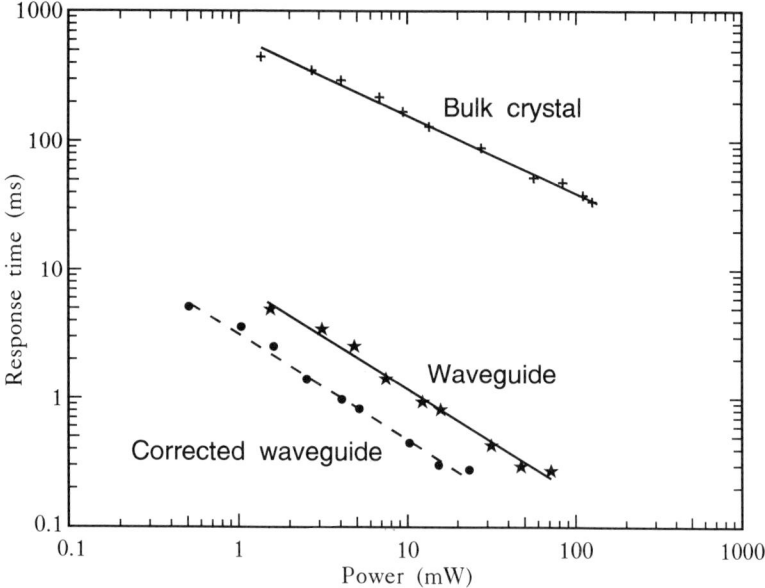

Fig. 7.15. Two-beam coupling response time versus the incident power for waveguide (*) and bulk (+) crystal regions at $\lambda = 488$ nm; the dashed line indicates the waveguide data corrected for measured waveguide losses of 14 dB/cm.

gain had been reversed in the waveguide compared to that in the bulk crystal. This provided additional confirmation that the photorefractive beam coupling effects were indeed those present in the waveguide, and were not just due to scattered light from the bulk or inadvertent bulk crystal coupling. The cause of the change in the gain direction is currently uncertain, but is likely to be due to a change in the dominant charge carrier species – from holes to electrons (Ducharme and Feinberg, 1986), brought about by the implantation process.

The photorefractive response time, τ, was measured for various total incident beam intensities ($I_1 + I_2$) in both the waveguide and the bulk. Figure 7.15 shows the measured response time (τ) versus the total input intensity at 488 nm, and for direct comparison the actual input power is also shown. It can be seen that a considerable effective speed-up has been achieved, as anticipated. At the moderate input power of 50 mW a response time of 200 μs was obtained in the waveguide, whereas the corresponding response time at the same input power in the bulk was only 25 ms, a factor of more than 10^2 slower. For conversion of the data in Figure 7.15 to equivalent average irradiance, the beam area for

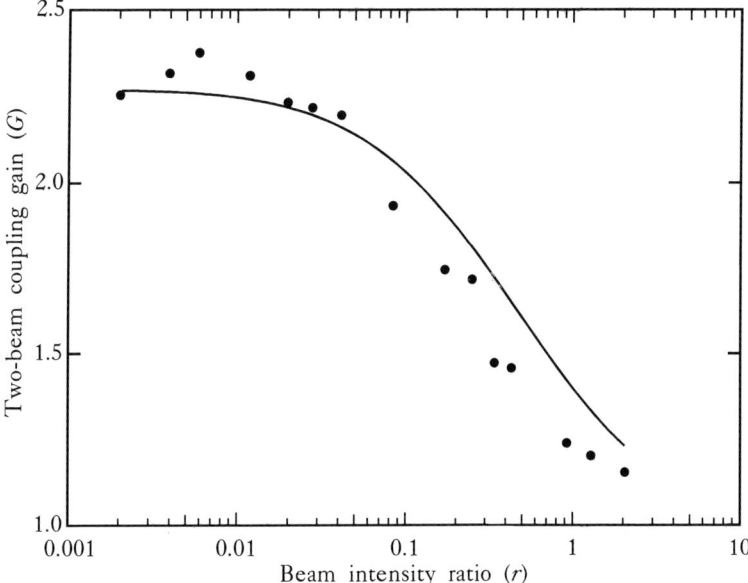

Fig. 7.16. Two-beam coupling gain G as a function of the beam intensity ratio r in the waveguide region at $\lambda = 568$ nm. The solid curve represents the calculated profile from standard two-beam coupling theory.

bulk and waveguide measurements was estimated to be $\sim 1 \times 10^{-2}$ and $\sim 8 \times 10^{-6}$ cm^2 respectively. The maximum irradiance obtained in the waveguide experiments, therefore, with correction for inferred losses, was ~ 1.4 kW/cm^2. The effective two-beam coupling gain was also measured at a wavelength of 568 nm as a function of the incident beam ratio. The experimental results are shown in Figure 7.16. The interaction length was estimated to be ≤ 0.5 mm, from calculations of the area and geometry of the beams in the waveguide.

7.5 Future and related applications

This chapter has emphasised the near-market applications of ion implantation for optical waveguide devices, because one assumes not only that these will soon be realised commercially, but also that the scale of the market is enormous. We have not specifically discussed some of the other attributes of the optical effects of ion implantation such as changes in $\chi^{(3)}$, optical storage, anti-reflection coatings, mirrors, sun glasses, security devices or improved phosphors, either because they are less well developed or are less general in the scale of their development. Some of the ideas are discussed in the patent literature and we sincerely hope that this book will act as a stimulus to produce many more applications now that one can confidently use ion beams to control so many optical properties.

References

Babsail, L., Lifante, G. and Townsend, P.D. (1991). *Appl. Phys. Lett.*, **59**, 384.
Baumert, J.C., Gunter, P. and Melchior, H. (1983). *Opt. Comm.*, **48**, 215.
Biaggio, I, Kerkoc, P., Wu, L.S., Zysset, B. and Gunter, P. (1992). *J. Opt. Soc. Am.*, **B9**, 507.
Bremer, T., Heiland, W., Hellermann, B., Hertel, P., Kratzig, E. and Kollewe, D. (1988). *Ferroelectr. Lett.*, **9**, 11
Brocklesby, W.S., Field, S.J., Hanna, D.C., Large, A.C., Lincoln. J.R., Shepherd, D.P., Tropper, A.C., Chandler, P.J., Townsend, P.D., Zhang, L., Feng, X.Q. and Hu, Q. (1992). *Opt. Mat.*, **1**, 177.
Chandler, P.J. and Lama, F.L. (1986). *Optica Acta*, **33**, 127.
Chandler, P.J., Lama, F., Townsend, P.D. and Zhang, L. (1988). *Appl. Phys. Lett.*, **53**, 89.
Chandler, P.J., Zhang, L. and Townsend, P.D. (1989). *Appl. Phys. Lett.*, **55**, 1710.
Chandler, P.J., Field, S.J., Hanna, D.C., Shepherd, D.P., Townsend, P.D., Tropper, A.C. and Zhang, L. (1989). *Elec. Lett.*, **25**, 985.
Chandler, P.J., Zhang, L. and Townsend, P.D. (1992). *Solid State Phenomena*, **27**, 129. also in *Trends in Ion Implantation*, M. van Rossum (ed.). TransTech Publications, Heverlee, Belgium.
Clarkson, W.A. and Hanna, D.C. (1989). *J. Mod. Opt.*, **36**, 483.
Davis, G.M., Zhang, L., Chandler, P.J. and Townsend, P.D. (1993). *IEEE Photonics Technology Lett.*, **5**, 430.
Ducharme, S. and Feinberg, J. (1986). *J. Opt. Soc. Am.*, **B3**, 283.
Faik, A.B., Chandler, P.J. Townsend, P.D., and Webb, R. (1982). *Rad. Effects*, **98**, 233.
Field, S.J., Hanna, D.C., Shepherd, D.P., Tropper, A.C., Chandler, P.J., Townsend, P.D. and Zhang, L. (1990). *Elec. Lett.*, **26**, 1826.
Field, S.J., Hanna, D.C., Large, A.C., Shepherd, D.P., Tropper, A.C., Chandler, P.J., Townsend, P.D. and Zhang, L. (1991a). *Elec. Lett.*, **27**, 2375.
Field, S.J., Hanna, D.C., Large, A.C., Shepherd, D.P., Tropper, A.C., Chandler, P.J., Townsend, P.D. and Zhang, L. (1991b). *Opt. Comm.*, **86**, 161.
Field, S.J., Hanna, D.C., Shepherd, D.P., Tropper, A.C., Chandler, P.J., Townsend, P.D. and Zhang, L. (1991c). *IEEE J. Quant. Elect.*, **27**, 428.
Field, S.J., Hanna, D.C., Large, A.C., Shepherd, D.P., Tropper, A.C., Chandler, P.J., Townsend, P.D. and Zhang, L. (1991d). *Opt. Soc. Am. Proc. ASSL 91*, **10**, 353.

References

Field, S.J., Hanna, A.C., Shepherd, D.P., Tropper, A.C., Chandler, P.J., Townsend, P.D. and Zhang, L. (1991e). *Opt. Lett.*, **16**, 481.
Field, S.J., Hanna, D.C., Large, A.C., Shepherd, D.P., Tropper, A.C., Chandler, P.J., Townsend, P.D. and Zhang, L. (1992). *Opt. Lett.*, **17**, 52.
Fluck, D., Binder, B., Kupfer, M. Looser, H., Buchal, Ch. and Gunter, P. (1992a). *Opt. Comm.*, **a**, 304.
Fluck, D., Moll, J., Gunter, P., Fluster, M. and Buchal, Ch. (1992b). *Elect. Lett.*, **28**, 1092.
Glavas, E., Zhang, L., Chandler, P.J. and Townsend, P.D. (1988). *Nucl. Inst. Methods.*, **B32**, 45.
Gunter, P. (1974). *Opt. Comm.*, **11**, 285.
Gunter, P. (1979). *Appl. Phys. Lett.*, **34**, 650.
Gunter, P. and Huignard, J.P. (1988). *Photorefractive Materials and their Applications II*, (Springer-Verlag, Berlin).
Hanna, D.C., Jones, J.K., Large, A.C., Shepherd, D.P., Tropper, A.C., Chandler, P.J., Rodman, M.J., Townsend, P.D. and Zhang, L. (1993a). *Opt. Comm.*, **99**, 211.
Hanna, D.C., Jones, J.K., Large, A.C., Shepherd, D.P., Tropper, A.C., Chandler, P.J., Townsend, P.D. and Zhang, L. (1993). *Proc. CLEO 93*, p.626 (Opt. Soc. Am., Washington).
Ito, H. and Inaba, H. (1975). *Opt. Comm.*, **15**, 104.
Kersten, R.Th. and Boroffka, H, (1976). *Opt. and Quant. Elect. Lett.*, **8**, 263.
Lifante, G. and Townsend, P.D. (1992). *J. Mod. Opt.*, **39**, 1353.
Naik, I.K. (1983). *Appl. Phys. Lett.*, **43**, 519.
Reed, G.T. and Weiss, B.L. (1986). *Nucl. Inst. Methods*, **B19/20**, 907.
Reed, G.T. and Weiss, B.L. (1987). *Elect. Lett.*, **23**, 792.
Risk, W.P. (1988). *J. Opt. Soc. Am.*, **b5**, 1412.
Shepherd, D.P. et al., (1993a). *Elect. Lett.*, to be published.
Shepherd, D.P., Hanna, D.C., Jones, J.K., Large, A.C., Tropper, A.C., Chandler, P.J., Kakarantzas, G., Townsend, P.D., Zhang, L., Chartier, I., Ferrand, B. and Pelenc, D. (1993b). *Proc. Int. Symp. on Optoelec. Mat., Honolulu, Nov. 1993*, (Amer. Ceram. Soc., Washington).
Strohkendl, F.P., Gunter, P., Buchal, Ch. and Irmscher, R. (1991a). *J. Appl. Phys.*, **69**, 84.
Strohkendl, F.P., Fluck, D., Buchal, Ch., Irmscher, R. Gunter, P. (1991b). *Appl. Phys. Lett.*, **59**, 3354.
Uematsu, Y. (1974). *Jap. J. Appl. Phys.*, **13**, 1562.
Youden, K.E., James, S.W., Eason, R.W., Chandler, P.J., Zhang, L. and Townsend, P.D. (1992). *Opt. Lett.*, **17**, 1509.
Zhang, L., Chandler, P.J. and Townsend, P.D. (1990). *Ferroelectr. Lett.*, **11**, 89.
Zhang, L., Chandler, P.D. and Townsend, P.D. (1991). *Nucl. Inst. Methods*, **B59/60**, 1147.
Zhang, L., Chandler, P.J., Townsend, P.D. and Thomas, P.A. (1992a). *Elect. Lett.*, **28**, 650.
Zhang, L., Chandler, P.J., Townsend, P.D., Alwahabi, Z.T. and McCaffery, A.J. (1992b). *Elect. Lett.*, **28**, 1478.
Zhang, L., Chandler, P.J., Townsend, P.D., Alwahabi, Z.T., Pityana, S.L. and McCaffery, A.J. (1993). *J. Appl. Phys.*, **73**, 2695.

Index

Advantages of implantation 7
Al_2O_3 (see sapphire)
alkali halides 57, 77, 83, 119, 145
amorphisation 39, 46, 82, 103
argon (solid) 135

$BaTiO_3$ 272
$Bi_4Ge_3O_{12}$ 137, 196, 219, 251
Boltzmann transport equation 66

CaF_2 122, 143
CaO 145
cathodoluminescence 146
channelling 55
channel waveguides 248, 263
colloids 105, 237, 240
colour centres 49, 85, 93
compositional changes 17, 52, 130, 136, 200, 211, 220, 235, 238
computer simulations 62
crystallographic effects 41, 51, 75, 199

damage distributions 24, 35, 54, 93
defect annealing 47, 49, 201, 220, 233, 240
defect complexes 49, 90, 198
differential cross-section 29
diffusion 16, 39, 51, 53, 197, 211
diffusion doping 7, 152, 239
divacancies 95, 137
displacement energy 36, 39
double barrier guides 181, 253

electrical properties 6
electron spin resonance (ESR) 70, 103
ENDOR 103
energy transfer 12, 24
electronic energy transfer 14, 24, 31, 38
excitons 119, 132

F type centres 85, 88

frequency doubling (SHG) 264
fission tracks 36

garnets 222, 258
glass waveguides 233

high dose effects/amorphisation 103

implant temperature 40, 117, 207
information storage 11
in situ luminescence 119, 135
ion beam processing 18
ion ranges 24, 34
isotopic effects 19, 100, 103

$KNbO_3$ 218, 250, 266
KTP 230, 270

Lasers 10, 137, 222, 255, 260, 263
LBO 231
LiF 90, 93, 145
$LiNbO_3$ 41, 53, 104, 130, 181, 196, 207, 239, 254
loss in waveguides 151, 189, 220
luminescence 115

MARLOWE 65
missing modes 166, 209
molecular beam effects 95
molecular dynamics 65

NaF 136
niobates (general) 207, 215
non-crystalline materials (glass) 233
non-linear materials 228, 247, 265
nuclear collisions 12, 24, 27

optical absorption 67, 109, 189
optical barrier waveguides 168, 196, 247
optical damage 213, 251, 272

Index

optical properties 8, 10
oscillator strength 101

pattern definition 2
phase precipitation 108
photoluminescence 135
photorefractive effects 213, 251, 272
point defects 14, 49
PRAL 66

quartz 37, 50, 102, 140, 180, 188, 196, 202, 265

radiation dosimetry 145
radiation enhanced diffusion 14, 18
range distributions 24, 34, 147
refractive index changes 167, 175, 201
reflectivity analysis 169, 179, 183
Rutherford backscattering 70

sapphire 45, 77, 117, 127, 137
secondary electron emission 18, 54
second harmonic generation 247, 265
semiconductor synthesis 136
silica (SiO_2) 13, 26, 52, 61, 101, 109, 123, 139, 140, 186

silicate glasses 45, 73, 100, 108, 237
silicon 49, 185
simulations (computer) 62, 67
spectroscopy of waveguides 259
sputtering 24, 57
strange modes 212
stress 45, 75, 143
surface effects and contaminants 10, 133
SUSPRE 66
synergistic effects 14, 125, 129, 219

tantalates 217
thermal effects 16
thermoluminescence 140
TRIM 62

waveguide lasers 137, 260, 263
waveguides 10, 151, 196, 248
waveguide analysis and mode theory 151, 167
waveguide coupling 160, 163
waveguide losses 189, 190, 192

YAG 137, 222

Zircon 42